T0212767

Library of Ethics and Applied Philosophy

Volume 35

More information about this series at http://www.springer.com/series/6230

Thomas Schramme

Editor

New Perspectives on Paternalism and Health Care

 Springer

Editor
Thomas Schramme
Department of Philosophy
Hamburg University
Hamburg, Germany

ISSN 1387-6678   ISSN 2215-0323   (electronic)
Library of Ethics and Applied Philosophy
ISBN 978-3-319-35473-6   ISBN 978-3-319-17960-5   (eBook)
DOI 10.1007/978-3-319-17960-5

Acknowledgements

The essays in this volume were discussed at the Spring School 'New Perspectives on Medical Paternalism' at the University of Hamburg in March 2012. This was a fantastic opportunity for young academics, most of them with a background in philosophy, to discuss issues in paternalism in a very intense – as some have put it, camp-like – atmosphere. In addition to these participants, who were selected by peer reviewers on the basis of applications to participate, there were also a number of senior scholars invited to the conference. Some of their papers are published here for the first time; some others are reprinted from different, scattered sources.

There are many people who have helped to publish this volume and whom I would like to thank. First of all, the authors have shown great team spirit and perseverance in keeping this a collective and coherent publication against the odds. Andrew Fassett expertly proofread several of the papers and translated two essays. Fallet helped in preparing the manuscript. Finally, the German Federal Ministry of Education and Research provided necessary, and indeed generous, funding.

Contents

Contributors

Diana Aurenque has completed her PhD in philosophy at the University of Freiburg, Germany, after studies in Santiago de Chile. From 2011 until 2014, she was a researcher and lecturer at the Institute of Ethics and History of Medicine (Habilitation), University of Tübingen. Since 2015 she is assistant professor (tenure track) in philosophy and ethics at the University of Santiago in Chile (USACH). Her areas of research in bioethics include the investigation of theoretical foundations of medical ethics, philosophy of medicine, ethical issues on reproductive medicine (particularly the ethics of PGD) as well as questions about the legitimacy of some controversial treatments in infants (intersex, religious circumcision, etc.).

C.R. Blease is Government of Ireland postdoctoral research fellow in philosophy at the University College Dublin. She completed her PhD – entitled 'Paul Churchland's Arguments for Eliminative Materialism' – in 2008 at Queen's University, Belfast, and has held teaching and research positions at Queen's University and Ruhr University, Bochum. Her current research interests are in the philosophy and cognitive science of psychiatry, in particular the relationship between lay concepts of mental disorders and scientific classifications. She also researches the nature of 'culture-bound syndromes' and is particularly interested in classifications of depression and different cultural expressions of 'disease'. Her research bent is strongly interdisciplinary.

Lorenzo del Savio is postdoc fellow at Universitätsklinikum Schleswig-Holstein, working on ethical, social and regulatory aspects of citizen science in biomedicine and health care. After his undergraduate studies in philosophy, he got a PhD at the Università di Milano, European School of Molecular Medicine, discussing a thesis on the ethics of public health interventions against unhealthy nutrition. His current academic interests include biomedical ethics and policy and democratic theory.

Dominik Düber studied philosophy, German philology and history of Eastern Europe at the University of Cologne and finished his studies with a Magister thesis on 'The Conception of Justice in Health by Norman Daniels'. He is currently working as a research fellow at the Centre for Advanced Study in Bioethics (University of Münster, Germany) and has submitted a PhD thesis called 'The Paternalism Argument Against Perfectionism'.

Gerald Dworkin is a distinguished professor emeritus at the University of California, Davis, and is part of the faculty there since 1996. Before that he has taught at Harvard, MIT and the University of Illinois, Chicago. From 1990 to 1997, he was the editor of *Ethics*. His areas of specialization are moral, political and legal philosophy and medical ethics. Some of his most recent publications are 'The Limits to Criminalization' in *Oxford Handbook on Philosophy of Criminal Law*, eds. John Deigh and David Dolinko (Oxford University Press, 2011), and 'Harm and the Volenti Principle', *Social Philosophy and Policy*, January 2012.

Jessica Flanigan is an assistant professor of leadership studies and philosophy, politics, economics and law at the Jepson School of Leadership Studies, University of Richmond. Her research focuses on medical ethics and public health. Flanigan's work has appeared in journals such as *Public Health Ethics, HEC Forum, Public Affairs Quarterly,* and *The Journal of Medical Ethics.*

Kalle Grill is senior lecturer at Umeå University, Sweden. He was previously a Marie Curie research fellow at Uppsala University and visiting/honorary research fellow at Keele University and the University of Birmingham School of Health and Population Sciences. He received a PhD in philosophy from KTH Royal Institute of Technology in 2009. His research interests are in moral and political philosophy, with special interests in paternalism and anti-paternalism, as well as moral reasoning, population axiology, public health and human relationships.

Clemens Heyder studied philosophy and history at the universities of Leipzig and Basel where he focused on biomedical ethics and graduated in the Master's programme 'Medicine, Ethics and Law'. Currently he is working on a PhD thesis about the ethical aspects of egg donation and especially on the arguments regarding the ban on it.

Douglas Husak is distinguished professor of philosophy at Rutgers University, New Brunswick. He holds both a PhD and a JD from Ohio State University, and his main interest is in legal philosophy. His specific interests are in the area of criminalization and the moral limits of the criminal sanction. He has published numerous books, chapters in books and articles about legal philosophy, most recently *The Philosophy of Criminal Law: Selected Essays* (2011) and *Overcriminalization* (2008). He is editor in chief of *Criminal Law and Philosophy* and a past editor in chief of *Law and Philosophy*.

Stefan Huster is professor at the Ruhr University of Bochum and chair of public law, social and health law and philosophy of law. He is, among holding numerous posts and memberships, coeditor of the journal *Medizinrecht* (MedR), coeditor of the book series *Bochumer Schriften zum Sozial-und Gesundheitsrecht* (Baden Baden), and member of the editorial board of the journal *Rationality, Markets, and Morals*. His main areas of research are constitutional law and ethics in public and social law, as well as in medicine. Another area of interest is religion and the neutrality of the state.

Roxanna Lynch currently works as a research assistant at the University of Bristol on a project that examines the ethical issues associated with dialysis in patients with dementia. She holds a PhD from Swansea University. Her thesis focussed on an analysis of the concept of care and an application of her original definition of that concept to various issues within bioethics.

André Martens graduated at the University of Hamburg. His MA dissertation was concerned with objectivist and constructivist accounts in the philosophy of psychiatry. In 2010/2011 he spent an academic year as a visiting student at the University of Oxford including tutorials in medical ethics, meta-ethics and the philosophy of psychiatry. He is currently finishing a BSc in Psychology.

Norbert Paulo is a junior faculty member [Universitätsassistent] in legal and social philosophy at the University of Salzburg, Austria. He recently completed his dissertation, 'Methods in Applied Ethics – A View from Legal Theory', at Hamburg University's Department of Philosophy. From August 2012 to January 2013, he was a visiting researcher at the Kennedy Institute of Ethics in Washington, D.C. Until 2009 he studied law and philosophy at Hamburg University and the University of East Anglia in Norwich. His research interests include ethics, applied ethics, jurisprudence, political philosophy and practical reasoning.

Bettina Schöne-Seifert is a professor of medical ethics at the University of Münster. She was a member of the German National Ethics Council and has been a member in several important scientific committees. Her areas of specialization are numerous and include ethical problems in modern medicine such as reproductive medicine, assisted suicide, research ethics, distributive justice in the health-care system, stem cell research, enhancement and ethics in psychiatry. Other areas of her research are theoretical questions and questions of justification in ethics, as well as ethical problems in relation to anthropology and neurobiology.

Thomas Schramme is professor of philosophy at the University of Hamburg. His main research interests are in political philosophy, ethics and the philosophy of medicine. He has published in journals such as *Ethical Theory and Moral Practice*, *Environmental Politics, and Bioethics*. He is currently writing a monograph on paternalism.

Kristin Voigt is an assistant professor at McGill University, jointly appointed in the Department of Philosophy and the Institute for Health and Social Policy.

Her research interests include egalitarian approaches to social justice and their implications for social policy, particularly education and health policy, conceptions of health inequality, the use of incentives to improve health outcomes, (childhood) obesity and smoking and tobacco control.

James Wilson is senior lecturer in the Department of Philosophy and director of the Centre for Philosophy, Justice and Health at University College London. He is also a joint coordinator of the International Association of Bioethics' Philosophy and Bioethics Network (INPAB). His main research interests are in public health ethics, the philosophy of intellectual property and political philosophy. Much of his work has focused on how philosophy can and should inform public policy and regulations in health.

Chapter 1
Introduction

Thomas Schramme

Paternalism can be briefly described as interference with choices or actions, which is targeted against the will but for the own good of the person interfered with. For a while, paternalism in health care was not only a common practice but also a major topic in the ethics of health care. Indeed, medical paternalism was arguably one of the main drivers of medical ethics, more specifically of the notion of respect for patient autonomy. In the early days of bioethics, paternalism, especially paternalism by doctors or medical personnel, was therefore a key issue and was discussed thoroughly (for fairly recent monographs, see Häyry 1991; Tännsjö 1999; cf. also Nys et al. 2007). This debate has significantly decreased in intensity, not least because the principle of respect for autonomy has prevailed and accordingly paternalistic measures within medicine have lost currency. Indeed, Western medicine today seems to be a spearhead of an anti-paternalist stance, where patients, as long as they are deemed competent, may do or allow to have done to themselves all kinds of grave harm, including their own death. Insofar the debate on the normative status of paternalism in medicine has been replaced by a debate on the proper understanding of patients' competence to decide and to consent (cf. Donnelly 2010). As a general rule, competent patients' decisions may not be overridden for their own good, whereas ignoring the will or desires of incompetents may be justifiable. In terms of the philosophical debate this means that "hard" paternalism (regarding competent patients) is never justified, whereas "soft" paternalism (regarding incompetents) is often justified (for these distinctions, see Feinberg 1986, 12ff.). All this seems straightforward.

However, more recent developments in health care have led to new issues regarding paternalism, which are still to be explored. There is for instance the increasing significance of public health measures. This area poses different issues than the well-known discussion about interpersonal paternalism (Callahan 2000; Coggon

T. Schramme (✉)
Department of Philosophy, Hamburg University, Hamburg, Germany
e-mail: Thomas.Schramme@uni-hamburg.de

© Springer International Publishing Switzerland 2015
T. Schramme (ed.), *New Perspectives on Paternalism and Health Care*, Library of Ethics and Applied Philosophy 35, DOI 10.1007/978-3-319-17960-5_1

2012). Public health policies concern most or all citizens, and interventions into choices are often aiming at the precursors of actual actions and are therefore less tangible. For instance, in case of so-called "nudges", paternalistic intervention may take the form of incentives for healthy behavior, or of changing the environment so that people are more likely to make "good" choices. Then there is the provision and even commercialization of medical services, such as assisted reproduction or assisted suicide. These lead to possible new forms of paternalism, where the party interfered with might not be the beneficiary. Whether this so-called "impure" or "indirect" paternalism is normatively less problematic than traditional interpersonal medical paternalism is still to be seen. Also, paternalism in psychiatry has not been targeted within the scholarly debate to the same degree as in general medicine. This is unfortunate, because psychiatry and psychiatric patients are special in several significant respects, most importantly in requiring a more complex account of the capacity to consent, and of autonomy more generally. It is certainly wrong to assume that psychiatric patients lack autonomy simply in virtue of their illness, hence it is not always straightforward which cases in psychiatric practice are instances of hard, as opposed to soft, paternalism. Finally, there are some more theoretical aspects of paternalism in health care that have not been on top of the agenda, but require thorough consideration. Some such issues relate to legal concerns, especially the criminalization of behavior (Husak 2008; Von Hirsch and Simester 2011). As already mentioned, legal paternalism poses other normative issues than interpersonal paternalism, not only because of its more broad-brush approach to influence choices, but also because the agent of such an intervention is not individual medical personnel but the state. The tasks and legitimate remits of a government are certainly different from those of an individual doctor, hence state paternalism in aid of medical aims poses new normative questions. In light of these new developments in the practice of health care it is not only advisable to develop new perspectives on paternalism and health care but also to revisit the very concept of paternalism and entrenched normative stances such as a liberal anti-paternalism. This volume therefore makes an effort in moving forward the debate on new and fairly unnoticed issues regarding paternalism in different areas of health care.

If we keep in mind the quite different areas of health care that have already been mentioned it seems less likely that we will end up with a straightforward general normative position, for instance that hard paternalism can never be justified. Indeed, to ask what is generally wrong with paternalism seems misguided. Different answers might be possible in different contexts, especially relative to various intervening institutions and their proper responsibilities or to different forms of intervention, for instance incentives and disincentives as opposed to legal duties and bans (cf. Grant 2012). Nevertheless it seems to be an important task to find out as to why we generally see a burden of justification on the side of the party willing to intervene paternalistically. This is the task of a (context-sensitive) justification of particular paternalistic measures. But before we take a closer look at the issue of justification of paternalism, it might be helpful to sort out, in more detail, what paternalism is and which forms it might take. Once we have a clearer understanding of the notion of paternalism and its possible justifications, we are in a better position to assess concrete areas of health care and policy, where paternalistic measures are considered or already practiced.

The remainder of this introduction hence follows the structure of the book and it briefly introduces the different sections. In virtue of pursuing this task it also summarizes the contents of the individual essays.

1.1 Paternalism and Anti-paternalism: Conceptual and Theoretical Issues

It is important, first of all, to be clear about the difference between conceptual and justificatory issues. Since paternalism is such a contested practice, at least in modern liberal states that highly value individual liberty, sometimes scholars in the debate on paternalism, and practitioners as well, tend to transfer normative issues into conceptual ones. But we cannot solve our normative disagreements by language policing. It doesn't matter to a hard-nosed opponent of legal paternalism, for instance, if politicians deem certain bans, say on drug consumption, a matter of social security, and hence claim not to be paternalists, in virtue of not aiming at the individual good of people interfered with.

Even though the important questions about the normative status of different forms of paternalism are not to be decided by conceptual analysis, it is still helpful to sort out the concept of paternalism, if only to be more cognizant about the issues at hand (cf. also Kleinig 1984; VanDeVeer 1986; Kultgen 1995). Gerald Dworkin and Dominik Düber therefore aim at a coherent definition of the concept of paternalism. Dworkin first distinguishes between dimensions and normative constraints before aiming at a definition of paternalism. One dimension of paternalism is the object to which the predicate "paternalistic" is applied. Individual acts or policies can be paternalistic, but also motives and persons, or even institutions. There are also differentiations possible regarding motives, outcomes, and reasons of actions. The normative constraints Dworkin discusses are, for instance, whether a paternalistic intervention necessarily requires a violation of autonomy of the person interfered with – where autonomy is here understood as a right. He also delineates acts and omissions, and acting against the consent of someone versus acting without considering their consent. Obviously, how we understand the concept of paternalism may pre-empt normative arguments, and it is therefore important to be clear about these normative constraints and how they themselves are laid out.

Dworkin has himself put forward a definition of the concept of paternalism that has become very influential. It was published in the Stanford Encyclopedia of Philosophy, which is available on the World Wide Web. He reiterates it in his contribution to this volume, yet he also lists other definitions from the literature that show the contested nature of the concept.

X acts paternalistically towards Y by doing (omitting) Z if and only if:

(1) Z (or its omission) interferes with the liberty or autonomy of Y
(2) X does so without the consent of Y
(3) X does so only because X believes Z will improve the welfare of Y (where this includes preventing his welfare from diminishing), or in some way promote the interests, values, or good of Y. (Dworkin (Chap. 2), this volume)

Düber critically engages with Dworkin's analysis. In doing so he adds content to the three criteria mentioned in the definition and ends up in general agreement with Dworkin, but stresses that coercion should be understood in a broad sense, as including compulsion that is not mediated by the will of a person interfered with and hence can incorporate cases where something merely happens to a person. This is different from other cases of coercion that are based on threats and other influences on the will, which do leave the person with a choice, if occasionally only one rational choice. Both kinds of coercion, which have in the tradition been called *vis absoluta* and *vis compulsiva*, ought to be included in a morally neutral definition of paternalism, according to Düber. The criterion of lack of consent should be interpreted in a narrow sense, as focusing only on actual consent. Finally, the benevolence criterion is supposed to refer to the intentions of intervening persons, not to the outcome of their actions.

Anti-paternalism has been an influential position, especially in the liberal camp. When discussing the normative justification of paternalism, anti-paternalists consider certain considerations or facts to be decisive and others not to count as reasons in a reasoned exchange. Especially principled aspects, such as rights of people, prepare the normative ground for such a debate about the normative status of paternalism, and can therefore work like filters. Kalle Grill scrutinizes this metatheoretical aspect. He compares the filter approach to a more dominant action-focused approach, which rejects paternalism on grounds of paternalistic actions bearing certain features that are found objectionable. His examples for such an action-focused viewpoint are especially the theories put forward by Seana Shiffrin and Peter de Marneffe. Grill points out difficulties of the latter approach that an alternative filter approach can avoid. A major problem for the action-focused approach is the need to sort out actions into paternalistic and non-paternalistic ones. This is a very difficult task, as we have already seen when discussing the wide variety of definitions of the concept of paternalism. The task is not made easier by the mixed motives of intervening parties we often find in reality. The filter approach to anti-paternalism, in contrast, simply needs to deny that the good of a person should count as a reason for interfering with the person's liberty. Although this substantive element needs to be justified, of course, it can then in a simple, straightforward way deny certain paternalistic justifications.

1.2   Justifying and Rejecting Paternalism

People can be in favor of paternalism in certain areas, such as road safety, but opposed to paternalism in other areas. They can also be in favor of paternalism for many different reasons. They might believe that human beings are prone to making stupid decisions, which should generally be prevented; they might also hold that we are not equally good at making wise decisions and that therefore paternalism might prevent significant inequalities resulting from individual choices. It is also sometimes assumed that paternalistic measures are justified if the person interfered with would have consented to such an intervention, had she been fully competent.

These variable aspects when discussing the normative status of paternalism should make us pause in thinking that paternalism can be justified or rejected *tout court*. It obviously depends on the case and context we are considering, and its circumstances. Still, it does not seem difficult to offer a provisional argument against any form of paternalism that might at least shove the burden of justification on the side of the paternalist: People want to make their own choices. This is, because they can express their individuality, which includes deep value commitments etc., in certain choices. This value of one's own choice is not merely extrinsic in that it is a requirement of desired states of the world. If that were the case, we would often not want to make these choices ourselves, but leave it to relevant experts. But even where we fail in achieving what we wanted and might end up in dire straits because of our bad choices (relative to our ends), we still often insist that it was valuable choice, because it was our choice and we have expressed ourselves in that choice, however flawed it was. Obviously, making stupid decisions is part and parcel of the way human beings determine themselves. Indeed, this is what makes them human. So paternalism is prima facie bad because it undermines human individuality.

Douglas Husak takes his cue from a relatively clear case of justified paternalism. This is the case of a father interfering with his young son's choice of ice cream over vegetables. He uses this prototypical case to sort out different criteria that might be referred to when justifying paternalistic intervention: The severity of the intrusion, which is minor in the case at hand; the aim of the intervention, which is supposed to be valuable; the means chosen, which ought to be effective; the level of competence to decide of the person interfered with and finally, the relationship of the involved parties. These criteria help us to understand better why certain forms of paternalism, especially criminalization of behavior and interference with competent adults, are so hard to justify.

Husak then considers whether consent of a person might deem an otherwise paternalistic intervention non-paternalistic. He believes that persons who consent to an intervention are not treated paternalistically. But this claim can be challenged and Husak therefore scrutinizes the idea of consent in relation to paternalistic intervention. For instance, consent may be given at different times, before or after an intervention. Future consent has often been deemed a valid reason for paternalistic action, but according to Husak it is flawed, for instance because it is unclear whether we can at all consent retrospectively. We might welcome certain consequences, so much is true, but that does not establish consent to an event that happened in the past. Also, in case of prior consent, such as living wills, the current expressions of consent are decisive, according to Husak, as people can change their minds. To be sure, it might be justifiable to treat a person paternalistically, who has before consented to this intervention, but it is then not the consent that is a reason for acting paternalistically. So again consent does not seem to be playing any role in justifying paternalism. Since this result seems counterintuitive, even to Husak himself, he considers other, more indirect, roles for consent, for instance because it might bear evidence as to whether the other mentioned criteria of justification are fulfilled or not. In general, paternalism is justified when it is reasonable, and the criteria should guide us in achieving this aim.

Kristin Voigt examines a particular argument offered in favor of paternalism that was provided by Richard Arneson. It is claimed that paternalistic intervention can prevent poor decision-makers from making choices that would put them in a disadvantaged position relative to more able choosers. Hence the rationale for paternalism in this case is egalitarian. Egalitarianism is a common position in political philosophy and hence would offer a springboard for many liberals, who are otherwise often opposed to paternalism. Voigt criticizes this argument on several accounts, for instance she takes it to task for providing a clearer distinction between voluntary and involuntary choices. For this, we would need some information as to how far decision-making capacities are themselves due to bad luck or unfair circumstances, and hence people cannot be held responsible for them. Another concern Voigt raises is a relational aspect: By assuming impaired capacity for choosing prudently – Arneson even uses the notion "prudential disabilities" – we might show disrespect to people interfered with.

The egalitarian argument in favor of paternalism focuses on distributional effects of people's choices and offers a neglected consideration when discussing the normative status of paternalism. It is easily applicable especially to issues of public health, where health inequalities are often aggravated by attempts to steer behavior in the desired direction without intervening too drastically, for instance by introducing incentives. Often, the aimed at healthy behavior is picked up by people who are already members of advantaged socio-economic groups. So egalitarian concerns can add to the relevant arguments when aiming at a justification of paternalism, although they can obviously not be sufficient.

Thomas Schramme discusses a peculiar form of paternalism that has been neglected in the debate. When someone is prevented from doing something, occasionally the reason might be to prevent harm to others. Now, this would not yet be an example of paternalism, but of any harm prevention that is perfectly in line with liberal concerns. But if the putative harm of one party is actually desired by the person whose welfare is under consideration, for instance when the person wants to be killed by another person and cannot do so herself, then the two parties are somehow linked in their interests. Seen in this way, it becomes more obvious that we have both elements of paternalism, infringement with liberty and providing benefit, applied to a kind of team, if not in just one person. These multiple-party cases are called indirect paternalism (sometimes impure paternalism). It is arguable whether these raise different normative concerns than "normal" cases of paternalism. Schramme indeed argues that indirect paternalism seems easier to justify, mainly because we normally do not have an entitlement to having a service we seek provided. So to prevent a person A to help another person B in harming B, even though B wants to be helped by A in carrying out this harm, does not seem to be as straightforwardly problematic as to prevent B directly from self-harming. Yet, as Schramme points out, there are certain conditions that undermine the justification of indirect paternalism after all. One is to do with the contested nature of what is actually to be considered as an instance of harm. Another aspect is the actual need for the sought service. If people cannot achieve ends that are not intrinsically immoral or otherwise defective on their own, but require the support of others, then they are not really free – in the sense of being able – to carry out the desired action. If these conditions are met, Schramme argues, then we should allow serviced self-harm and prevent indirect paternalism.

Roxanna Lynch considers whether there is a tension between a caring relationship and hard paternalism. In the actual practice of health care it is not easy to consistently hold a principled anti-paternalist stance, because the situations are complex and evaluation of the case may strongly depend on the context. In assessing paternalism in caring relationships she focuses specifically on aspects of communication between carer and the recipient of care. Lynch considers threats to caring relationships, such as epistemic injustices and imbalance of power. Epistemic injustice occurs, for instance, when a patient's testimonial credibility is undermined. Imbalance of power can more easily occur if there is a known increased vulnerability of either party. In general, paternalism does not seem special in this context by posing unique threats to caring relationships. Rather, it adds further risks of undermining them, because it may prevent good communication. Lynch identifies a specific danger in case of vulnerable persons, who often lack confidence in opposing paternalistic intentions. Still, this is not a principled argument against paternalism but an attempt to reason that we ought to be wary of negative consequences of paternalism on caring relationships.

A more principled argument against paternalism is put forward by Norbert Paulo. He stresses the importance of individual rights, which seem to restrict the potential justifications of paternalistic intervention. Hard paternalism, according to Paulo, undermines the very idea of having rights. Therefore it cannot simply be assumed that our reasons for hard paternalism might be balanced against the right to individual liberty of the person who is interfered with, at least provided our usual understanding of what a right is. Only rights and interests of others can limit rights, but these issues are not forthcoming in the cases of preventing harm to the person interfered with. Paulo also introduces the common philosophical underpinnings of rights, the will theory and the interest theory. He then focuses more on an applied context and shows that principlism, the well-known theory of medical ethics developed by Tom Beauchamp and James Childress, is incompatible with either of these rights theories. In sum, his perspective on rights as structural limits to certain intrusions leads him to a firm anti-paternalist stance.

1.3 Paternalism in Psychiatry and Psychotherapy

The debate on paternalism in psychiatry has been fuelled in the last few years by several legal and political decisions in several countries. For instance, the presumption of capacity, as included in the Mental Capacity Act 2005 in the United Kingdom, has led to a new way of perceiving the decision making capacities of persons with a mental illness. For a long time, there was in psychiatric practice and often in theoretical debate as well rather a contrary presumption, which contributed to the already mentioned dearth of scholarly contributions to the ethics of psychiatric paternalism. In Germany recent court rulings introduced a legal distinction between detainment and medical treatment. According to these decisions, psychiatric patients may not be treated against their will, though they may be detained. The purpose was to address the situation of forensic patients, who might pose a threat to others, but the rulings applied to other psychiatric patients as well, so treatment against the will was also generally ruled out in case of possible self-harm. Hence, a particular form of paternalism, which is based on the therapeutic intent of not simply preventing harm but also fighting its causes, was effectively made subject to extremely high legal standards. All this has led to an increased interest in a debate on psychiatric paternalism.

Bettina Schöne-Seifert sorts out the different aspects of paternalism as they apply to psychiatry. She explicitly mentions that soft (sometimes called weak) paternalism can be problematic, even though the person interfered with is, in these cases, not capable to make autonomous decisions. The problematic nature of these cases is often overlooked, as the main ethical problem with paternalism is identified in its overriding autonomous decisions. It seems that where there is no such capacity, it is justified to disregard the decisions actually made by patients. But this is too casual, particularly in the context of psychiatry, because psychiatric patients are in a special situation, first because they do not generally lack capacity to decide autonomously, like small children or comatose patients, but fluctuate in regard to the necessary capacities. Second, they are usually in a very vulnerable position in relation to their caretakers, and any overriding of their will, however defective it may be at times, undermines trust in the benevolent motivation of psychiatric personnel.

Schöne-Seifert discusses several attempts to justify weak paternalism and thereby focuses especially on the notion of autonomy, as it is necessary to clarify what we mean by it and where to locate the threshold of necessary abilities. After all, the delineation between autonomous and heteronomous patients also marks the difference between hard and soft paternalism. For her, there are a number of abilities that together secure the required competence for autonomous action, the abilities to understand, process information, evaluate and to guide oneself. Schöne-Seifert's more general considerations are finally transferred to concrete psychiatric cases. She also makes suggestions for future changes and topics for debate in psychiatric practice.

Charlotte Blease addresses the context of psychotherapy, more specifically psychodynamic psychotherapy. An interesting feature of psychotherapy is that at least partly its therapeutic success seems to depend on the placebo effect, i.e. roughly the therapeutic effect that is based on the simple unjustified belief that a certain procedure will help. One important aspect of paternalism is that an intervener assumes (comparatively better) knowledge about the good for another person – otherwise there would be no point in intervening with a decision or action. But it would be ethically unjustified to simply keep a person in the dark about the facts of a situation. Better knowledge about the good must not be based on misinformation of another person. So a psychotherapist has to release all relevant information regarding a therapeutic option in order to obtain valid consent. Yet, information about the placebo effect in psychotherapy is not standardly given. Indeed, the very effect of placebo seems to rely on the ignorance of the client or patient regarding its status, so to use the placebo effect necessarily seems to require deception.

In pursuing this ethical problem, Blease carefully analyzes the level of deception that is involved in placebo and also its significance. She also discusses the concept of placebo and its effects. According to her analysis several aspects of psychotherapy are placebogenic, i.e. causing the placebo effect. The disclosure of the status of placebo in psychotherapeutic practice therefore seems to be required to obtain proper informed consent. Otherwise it might be deemed a case of paternalistically withholding information.

André Martens examines a very important case of psychiatric disorder that poses difficult problems for the justification of paternalism. Patients with anorexia nervosa have peculiar beliefs about their physical appearance and weight. Their desire to be thin might go so far that their life is endangered. In these situations, there seems to be a good case for compulsory treatment. A major issue regarding these patients, which is discussed thoroughly by Martens, is their decision-making competence. On the one hand, they seem to make factual errors, on the other hand they seem to have all the intellectual capacities usually regarded necessary for autonomous decision-making. Martens therefore widens the focus to emotional and evaluative aspects of such capacity. He refers to the notion of pathological values that was introduced by Jacinta Tan, and he adds the idea of compulsive values, i.e. values that are due to compulsion and not freely chosen. Here, not the process of value genesis but its form is important, as it is immune to revision and checks against evidence to the contrary. Yet, Martens is also explicit in denying this to be already enough to justify soft paternalism, as the latter requires knowledge of superior values, which is not secured merely by rejecting compulsory values.

As can be seen from the three papers in this section, psychiatry and psychotherapy pose contexts for the debate on paternalism that are quite rich and complex. It will be interesting to see whether in the future there will be a more concerted effort in medical ethics and the ethics of psychiatry to clarify these issues.

1.4 Paternalism and Public Health

Public Heath has become increasingly important as a way to address disease in the population. Curative medicine tackles existing health problems and it might or might not be successful in restoring health. Medicine might also aim at preventing disease. Yet, medicine traditionally conceived aims at the individual organism, even in its preventive efforts. Public health instead focuses on all kinds of aspects that might cause and influence the development of disease, especially the so-called social determinants of health. These might be living conditions, environmental pollution, lifestyles, the level of stress experiences or indeed any other causal contribution to changes in human organisms. In virtue of its wide remit, public health necessarily leaves the traditional borders of medicine behind and becomes an interdisciplinary attempt to increase population health. In addition, it also focuses on the level of health inequality in society, as it can be shown that specific socioeconomic groups suffer significantly more often from particular diseases. This fact seems to point at undeserved disadvantages, i.e. according to many theories of justice, to cases of injustice. The means of influencing circumstances and lifestyles that public health uses are often certain policies and occasionally even legal measures. Obviously these measures can be paternalistic, and they can vary in degree regarding their penetration into the liberty of citizens.

In recent years, a new theory of paternalism has made its name and actually gained considerable impact on real politics. "Libertarian paternalism" tries to bring people to do what is good for them by intentionally designing the "choice architecture", i.e. the environment that forms the background of citizens' decisions. This might include setting defaults in specific insurances to opt-out instead of opt-in, or to lay out healthy food in a certain way so that it will be more likely chosen in cafeterias. All this is done in a way that allows people still to choose what is deemed less optimal, e.g. the unhealthy option. People are not coerced, but only "nudged" in the right direction (see Thaler and Sunstein 2008; cf. Conly 2012; Rebonato 2012; Coons and Weber 2013; White 2013).

Part of the justification for these interventions is that people are bad choosers due to irrational influences, such as biases or hyperbolic discounting, which involves underestimating the significance of the future. So nudges are sometimes supposed to bring about the choice that people actually want, but cannot make, because of their defective decision-making. The mentioned interventions have obvious applications to public health, and indeed some countries, such as the UK and the USA, have begun to use the underlying model to influence choices and even to build special agencies for this purpose. The UK, for instance, now has a "Behavioural Insights Team", also called "Nudge Unit".

James Wilson aims to ease the worries of the anti-paternalists as regards health policy. He makes clear that we should not call policies but justifications for policies paternalistic, so that we do not confuse conceptual and justificatory issues and avoid potential circularity. He also points out that all public policy includes some level of interference with liberty and lack of individual consent. So the related worry of the anti-paternalist should be consistently targeted at all public policy, which however would undermine its reasonableness. This argument establishes incongruence between public policy and the often-discussed problems of interpersonal paternalism, hence the debate on this form of paternalism should not simply be transferred to the political realm. In conclusion he recommends changing the focus from the terminological issue as to whether a particular policy is paternalistic to its justification. He regards a wide range of such policies, when they aim at public health, as justified and rejects a number of objections to the contrary.

Stefan Huster is concerned with the relation of individual responsibility and paternalism in health law. His examples are mostly drawn from the German legislature and legal practice. He stresses the high level of attention that is given to healthcare, in contrast to preventive medicine or public health. This does not seem justified, both for reasons of opportunity costs and of health justice. The issue of paternalism is related to individual responsibility insofar as an anti-paternalist stance apparently involves respecting individual freedom even where it leads to bad outcomes. For instance, if a fully informed person voluntarily leads an unhealthy lifestyle, we might think he or she should be held responsible for the results. Yet despite some level of rhetoric strengthening of the idea of individual responsibility in insurance law, there are only limited examples of application of using it as a criterion for distributing resources. In contrast, in public health, the rhetoric is different. People are not deemed responsible for their health but it is regarded to be an outcome of influences by social determinants. So citizens do not seem to be taken seriously enough. According to Huster, the justification of public health policies seems to rely, in the final analysis, on a concern of the state for the individual health of citizens, which would obviously lead to possible worries about state paternalism and is indeed tried to avoid by the German legislature at all costs in relevant cases, such as the regulation of smoking in public places. Huster finally considers solutions as to how we might be able to live with the two ways of thinking about individual responsibility in health care and public health. Still, we need to be more consistent in our perspective on individual responsibility. For instance, if we criticize state "healthism" and reject public health measures, we would also need to put greater weight on the consequences of individual autonomous choices in health care, and we would also need to live up to resulting social differences.

Jessica Flanigan asks whether social costs that are implied by public health problems can be used to support a justification of coercive paternalism. Such a rationale might put it into a more straightforward liberal perspective, as harm to others is a common justification for intervening into choices. Indeed, it might be added, mixed motives are quite common in real life examples of paternalism. However, Flanigan rejects social costs arguments on grounds of a thorough discussion of the reasons for the provision of healthcare. Either there is an entitlement to healthcare, which does not allow for an infringement, even where healthcare needs are caused by lifestyle choices, or healthcare is a simple benefit that we don't have a justified claim on. But both foundations cannot lead to a justification of paternalistic public health intervention: If healthcare is a right then we may not make it conditional on individual lifestyles or on the frequency and volume of its use. If healthcare is simply optional then it might be legitimate to make it conditional, but there would still be no justification of public health intervention either, because such a benefit might be refused by people who oppose paternalistic intervention. Obviously, this might lead to other unwanted consequences, such as letting people suffer for their choices who had opted out of the conditional provision of benefits, but this would only possibly undermine the idea of deeming healthcare a benefit instead of an entitlement. Flanigan concludes that paternalistic arguments need another foundation and cannot be supported by considerations regarding the costs of unhealthy lifestyles to others.

Lorenzo Del Savio focuses on the determinants of choice. Anti-paternalism – be it of a deontological or a consequentialist bent – usually deems autonomous choices valuable. Yet in fact people often make flawed decisions, and therefore even a liberty-based approach needs to take into account some welfare-related aspects of choice. This applies, as Del Savio posits, especially to broadly utilitarian approaches, such as John Stuart Mill's. Indeed, what liberals of any theoretical background seem to be interested in is not just choice per se, but in valuable choice. This, again, seems to require a presumption of some level of individual control in making decisions. After thoroughly reviewing the empirical data on impaired choice making, Del Savio concludes that it undermines a simple liberty model. Similarly, the ubiquity and unavoidability of at least some level of intervention puts a strong anti-paternalism into question. In consequence Del Savio sees a case for justifying paternalistic public health intervention.

Public health paternalism will very likely be an increasing matter of concern in the future, as population health becomes a more important issue on the political agenda, and as more interventions are considered. It is an intricate topic, as its normative assessment requires factual knowledge about the psychology of human decision-making, and also because some aspects of it do not align easily with common models of paternalism in the ethics literature. For instance, the intentional arrangement of the circumstances of individual choice by public health institutions for purposes of promoting health is not as clear-cut a paternalistic intervention as a legal ban on certain choices. So the various means of public health intervention call for a complex model of normative assessment, including different perspectives, such as political philosophy and jurisprudence.

1.5   Paternalism and Reproductive Medicine

Reproductive medicine is, maybe like no other area of health care, concerned with intimate and deep commitments of people, which form the basis of their decisions and values. If making one's own choices is valuable then making one's own choices in matters of reproduction is indeed of vital importance. It is maybe no surprise, then, that the issue of "reproductive autonomy" has been at the forefront of several debates in biomedical ethics within the last decades. This concern has expanded from the initial main aspect of women's right to abortion to several reproductive technologies, such as in-vitro-fertilization or preimplantation genetic diagnosis. Access to these technologies is now seen by many people as being implied by the right to reproductive autonomy. Now obviously, in virtue of its aspect of concerning issues that we hold dear to our hearts and of its being at the same time of some contested nature, it allows for some paternalistic concerns as well. The welfare aspect of paternalistic intervention might relate to the (neglected) interests of the willing users of reproductive technology, or indirectly to the interests of other parties, most notably the developing fetus.

Diana Aurenque starts with a case, which has caused some debate within medical ethics, where a deaf couple decided to use reproductive technology to choose a fertilized egg with the genetic disposition for deafness to ensure that their child will be raised to their own cultural setting. If a ban on such choices is considered, this might be seen as an instance of indirect paternalism (although Aurenque does not use this terminology), as the embryo needs the assistance of its parents to develop into a child, yet they are prevented from choosing it on grounds of protecting the interests of the child. An interesting additional feature of this case is that in virtue of its undeveloped nature embryos cannot make choices themselves, hence the party that is supposed to benefit from a ban is incompetent and the case could be considered a matter of soft paternalism. Yet, as Aurenque discusses, it is not straightforward to even see deafness as a matter of impaired welfare that a person needs to be protected against.

Aurenque also assumes that paternalism and autonomy are not opposing ideas, but that paternalism indeed allows for promoting autonomy. She pursues this thesis on the basis of Emmanuel Lévinas's idea of "paternity". His account calls for taking responsibility for other persons, but also for seeing our limits in understanding other persons. This might provide a kind of bridge between respecting autonomy and accepting justified occasions for paternalism.

Clemens Heyder takes his cue from the case of the German Embryo Protection Act, which prohibits certain reproductive technologies and related practices, such as egg donation. He discusses such a ban from the perspective of the liberal framework, developed by John Stuart Mill, which requires a justification on any restriction of individual liberty. Heyder first clarifies the concept of reproductive autonomy and then considers the case of egg donation from the involved perspectives of the parents, the potential child, and society. Altogether he does not see sufficient reasons for intervening into reproduction in this case, since there is no harm involved that might be avoided. In order to avoid this conclusion we would have to give up liberalism as our general framework for issues of social morality, which could obviously ask too much of us.

Surely the papers in this volume will not be sufficient to clarify the several thorny theoretical and practical conundrums surrounding paternalism in different contexts. Hopefully they will nevertheless ignite new research and critical engagements with the proposed arguments.

References

Callahan, Daniel (ed.). 2000. *Promoting healthy behavior: How much freedom? Whose responsibility?* Washington, DC: Georgetown University Press.

Coggon, John. 2012. *What makes health public? A critical evaluation of moral, legal, and political claims in public health.* Cambridge, UK/New York: Cambridge University Press.

Conly, Sarah. 2012. *Against autonomy: Justifying coercive paternalism.* Cambridge: Cambridge University Press.

Coons, Christian, and Michael Weber (eds.). 2013. *Paternalism: Theory and practice.* Cambridge: Cambridge University Press.

Donnelly, Mary. 2010. *Healthcare decision-making and the law: Autonomy, capacity and the limits of liberalism.* Cambridge/New York: Cambridge University Press.

Feinberg, Joel. 1986. *Harm to self,* The Moral Limits of the Criminal Law, vol. 3. New York: Oxford University Press.

Grant, Ruth. 2012. *Strings attached: Untangling the ethics of incentives.* Princeton: Princeton University Press.

Häyry, Heta. 1991. *The limits of medical paternalism.* London/New York: Routledge.

Husak, Douglas. 2008. *Overcriminalization: The limits of the criminal law.* New York: Oxford University Press.

Kleinig, John. 1984. *Paternalism.* Totowa: Rowman & Littlefield.

Kultgen, John. 1995. *Autonomy and Intervention: Parentalism in the caring life.* Oxford: Oxford University Press.

Nys, Thomas, Thomas Nys, Yvonne Denier, and Toon Vandevelde (eds.). 2007. *Autonomy & paternalism: Reflections on the theory and practice of health care.* Leuven/Dudley: Peeters.

Rebonato, Riccardo. 2012. *Taking liberties: A critical examination of Libertarian paternalism.* Houndmills/Basingstoke/Hampshire/New York: Palgrave Macmillan.

Tännsjö, Torbjörn. 1999. *Coercive care: The ethics of choice in health and medicine.* London/New York: Routledge.

Thaler, Richard H., and Cass R. Sunstein. 2008. *Nudge: Improving decisions about health, wealth, and happiness.* New Haven: Yale University Press.

VanDeVeer, Donald. 1986. *Paternalistic intervention. The moral bounds on benevolence.* Princeton University Press: Princeton.

Von Hirsch, Andreas, and A.P. Simester. 2011. *Crimes, harms, and wrongs: On the principles of criminalisation.* Oxford: Hart Publishing.

White, Mark D. 2013. *The manipulation of choice: Ethics and libertarian paternalism.* New York: Palgrave Macmillan.

Part I
Paternalism and Anti-paternalism: Conceptual and Theoretical Issues

Chapter 2
Defining Paternalism

Gerald Dworkin

> *There are three concepts all of them vague. Imagine 3 solid pieces of stone. You pick them up, fit them together and now find they make a ball. What you've now got tells you something about the 3 shapes. Now consider you have 3 balls of, or lumps of soft mud or putty – formless. Now you put them together and mold out of them a ball.*
>
> (Wittgenstein, from Bouwsma, *Wittgenstein: Conversations, 1949–1951*)
>
> *The merit of any definition . . . depends upon the soundness of the theory that results; by itself, a definition cannot settle any fundamental question.*
>
> (Rawls, *A Theory of Justice*)

Any definition of a concept is subject to various criteria for a good definition in the context at hand. Unless we are simply stipulating how we shall be using the word – and even then questions will arise about why we picked that word to use for this stipulation – there will be some, usually implicit, ideas of what makes for a good definition. In addition to trivial ones – such as consistency – there will be a set of problems that the definition will be used to clarify or, if possible, resolve. There will be a set of constraints – weak or strong – on how the word is currently being used. There will be a context – perhaps one of personal ethics or perhaps one of current law – in which the concept finds a place. There will be some conceptual or normative issues that will be used to assess the usefulness or correctness of the definition.

There may be stipulated criteria, e.g., that the concept should not settle some particular normative matter, thereby avoiding what Hart has called the "definitional stop." For example, if one defines terrorism as the morally illegitimate use of violence on innocent persons then the question of whether terrorism is ever morally legitimate has been settled by the definition. Sometimes, this is not an objection to

G. Dworkin (✉)
Department of Philosophy, University of California Davis, Davis, CA, USA
e-mail: gdworkin@ucdavis.edu

© Springer International Publishing Switzerland 2015

T. Schramme (ed.), *New Perspectives on Paternalism and Health Care*, Library of Ethics and Applied Philosophy 35, DOI 10.1007/978-3-319-17960-5_2

a definition. If we define "murder" as the wrongful taking of human life then although we cannot raise the question of whether murder is ever right, we can raise the question of whether some killing of a human being is murder. I am going to begin by canvassing a wide variety of definitions of paternalism which may have been developed in quite different contexts for quite different purposes. It is helpful both to see how wide the variety is and to see the various dimensions along which the definitions vary.

The first crucial dimension is what the term is predicated of. People can be paternalistic. Reasons can be reasons of paternalism. Motivations can be paternalistic. Institutions can be paternalistic. Acts can be paternalistic. Policies can be paternalistic. It may be that acts are primary in some definitions with the other elements being defined in terms of acts, or the order might be the reverse. But I shall be concerned primarily with the notion of a paternalistic act or the notion of a paternalistic policy.

2.1 Dimensions

2.1.1 Outcomes vs. Motives

A paternalistic act may be defined in terms of the outcomes it produces. If a state enacts legislation requiring boaters to wear life jackets, and if wearing life jackets is beneficial to the interests of boaters, then this is an act of paternalism. The alternative view is that whether an act is paternalistic or not cannot be determined without reference to the reasons for which the state acts. Two acts may have the same outcome, an improvement of B, yet only one counts as paternalistic.

2.1.2 Actual vs. Hypothetical Motives

An act may be defined as paternalistic in terms of the reason for which A acts. If she has more than one reason there is an issue of how to specify the relation between her various reasons and the characterization of the act. If we are considering a piece of legislation which is passed by many voters, with differing reasons, the issue is even more complex. But in both cases it is the actual reasons which must be considered.

The alternative view is that the reasons which count in determining whether an act is paternalistic are the hypothetical reasons which could motivate or justify the act. So a doctor's lying to a patient about his terminal condition is paternalistic because considerations of the patient's welfare would have led to his lying (even if his current act was motivated by a desire to avoid a long discussion after a tiring day).

2.1.3 *Motives vs. Reasons*

As opposed to what explains the act, the motives for which the agent acts, the important question may be whether there are reasons which are sufficient to justify the policy which are of the appropriate kind. The agent may have acted, say, to promote the interests of third-parties; but if there are reasons which refer to B's welfare, and such reasons are sufficient to justify the act, then the act is paternalistic.

2.2 Normative Constraints

2.2.1 *A's Act Must Violate B's Autonomy vs. A's Act Need Not Be a Violation of B's Autonomy*

Consider the case where a husband hides his sleeping pills because he fears that his wife may find them and use them to commit suicide. A definition of paternalism might classify the husband's act as non-paternalistic on the grounds that what A does is not a violation of a sphere of autonomy of his wife. She has no right that he keeps his sleeping pills in clear view. Or, one might consider the act as paternalistic even while conceding this point. In some sense he substitutes his judgment for hers in the belief that his judgment is better than hers. Obviously, interference with autonomy is not a sufficient condition for an act's being paternalistic. I interfere with your autonomy when I steal your bicycle but that is not a case of paternalism. The issue is whether it is a necessary condition. Can I act paternalistically towards B even if I do not in any way violate his autonomy? I speak of "autonomy" rather than, say, "liberty" because paradigm cases of acting paternalistically, such as a doctor lying to his patient, might not be considered infringements of liberty.

2.2.2 *Acts vs. Omissions*

Is paternalism defined only over the range of actions, or can it include failures to act as well? If I push you out of the way of a car to avoid injury, this could be thought to be paternalistic.

Suppose I do not push you out of the way (when I could), envisaging minor bruises and scrapes, so that you will miss a business meeting where I believe you will make a seriously mistaken deal. Is that paternalistic?

2.2.3   Against Your Consent vs. Failing to Consider
Whether You Consent or Not

If I act knowing that you do not (or would not) consent to what I am doing, I act against your consent. If I act not knowing whether you consent (or would consent) or not, I act without your consent. A related question is whether the test is objective or subjective, i.e., whether the issue is whether you have consented (whether or not I know this) or whether I believe you have consented (whether or not you have).

2.2.4   Interests at Stake

The most common definitions of paternalism make reference to the welfare/interest of the person whose autonomy is being limited. The broadest definition I have encountered is that of Seana Shiffrin who defines paternalism in such a way that the beneficiary of the action may be, and may be intended to be, someone other than the person towards whom we are acting paternalistically.[1] I will return to this definition later.

2.2.5   Physical Welfare vs. Moral

The issue here is whether the harm to be avoided is psychological or physical, such as death or torment, or is moral such as being corrupted or degraded. Moral paternalism is to be distinguished from legal moralism. In the latter case the grounds for acting are that the conduct in question is wrong or evil but not that it harms the agent who acts in these ways. It is a distinct, substantive question of whether, for example, if your character is made worse by what you do, you are worse off, i.e., whether your wellbeing is diminished. Some philosophers such as Plato have asserted the truth of this view. Some philosophers such as Feinberg have denied it.

2.3   Definitions

Given the number of these dimensions, and the possibility of combining the dimensions in various ways, there are obviously a large number of definitions that are possible. Obviously not all possibilities have been seriously put forward for acceptance. But the variety is larger than might be thought and I will set out a number of proposed definitions to give the reader an idea of what such a variety might look like.

---

[1] Shiffrin (2000).

2.3.1   Definition A

X acts paternalistically towards Y by doing (or omitting) Z if and only if:

1. Z (or its omission) interferes with the liberty or autonomy of Y
2. X does so without the consent of Y
3. X does so just because doing Z will improve the welfare of Y (where this includes preventing his welfare from diminishing), or in some way promote the interests, values, or good of Y.[2]

2.3.2   Definition B

The second definition is from an economics paper by Sunstein and Thaler.[3] If an employer is "attempting to steer employees' choices in directions that will promote employees' welfare," he is acting paternalistically.[4] Thus, if an employer automatically enrolls an employee in the company's 401 k plan (with the option to withdraw at any time) this counts for them as paternalistic. Elsewhere, Sunstein claims that the anti-paternalist position is "incoherent, simply because there is no way to avoid effects on behavior and choices."[5] This definition does not require any interference with liberty, or coercion, or infringement upon autonomy. If an action has effects on choices, and one intends those effects because they will enhance the welfare of the person being affected, then on their definition the action is paternalistic.

---

[2] See Dworkin (1972). Condition one is the trickiest to capture. Clear cases include threatening, bodily compulsion, lying, withholding information that the person has a right to have, and imposing requirements or conditions. But what about the following case? A father, skeptical about the financial acumen of a child, instead of bequeathing the money directly, gives it to another child with instructions to use it in the best interests of the first child. The first child has no legal claim on the inheritance. There does not seem to be interference with the child's liberty; nor on most conceptions the child's autonomy. Or consider the case of a wife who hides her sleeping pills so that her potentially suicidal husband cannot use them. Her act may satisfy the second and third conditions but what about the first? Does her action limit the liberty or autonomy of her husband? The second condition is supposed to be read as distinct from acting against the consent of an agent. The agent may neither consent nor not consent. He may, for example, be unaware of what is being done to him. There is also the distinct issue of whether one acts not knowing about the consent of the person in question. Suppose the person in fact consents but this is not known to the paternalizer.

[3] Thaler and Sunstein (2003).

[4] Thaler and Sunstein (2003), 177.

[5] Sunstein (2002), 195.

2.3.3 Definition C

X acts paternalistically towards Y by doing (or omitting) Z if and only if:

1. X aims to close an option that would otherwise be open to Y or X chooses for Y in the event that Y is unable to choose for himself
2. X does so, to some extent, in order to promote Y's good.[6]

2.3.4 Definition D

X acts paternalistically in regard to Y to the extent that X, in order to secure Y's good, as an end, imposes upon Y.[7]

2.3.5 Definition E

X acts (hard) paternalistically with regard to Y if:

1. X restricts Y's liberty
2. X does so primarily out of benevolence towards Y
3. X must disregard Y's contemporaneous preferences
4. X must either disregard whether Y engages in the restricted conduct voluntarily or deliberately limits Y's voluntary conduct.[8]

2.3.6 Definition F

Essentially the preceding with (4) replaced by: X's action must be a violation of a moral rule or X recognizes (or should) that his action towards Y needs moral justification.[9]

2.3.7 Definition G

A policy is paternalistic if it limits a person's liberty for her own good, or for the reason that it benefits her or improves her situation in some way. To limit a person's liberty "for her own good" is to limit her liberty for a certain kind of reason: that this policy will promote her welfare or improve her situation in some way.

---

[6] Clarke (2002).
[7] Kleinig (1983).
[8] Pope (2004).
[9] Gert and Culver (1979).

The government acts for a reason in limiting someone's liberty if and only if this policy cannot be fully justified if this reason is not counted in its favor and the government adopts it only because someone in the relevant political process takes or has taken this reason as sufficient to justify it. A policy is paternalistic, then, if it cannot be justified by nonpaternalistic reasons alone, and the government adopts it only because someone in the relevant political process takes some paternalistic reason as sufficient to justify it.[10]

2.3.8 Definition H

My last definition and the one I will concentrate on is that of Seana Shiffrin: X acts paternalistically towards Y by doing Z if:

1. Z is aimed to have an effect on Y or her sphere of legitimate agency
2. that involves the substitution of X's judgment or agency for Y's
3. is directed at Y's own interests or matters that legitimately lie within Y's control
4. undertaken on the grounds that compared to Y's judgment or agency with respect to those interests or other matters, X regards her judgment or agency to be (or as likely to be), in some respect superior to Y's.[11]

The feature that stands out in this definition is the absence of the idea that the action has to be directed at Y's own interests. It may be, but it also may simply concern matters that are legitimately within Y's control.

2.4 An Examination in Detail of One Case

Having seen the variety of ways of defining paternalism, and the various dimensions along which definitions differ, I want to explore this last definition in some detail. My purpose in doing so is to try and see what the nature of the dispute between various rival definitions amounts to. After all, if the whole process were merely stipulative – this is how I propose to use this word – then one would simply present the definition, perhaps point out some of its implications, and go on to the justificatory issues. But proponents do not follow this pattern. They seem to be defending their definitions against rivals. What could such a defense look like?

The most striking part of Definition H is clause (3). By expanding the scope of the justification to include matters other than Y's interests it allows for paternalism which does not provide a gain or avoid a loss to Y.

---

[10] de Marneffe (2006).
[11] Shiffrin (2000).

The first thing to note is that the entire discussion of paternalism takes place in the larger context of a discussion of the Unconscionability Doctrine (UD) in contract law. This doctrine enables a "court to decline to enforce a contract whose terms are seriously one-sided, over-reaching, exploitative, or otherwise manifestly unfair."[12] There is a normative dispute about the use of the doctrine. Liberals tend to favor it as a way of enabling poor people who are taken advantage of to get out of contractual obligations. But opponents of the doctrine, often of a libertarian or conservative bent, object to the doctrine on the grounds that it is an instance of paternalist behavior. Liberals respond, not by denying this characterization, but by arguing that under certain circumstances paternalism is legitimate.

Shiffrin's contribution to this debate is to dispute the common ground between the two sides by denying that the defense of the doctrine need be a paternalist one. She will argue that the characterization of the doctrine as paternalist "reflects some common but misleading thought about paternalism."[13] So her discussion of what paternalism is reflects three prior normative commitments: (1) the legitimacy of a particular legal policy; (2) her acceptance that, at least generally, paternalism is wrong; and (3) an account of what makes it wrong.[14]

To defend the policy she must show, given (2) that it is not an instance of (3) and the way to do this is to show that it is not an instance of paternalism. But to do this she must show that it is not an instance of paternalism in the sense that her opponent in the dispute accepts something as paternalistic. After all, it would be foolish to simply define something as paternalistic only if it is, say, an instance of coercion, argue that a court not upholding a contract is not an instance of coercion, and therefore claim that the UD is not an instance of coercion. For her opponent already concedes that the UD is not an instance of coercion, but argues nevertheless it is an instance of paternalism.[15]

As it turns out, however, Shiffrin doesn't need a new conception of paternalism to argue against the traditionalists. For the key to her normative argument is that she believes that there is a distinct motive which can justify not enforcing unconscionable contracts having nothing to do with protecting one of the parties. She argues that the state has a right not to be complicit in enforcing contracts that it believes to be immoral, because exploitative. Given that the traditionalist believes that only a certain range of motives makes an act paternalistic, and that the desire to not be

---

[12] Shiffrin (2000), 205.

[13] Shiffrin (2000), 207.

[14] For 2 and 3, "I agree with many of its opponents that paternalist doctrines and policies convey a special, generally impermissible, insult to autonomous agents" (Shiffrin 2000).

[15] It might be noted that in the debate as to whether or not Mill's saying that we should not allow the enforcement of contracts for slavery was consistent with his absolutist prohibition of paternalism, some have argued that not enforcing such contracts was not for Mill paternalistic on just such grounds.

complicit is not one of them, they could agree that there is a nonpaternalistic justification for the UD.[16] No new definition of paternalism is required.[17]

So, although her definition of paternalism takes place in the context of a normative disagreement it is, strictly speaking, not required for her normative argument. Nevertheless, the context does affect her discussion of how to characterize the concept because it opens the discussion into a general examination of the role of autonomy, and the relation of that notion to ideas of what she calls "accommodation."[18]

She is methodologically self-conscious about how to argue about competing characterizations of paternalism. She believes that the traditional idea of testing these against our linguistic intuitions is plausible but she thinks that it also should involve our normative intuitions. We want to arrive "at a conception of paternalism that fits and makes sense of our conviction that paternalism matters. That is, it seems worthwhile to assess what is central in our normative reactions to paternalism and to employ a conception of paternalism that complements and makes intelligible our sense of paternalism's normative significance."[19]

More importantly, she specifies this rather abstract idea of "complementing . . . and making intelligible" by saying that she is going to argue for her view, and against those of others, by claiming that alternative conceptions "deploy overly narrow criteria that draw somewhat arbitrary and unmotivated distinctions between cases; that is, they draw distinctions that do not seem to have much normative significance in light of what seems to be the driving force behind our aversion to paternalism."[20]

This is, then, a research program. First, try and spell out why we are concerned about paternalism. What is our interest in it? What is the normative point of classifying some modes of action as paternalistic in the first place? Second, in light of that knowledge formulate a characterization which includes (and excludes) acts on the basis that they are sufficiently similar with respect to the normative point of the notion. Argue against alternative notions because they either include acts which do not seem sufficiently similar, or exclude acts which do seem similar (so that their exclusion seems arbitrary).

Let us look at how the actual argument proceeds. If one looks at her definition the features that stand out as original include the clause which allows the interests of

---

[16] I made a similar point many years ago in arguing that there is a good reason to impose some kind of bright clothing on hunters, not to protect them from being shot by other hunters, but to protect those who might shoot them from the damaging psychological consequences of killing another person.

[17] Shiffrin recognizes this: "I should note, though, that the defense of the unconscionability doctrine that I will pursue does not depend upon my particular characterization of paternalism; it could be deployed with many other characterizations in mind" ("Paternalism, Unconscionability Doctrine, and Accommodation," 212).

[18] This is, very roughly, the idea of when we should tolerate the burdensome, other-regarding conduct of our fellow citizens. One example would be not to discriminate between smokers and non-smokers in setting health care premiums.

[19] Shiffrin (2000), 212.

[20] Shiffrin (2000).

others to be defining of paternalism, the absence of a clause referring to the action being counter to Y's will, the absence of a clause referring to interference with liberty or autonomy, and the presence of a clause referring to Y's sphere of legitimate agency. I do not propose to examine each of these inclusions or exclusions but to look at two of them to give the flavor of the argument.

Let us start with the most startling. Reference to the welfare or interest of Y is no longer a necessary condition. Nor, indeed, as she concedes, is anyone's welfare necessarily at stake. Indeed, from this broad perspective, a paternalist motive need not concern any person's welfare at all. Suppose an interlocutor raises his hand at a talk. He is called upon and just as he haltingly begins to articulate his point, an excited, sympathetic colleague loses self-control and interjects: "Isn't this a better way to put the point?" She goes on to drown him out while cleverly and eloquently articulating his point. She takes over his question because she feels she has a better command of it than he does. I think her taking command over his question for this reason makes her action paternalist, even if her motive is really that she wanted to see the point formulated properly and not that she wanted in particular to help him formulate the point or to make his point understood.[21]

She begins with an example: "Suppose a park ranger has the power to refuse permission to climb a steep, dangerous mountain path . . . Suppose the ranger says, 'Of course, you may take whatever risks you want to with your life, but I refuse permission because you might die and leave your spouse grief-stricken.' Such a refusal also seems paternalist."[22]

I assume the person to whom it seems paternalist is Shiffrin. My own reaction is to think of this as precisely the contrast class to paternalism. But her argument does not reduce to "It is," "It isn't." She even concedes that "most accounts of paternalism do not encompass this sort of behavior in their characterizations" but neither common usage nor what "most" philosophers say is the test.[23] Her argument is the following:

> But, it is unclear why we should draw a bright line here, separating the cases so sharply. For example, both of the cases [where the ranger refuses permission to protect the climber and the case above] involve an effort on the ranger's part to assert her will over a domain in which the ranger does not have (or even assert) legitimate authority on the grounds that her judgment is superior. *Both cases seem to involve the same sort of intrusion into and insult to a person's range of agency* . . . [W]e should have the same sort of normative reaction to the case in which the ranger forbids the climb from concern for the spouse [as to the case when she forbids it from concern for the hiker]. What concerns us about paternalism, narrowly construed, should spark the same concern about these closely related, similarly motivated cases.[24]

---

[21] Shiffrin (2000), 217.

[22] Shiffrin (2000). The example is an odd one because she starts by assuming the ranger has the power (right?) to refuse permission to take the path. But, then, in what sense is the decision within the hiker's legitimate sphere of control? Shiffrin seems to think that the area of control that is being interfered with is how the hiker should treat his wife. But not every way of treating his wife is within his sphere or control. Whether he beats her, for example, is not. Maybe, how he acts to cause her great grief is also not. I elaborate on this below.

[23] Shiffrin (2000).

[24] Shiffrin (2000), 217–218. Italics mine.

The argument here is essentially casuistical. Starting from a central case (acts motivated by concern for Y) we classify another case (concern for Y's spouse) as sufficiently similar so as to warrant being classified in the same way. But similarity requires a metric. Her proposal is to use as the metric whether the act warrants the same kind of normative reaction as the central case, i.e., the insult to Y which stems from X "taking over" some portion of Y's sphere of legitimate agency on the grounds that X's judgment is, in some respect, better than Y's.

This last clause is essential in order to avoid objections such as the following. "Look, if you accept such a broad notion of paternalism then why not count the following as paternalistic? X robs Y of his wallet because X wants the money." But here X does not justify his action by claiming that he knows better how to spend Y's money than Y knows how to spend his money.

It is true that her analysis must classify the following as paternalistic. X robs Y of his wallet because X thinks Y should spend his money on tsunami relief – which X proceeds to do. If one just appeals to intuition then this seems absurd. But her argument is that what is done to Y in this case is sufficiently similar (in the nature of the grievance that Y has against X) to central cases of paternalism (such as X stealing Y's wallet because he thinks that Y is manic and will spend the money unwisely).

How might one oppose a definition such as Shiffrin's? It seems to me there are the following possibilities. One can simply reject the strategy as the wrong one, or mistaken.

This is to argue that even if the normative objections are sufficiently like the ones in central cases, that is only a necessary condition not a sufficient one for similar classification. The objection to lying might be thought quite similar to the objection to coercion, i.e., the bypassing of the person's rationality to achieve ends which they cannot share. But one might still think there are sufficient differences (in this case the mechanism by which this is brought about) so that it is important to distinguish them.[25]

Or one could argue, as I have with respect to the concept of autonomy, that definitions are only evaluable relative to some problem or issue. Once the issue(s) has been identified then there may be constraints on the definition other than the normative ones stressed by Shiffrin. For example, the concept might have to be scalar rather than on–off.

One could concede the strategy to be correct but argue about the degree of similarity. One might think that overruling a person's judgment to benefit that person is to treat them like a child. Whereas to overrule a person to benefit others is to treat them as morally incompetent. And these are sufficiently different to warrant different classification.

Or consider the following case which (arguably) falls under her classification. You are about to walk by a drowning child. I, who cannot swim, pull out my gun and

---

[25] Of course, in these cases there is a descriptive core which is lacking in the case of paternalism. Paternalism does not specify means; it specifies motives plus some characterization of the effect of an action.

order you to rescue the child. Since (it might be argued) there is no duty to rescue, the issue of when and whether to rescue is within your legitimate sphere of agency. Since I believe that you are making the wrong decision (although not acting impermissibly) I intervene to overrule your judgment. Again, one might say that the shortcoming or deficiency that is overruled in standard cases is that of lack of prudence, whereas in this case it is lack of sufficient concern for the welfare of others, and there ought to be a marker of this difference which is incorporated into the classification of paternalism.

Again, one could agree about the criterion, agree about the proposed definition, but disagree about which cases fall under it because one disagrees about the application of some of the concepts in the definition. For example, one might disagree about what falls within the legitimate sphere of a person's agency. In the above example, one might say that under one description (whether I go swimming) rescue is within the proper sphere, but under another (not performing an "easy rescue") it is not.

One can also criticize the definition on the grounds that the central concept it invokes (substitution of judgment in matters within another person's legitimate sphere of control on the grounds that one believes that one's judgment is superior) is not linked in any essential way with the normative notion invoked (that of insult to the person).

Consider all the following cases:

The law requires you to wear a seatbelt when you drive because it believes that it is better for you to do so.
I refuse to play tennis with you after your recovery from a heart attack because I think it is too dangerous.
The law requires you to save for your retirement (Social Security).

Now all these cases could be claimed to be cases in which the descriptive condition is satisfied. But it is not at all clear that there is any insult to the person whose judgment is overridden. If we believe that you will be risking harm to yourself and we don't trust you to drive without requiring you to use seatbelts we do think that you are in some way defective but why is that insulting? When I correct your addition is that an insult? When I tell you to go to the emergency room with your fever is that an insult?

People make lots of mistakes in their practical judgments about what is best for them. This is something that those being paternalized can acknowledge without assuming that there is something fundamentally inadequate about their decision-making capacities. If we overrule their judgment we need only be referring to a common human condition, not some fundamental defect of the person.

It might be objected that the insult arises from the fact that someone else's judgment is being substituted for yours in an area where you have the right to make your own decisions (the legitimate sphere of control that Shiffrin invokes). But this begs the question against the advocate of paternalistic measures, for she will argue that the area in question cannot be within the sole authority of the agent since it is being claimed that the intervention is a legitimate one.

Note that in cases where the government intervenes on other-regarding grounds it is not assumed that this must be an insult to the person being restricted. Of course, a judgment is being made that the conduct is defective, and a judgment is also made that you would not recognize this on your own (or at least that some further incentive is needed to act on that recognition). But in the first place no global judgment is being made that you are a bad person. And even if it is why is this an *insult*, if true?[26]

These are all substantive criticisms of Shiffrin's account but they do not take the form of supposing that we can simply read off from the definition and the way it classifies cases that it must be wrong.

Paternalism is not a natural kind, and while ordinary usage has some force in evaluating the definition (the so-called "change the subject" argument) it is not definitive. I am also inclined to agree with her that the nature of the "insult" to another person is a useful way of categorizing forms of interference with liberty and autonomy. But any of these may be trumped by, for example, the fact that one is concerned with state coercion, and wants to draw the line on legitimacy between conduct that is regulated to protect others vs. conduct regulated to protect the person being coerced. If that is the issue being examined, then any definition which broadens paternalism so as to include interference for the sake of the protection of third parties will be rejectable. But that will only show that some other term needs to be invoked.[27]

References

Clarke, S. 2002. A definition of paternalism. *Critical Review of International Social and Political Philosophy* 5: 81–91.

Coons, Christian, and Michael Weber (eds.). 2013. *Paternalism: Theory and practice*, 25–38. Cambridge: Cambridge University Press.

de Marneffe, P. 2006. Avoiding paternalism. *Philosophy and Public Affairs* 34: 68–94.

Dworkin, G. 1972. Paternalism. *The Monist* 56: 64–84.

Gert, B., and C.M. Culver. 1979. The justification of paternalism. *Ethics* 89: 199–210.

Kleinig, J. 1983. *Paternalism*. Manchester: Manchester University Press.

Pope, T.M. 2004. Counting the Dragon's teeth and claws: the definition of hard paternalism. *Georgia State University Law Review* 20: 659–722.

Shiffrin, S.V. 2000. Paternalism, unconscionability doctrine, and accommodation. *Philosophy and Public Affairs* 29: 205–250.

Sunstein, C.R. 2002. The laws of fear. *Harvard Law Review* 115: 1119–1168.

Thaler, R.H., and C.R. Sunstein. 2003. Libertarian paternalism. *The American Economic Review* 93: 175–179.

---

[26] These points are made in de Marneffe (2006).

[27] This article first appeared in Coons and Weber (2013), and is reprinted by kind permission of Cambridge University Press.

Chapter 3
The Concept of Paternalism

Dominik Düber

3.1 Introduction

Since the publication of seminal articles concerning paternalism in the 1970s (e.g. Feinberg 1971; Dworkin 1972; Gert and Culver 1976; Beauchamp 1977), the debate focusses both on the conceptual question, which kinds of behaviour, action or regulation are to be understood as 'paternalistic' and the evaluative question regarding the moral status of paternalism. Often, both questions get confounded, e.g. taking soft paternalism to be morally acceptable or hard paternalism to be illegitimate. When we want to debate the moral status of the very different kinds of paternalism, it is helpful to identify a conceptual core of paternalism that includes all the actions and policies typically discussed under this label while excluding those that lack characteristic features. This delineation should remain neutral with regard to the moral legitimacy of such actions and policies. In what follows, I will try to shed some light on this conceptual question in developing a concept of paternalism that is designed to fulfill the function of identifying the kind of conduct that poses so many difficult moral questions, without deciding these questions right away. It should therefore be apt to serve as a basis for discussion between paternalists and anti-paternalists in order to identify the kinds of behavior which they are in dispute about. Although I will remain silent on the evaluative issue, the broad range of paternalism makes it likely that a pure paternalistic or anti-paternalistic stance cannot cope with all of its instances.

As a first approximation, 'paternalism' is the interference with a person for her own good. The great interest of moral and political philosophy in discussing the legitimacy of paternalistic actions seems to stem from two opposed findings. On the one hand we have the Harm Principle, which holds that the only justification for

D. Düber (✉)
Centre for Advanced Study in Bioethics, University of Münster, Münster, Germany
e-mail: dominik.dueber@uni-muenster.de

© Springer International Publishing Switzerland 2015
T. Schramme (ed.), *New Perspectives on Paternalism and Health Care*, Library
of Ethics and Applied Philosophy 35, DOI 10.1007/978-3-319-17960-5_3

interfering with a person's liberty is that her conduct harms someone else. Its most influential formulation is in the first chapter of John Stuart Mill's *On Liberty*:

> That principle is, that the sole end for which mankind are warranted, individually or collectively, in interfering with the liberty of action of any of their number, is self-protection. That the only purpose for which power can be rightfully exercised over any member of a civilized community, against his will, is to prevent harm to others. His own good, either physical or moral, is not a sufficient warrant. (Mill 1859/1977: 223)

A similar anti-paternalistic stance – at least with regard to state action – can be found in Immanuel Kant's *Theory and Practice*. In repudiation of Leibnizian ideas of perfectionist enlightened absolutism as inherited in Christian Wolff and his school (Moggach 2009), Kant claims:

> If a government were founded on the principle of benevolence toward the people, as a *father's* toward his children – in other words, if it were a *paternalistic government* (imperium paternale) with the subjects, as minors, unable to tell what is truly beneficial or detrimental to them, obliged to wait for the head of the state to judge what should constitute their happiness and be kind enough to desire it also – such a government would be the worst conceivable *despotism*. (Kant 1793/1974: 290f.)

It is this anti-paternalistic stance that finds wide approval across different lines of tradition in contemporary philosophy.

But *on the other hand*, even in Western societies, we find many public policies that seem to rely on a paternalistic rationale.[1] And it seems that at least some of these practices cannot be described as some ancient heritage to be overcome soon, but they are valued by the majority of people in these societies, liberal-minded or not. It looks like this tension between a widely shared – though not unconditional – devaluation of paternalism on the one hand and a number of valued paternalistic practices on the other explains the sustained or recurrent interest in paternalism.

Even if paternalism is prima facie wrong and we have reason to avoid it, we should distinguish between two questions that are often confounded by those who treat paternalistic implications as an objection against a particular theory or class of theories. This is not a new observation. Donald VanDeVeer noted more than two decades ago:

> Our ultimate focus in this study concerns the *justifiability* of paternalistic acts. We want to *evaluate* them. But when is an act correctly *described* as a 'paternalistic act'? The latter is a conceptual issue; the former is an evaluative one. In familiar discussions these two questions often get confused and, as a result, the inquiry or dispute gets muddled. [...] If we wish to avoid begging the moral question (by simply *assuming* or supposing an act is wrong in labelling it 'paternalistic') we need to identify a morally neutral definition of 'paternalistic act,' or 'paternalism.'" (VanDeVeer 1986: 16f.)

Following this plea, my sole concern in what follows is the conceptual question, i.e. which kinds of conduct or regulation should be described as paternalistic. The only implication for evaluating paternalistic acts is that this understanding makes it highly probable that some kinds of paternalism are justifiable while others are not.

---

[1] VanDeVeer (1986: 13 et seq.) provides a list of 40 examples of potentially paternalistic measures.

3.2 The Features of Paternalism

In what follows, I shall proceed in three steps. First, I will identify the features of paternalistic actions as they are developed by Gerald Dworkin in his entry in the *Stanford Encyclopedia of Philosophy*. In so doing, I will suggest some refinements or minor changes that are helpful in developing a more neutral concept of paternalism (Sect. 3.2.1). Secondly, I will defend this refined account against competing attempts that try to add other features or remove some of the features carved out in showing that these accounts miss the core of what we normally identify as paternalism (Sect. 3.2.2). In a final step, I will scrutinize one feature – the feature I will call the *coercion condition* – in more detail in order to specify a notion of coercion that is suitable for contexts of paternalism (Sect. 3.3).

3.2.1 Identifying the Features

The most recent proposal for defining paternalism by Gerald Dworkin proposes

> "[…] the following conditions as an analysis of *P acts paternalistically towards Q by doing (omitting)* Z:
>
> (1) Z (or its omission) interferes with the liberty or autonomy of Q
> (2) P does so without the consent of Q
> (3) P does so just because Z will improve the welfare of Q (where this includes preventing his welfare from diminishing), or in some way promote the interests, values, or good of Q." (Dworkin 2010; changed variables)

Although I do not agree with every detail of this definition, I think it is superior to most of the competing approaches – especially those brought forward in the debate on perfectionism – and provides a helpful starting point for identifying the features of paternalism. According to the definition, an action has to show three features for being paternalistic. Dworkin's first feature hints at what shall be called the *coercion condition*. Departing from Dworkin, the use of multifaceted terms like liberty and autonomy should be avoided, if possible. The reason for this is not only that there is a wide range of conceptions of liberty and autonomy, but also the danger of neglecting VanDeVeer's plea for a morally neutral definition of paternalism. Those approaches that have fewer scruples regarding the justifiability of paternalism often employ a different understanding of liberty and autonomy than those that hold stronger reservations. Joseph Raz, for example, holds the view that personal autonomy is impaired only if a person is deprived of morally worthwhile options (Raz 1986: V; Raz 1987: 316). Since paternalism aims at the good of a person, it is unlikely to deliberately deprive a person of worthwhile options and therefore of autonomy in Raz's sense. This leads to the first condition of paternalism which is not fulfilled in most instances of what normally would be called paternalism. Robert Nozick, on the other hand, strongly ties liberty and voluntariness to rights (Nozick 1974: 262).

Since not every instance of paternalism must violate moral or legal rights, such an understanding of liberty would not be very helpful either. Hence, it is desirable to avoid concepts like liberty and autonomy where possible.[2]

For that reason, I shall preliminarily call this element of paternalism the *coercion condition*. I am not attached to this particular term; alternatively it may be called compulsion. Anyway, 'coercion' should be understood in a wide sense, i.e. not in the narrow sense of contemporary discussions of coercion.[3] Coercion in this wide sense includes what Joel Feinberg (1986: ch. 23) calls coercion as well as what he calls compulsion and what in the tradition of Roman Law is called *vis compulsiva* and *vis absoluta* (du Plessis 1997: 4). I will discuss this issue in more detail in Sect. 3.3.[4]

I will call the second feature of paternalistic actions mentioned by Dworkin the *consent condition*. In determining the conceptual characteristics of paternalism, it is helpful to understand lack of consent as lack of *actual* consent. This implies that all benevolent interference with the actual will – or the operative preference – counts as paternalism. This keeps the range of phenomena subsumed under this term wide enough to include some cases that might bear ethical relevance, such as cases of 'weak' or 'soft' paternalism, i.e. cases in which the will of a person is "substantially nonvoluntary" (Feinberg 1986: 12). Joel Feinberg – following a critique of Tom Beauchamp (1977: 67f.) on his former approach (Feinberg 1971) – admitted that weak paternalism is no paternalism "in any clear sense" (Feinberg 1986: 12). But this is only plausible if one confounds the conceptual and the moral issues of paternalism. Whereas it may or may not be true that deficits of the will or belief of a person render these instances of paternalism ethically non-problematic, this does not imply that they should be excluded from the scope of paternalism hastily. It is far from clear why a defective belief should render a benevolent interference non-paternalistic (Kleinig 1983: 8–10).

Typical examples of a defective will are weakness of will (akrasia), defective beliefs, and psychological or intellectual incompetence (cf. Feinberg 1986: ch. 22, 25 and 26). A discussion of the moral status of instances of benevolent interference in a defective will should not be avoided through conceptual presuppositions.[5]

With this reading at hand, the consent condition includes two types of cases. Cases in which an action is carried out against the known will of the paternalized person and cases in which the – known or unknown – will of the paternalized person

---

[2] It is, of course, possible to discuss the legitimacy of particular kinds of actions (that appear paternalistic) in two ways: One could start from a morally neutral definition of paternalism and then check which of those actions can be justified. This is the route I propose by working on the first step in this paper. The other possibility is to hold a general anti-paternalistic stance and then discuss various notions of liberty and autonomy and various notions of 'voluntary will' in order to see which actions can be justified, which in turn is identical with being non-paternalistic.

[3] See Anderson (2009) for an overview.

[4] Another terminological possibility would call this the *interference condition* and understand "interference" as an umbrella term covering compulsion and coercion.

[5] For a critique of soft paternalistic procedures for establishing the voluntariness of living organ donation in Germany, see Fateh-Moghadam (2010) and Gutmann and Fateh-Moghadam (2014).

is declared as irrelevant. This reading clearly excludes cases of actual consent to a benevolent coercive measure. Thereby, we gain a clearer understanding according to which an action is paternalistic. The action is paternalistic insofar as it aims at the good of those who do not consent. Two examples elucidate how this can become unclear. Simon Clarke criticises VanDeVeer and Archard for integrating a consent condition into their notion of paternalism:

> Both definitions are, however, too narrow. Laws requiring drivers and passengers to wear seatbelts while in motor vehicles may not be contrary to the operative preferences of many of those drivers and passengers. (Clarke 2006: 86)

Clarke's observations that some drivers might support the obligation of wearing a seatbelt only shows that it is important to take the intentions of the paternalizing agent into account. If the (potential) paternalizer – the state in this case – only aims at those people who wear seatbelts anyways (maybe for rendering their risk-avoiding behaviour less uncool), this is not an instance of paternalism. But if it aims at protecting those people who would not wear seatbelts otherwise, it is an instance of paternalism. If the state affirms both rationales, it is a *mixed case*. In such a mixed case the benevolent rationale can be a main, a equipollent, or a minor intention and it can be sufficient or only a contributory for carrying out the partly paternalistic action.[6]

The proposed solution also holds true for the cases discussed by Douglas Husak (2003: 387–399), who claims that most analytical work done for clarifying the concept of paternalism is not suitable for treating paternalism in law, since this analytical work regularly takes paternalism between two persons as its starting point, whereas law always aims at a group or even at the whole society. In this group, it is likely to find some people supporting this law, while others reject it. As already mentioned, I treat such a law as paternalistic insofar as it (benevolently) aims at influencing the behavior of those people who do not support it. Husak now fears that this will be problematic, for it is difficult to establish what the rationale of a particular law really is. This problem stems from the great number of members of parliament, who may have different intentions in affirming a law, as well as from the continuity of law, which implies that the rationale for supporting a law may vary over time. Husak is obviously right in making these observations. But this does not disable our reviewing the justification of a law as if it were a paternalistic law. If we want to find out whether a potentially paternalistic law is justified, we have to check, if it can be justified on paternalistic grounds. But for doing this, we can – and

---

[6] This, by the way, seems to account for the worries, Kalle Grill utters in "The Normative Core of Paternalism" regarding the treatment of intention as qualifier on what actions should count as paternalistic (Grill 2007: 444–448). He repudiates definitions of paternalism that incorporate "the reason component of paternalism as a qualifying condition on what actions count as paternalistic" (Grill 2007: 448), because he confuses an action being partly paternalistic with being partly (morally) justified (Grill 2007: 447). Whereas the latter sounds peculiar, the former bears no problems. This confusion may come from the ambiguity of the term 'justified' that is described by Kleinig (1983: 10). The term can either be understood as a success-word in being morally justified or can simply refer to the intentions someone has in performing a particular action.

presumably have to – apply a concept of paternalism like the one defended here. This means, we have to look whether the law in question can be justified, if it is reconstructed as a benevolent interference with the will of the beneficiaries. If it cannot be justified that way, it cannot be justified on paternalistic grounds. In this case, it may be legitimate for other reasons, for example as a rationale of avoiding harm to others, or it may be completely illegitimate. This, then, is no longer a debate concerning paternalism. Hence, it might be difficult to classify a law as paternalistic, if one simply has an interest in classifying laws for its own sake. But if the classificatory work is done with an eye on justificatory issues, the problem vanishes.

Additional to this, Husak correctly describes that laws can apply to a wider group of people than the beneficiaries addressed in the benevolent rationale. If this is the case, the justificatory check requires – similar to the justification in cases of indirect paternalism (Feinberg 1986: 9) – two steps; first, checking whether the law can be justified insofar it is reconstructed as a paternalistic law; and second, checking whether the effects on the group not included in the benevolent rationale can be accepted as reasonable trade-offs with regard to the benevolent aim.[7]

I will call the third feature that Dworkin mentions in his definition ("P does so just because Z will improve the welfare of Q or in some way promote the interests, values, or good of Q.") the *benevolence condition*. I want to clarify my understanding of this element, since it is not clear whether the intentions of the paternalizing agent or the consequences are decisive for an action to be paternalistic. Dworkin's formulation "P does so just because Z *will* improve the welfare of Q" (my italics; DD) may be understood as a condition of success. If one is concerned with the justification of paternalistic actions, and not with potential voluntariness of the paternalized person, it seems more plausible to include actions of the paternalizing agent in which he aims at a good for the paternalized person, but for some reason fails to realize it. Therefore, it is helpful to look at the *intentions* of the paternalizing agent and not at his success. When someone performs a poietic or target-oriented action, there are always insecurities regarding his success. The likeliness of failure can sometimes be anticipated by people with normal epistemic capacities, sometimes it cannot. If it can be anticipated that the intended good cannot be realized with the means the paternalizing agent employs, we would not criticize him for benevolently interfering with the will of the paternalized person, but we would criticize him for neglecting his epistemic virtues in not anticipating that he will miss the intended good. But this neglect is not a reason not to treat these cases as instances of paternalistic actions. To the contrary, this would exclude the well known critique of paternalism that claims that paternalistic actions will necessarily miss the intended good, since the subject always knows best what is his own good (Buckley 2009: ch. 5). Therefore, the benevolence condition should be understood as requiring the paternalizing agent to have a benevolent intention.

---

[7] Husak (2003: 402–404) suggests that the acceptability of such trade-offs can be assessed by balancing the value of the losses of the one group against the gains of the other. Whether this is the right measurement has to be discussed when it come to justificatory issues.

But there are, of course, two possibilities why a paternalistic action may fail to succeed in realizing the intended good. On the one hand, it is possible that the person performs the paternalistic action the way he intended to, but his action is not suitable for producing the good. This should be called 'paternalistic action' nevertheless. But on the other hand, the paternalizing agent can fail to perform the paternalistic action itself. The paternalizing agent could, for example, try to hide a friend's cigarettes in order to save him from the negative impacts of smoking, but accidentally take someone else's cigarettes. Such an action fails to succeed not in virtue of being inappropriate for its end, but because the intended action is not performed at all. In this case, it makes sense to speak of an *attempted paternalistic action*, since the agent acted neither paternalistically to his friend, nor to the other person who, at best, accidentally benefits from the agent's doing.

Two other types of cases are excluded by such an understanding of the benevolence condition. First, it does not make sense to say that an action has *paternalistic effects* (without benevolent intentions). If the content of the intention is not benevolent, there is no instance of paternalism, even if an interference has positive consequences by chance (like in the cigarette-case) or as a foreseeable approved side-effect. Second, an action is not an instance of paternalism if it aims at the well-being of third parties. This, as already mentioned, does not exclude mixed rationales that aim at the good of the affected person and of third parties. Such an action is paternalistic insofar as it aims at the good of the paternalized person (Feinberg 1986: 8).

3.2.2 Objections

It is assumed that the three features mentioned so far are each necessary and – taken together – sufficient for an action to be paternalistic. As this assumption is not accepted by all of the disputants, I shall have a brief look at systematically relevant alternatives. Whereas the necessity of a benevolence condition is commonsensical, there are attempts to deny the necessity of either the coercion condition or the consent condition or of both of them.

Simon Clarke, for instance, thinks it is necessary to *discard the consent condition*. His relatively thin definition of paternalism reads as follows:

> P behaves paternalistic towards Q: (1) only if P aims to close an option that would otherwise be open to Q, or P chooses for Q in the event that Q is unable to choose for himself; and (2) to the extent that P does so in order to promote Q's good. (Clarke 2002: 81; changed variables)

Referring to a consent condition employed by David Archard (1990: 36, 39), he claims:

> However, this extra condition is not needed. The case where P believes that Q shares his view of the correctness of interference may not be a case of paternalism. But this, I suggest, is because P's reason for acting is to give Q what Q desires. If so, then insofar as this, rather

> than to bring about Q's good, is P's motivation, the intervention is indeed not paternalistic. My definition accounts for this, since in that case the second condition, that the behaviour is intended for Q's good, is not satisfied. (Clarke 2002: 89; changed variables)

This seems to be an ad hoc solution in favour of Clarke's definition. *Either* Clarke has to stipulate that P in these cases always is motivated to give Q what Q desires. But this would presuppose what is in question. *Or* Clarke has to admit that cases, in which P's motivation for interfering is Q's good, even though P knows about Q's corresponding will, are instances of paternalistic actions, which would be peculiar and is exactly the consequence Clarke tries to avoid. This finding supports the necessity of a consent condition.

Jonathan Quong tries to develop a plausible definition of paternalism in the context of his argument against perfectionism in political philosophy. He admits that perfectionist policies can rely solely on non-coercive means, but claims this not to be sufficient for avoiding paternalism:

> [...] I argue that liberal perfectionism, despite claims to the contrary, remains a paternalistic doctrine. Although perfectionist policies can be pursued by non-coercive means – something favored by most contemporary perfectionists – perfectionist policies cannot avoid being paternalistic since they imply a negative judgment about citizens' capacities to make good decisions and run their own lives. (Quong 2011: 9)

With this idea in the background, he develops a definition of paternalism that needs *neither a coercion condition, nor a consent condition*:

> I offer the *judgmental definition*, where paternalism is defined as an act where:
>
> 1. Agent P attempts to improve the welfare, good, happiness, needs, interests, or value of agent Q with regard to a particular decision or situation that Q faces.
> 2. P's act is motivated by a *negative judgment* about Q's ability (assuming Q has the relevant information) to make the right decision or manage the particular situation in a way that will effectively advance Q's welfare, good, happiness, needs, interests or values. (Quong 2011: 80; changed variables)

This 'judgmental definition' goes too far in broadening the notion of paternalism. Almost every instance of providing help would be an instance of paternalism in this reading. In cases of help, the person who receives help (Q), is often in a worse position to take care of her good than the person who provides help (P). If this is part of P's motives, then the second condition in Quong's definition is fulfilled. If, in addition to that, P's motivation for helping Q is Q's good, then the first condition is fulfilled as well. This seems to hold true for many instances of helping, as well as for most instances of state action, i.e. whenever the state is used as means for realizing those purposes an individual person is unable to. Therefore, it is easy to see that the judgmental definition of paternalism would cover a vast amount of cases that neither suggest any moral uncertainty nor would normally be treated as paternalism. It may be convincing to assume that paternalistic actions are regularly accompanied by a negative judgment regarding the abilities of the paternalized person in supporting her good, but this does not warrant discarding the coercion condition or the consent condition.

The definition of paternalism by Donald VanDeVeer, which *abstains from a coercion condition*, bears a similar, but less severe problem.

P's doing or omitting some act Z to, or toward, Q is paternalistic behavior if and only if

(1) P deliberately does (or omits) Z and
(2) P believes that his (her) doing (or omitting) Z is contrary to Q's operative preference, intention, or disposition at the time P does (or omits) Z [or when Z affects Q – or would have affected Q if Z had been done (omitted)] and
(3) P does (or omits) Z with the primary or sole aim of promoting benefit for Q [a benefit which, P believes, would not accrue to Q in the absence of P's doing (or omitting) Z] or preventing harm to Q [a harm which, P believes, would not accrue to Q in the absence of P's doing (or omitting) Z]." (VanDeVeer 1986: 22; changed variables)

I think this definition is helpful altogether, since it makes reference to the operative preference of the paternalized person as well as to the intention of the paternalizing agent. But it should not discard the coercion condition, since paternalism is normally linked to having some kind of intervening character and therefore should have some kind of influence on a person's options or opportunities of action. Nevertheless, there will be few actions that fulfil VanDeVeer's definition and avoid having any intervening character.[8]

The examples taken from Clarke, Quong and VanDeVeer speak in favour of taking all the three features as necessary for an action to be paternalistic. Nevertheless, there have been other proposals for defending other conditions as necessary for paternalism, which, if true, would make the conjunction of the three features not sufficient for an action to be paternalistic. Bernard Gert and Charles Culver claimed that it is part of paternalism to violate a moral rule.

We believe that an essential feature of paternalistic behavior toward a person is the violation of moral rules (or doing that which will require such violations), for example, the moral rules prohibiting deception, deprivation of freedom or opportunity, or disabling. (Gert and Culver 1976: 48)

With regard to this assumption, it seems natural to follow Gerald Dworkin's reply:

On the other hand, the attempt to broaden the notion by including any violation of a moral rule is too restrictive because it will not cover cases such as the following. A husband who knows his wife is suicidal hides his sleeping pills. He violates no moral rule. They are his pills and he can put them wherever he wishes. (Dworkin 1983: 106.)

Therefore, the violation of a moral rule should not be included in the notion of paternalism.

In addition to that, following Quong's idea of a negative judgment, one could be inclined to include a superiority condition

which restricts paternalism to such actions as are performed by an agent who considers herself in some way superior to the person(s) interfered with. (Grill 2012: Sect. III)

---

[8] This, of course, is something else than Dworkin's success condition criticized above.

as a necessary condition for paternalism. But even if paternalistic actions are often or even always accompanied by negative judgments or superior self-understandings, I see no reason why an action that does not show this feature but the other three should not be treated as paternalistic. However, it would need such a reason for claiming that the three features taken together are not sufficient for paternalism. Therefore, I think we can assume the three features to be each necessary and jointly sufficient for labelling an action 'paternalistic'.

3.3   The Notion of 'Coercion' in Contexts of Paternalism

As mentioned earlier when introducing the coercion condition – or whatever its appropriate name may be – the notion of coercion in contexts of paternalism should not be understood in the narrow sense that is prevalent in contemporary analytic philosophy.[9] The wider notion introduced in what follows is solely meant to be useful for issues regarding paternalism. This means that it shall be understood as a context-relative notion of coercion. This reading of coercion is not innovative in any sense, but simply unifies two different meanings of coercion or force, well known since ancient philosophy.

3.3.1   Vis Absoluta (*Coercion Not Mediated by the Will*)

The first notion of 'coercion' is what in the tradition of Roman Law is called "*vis absoluta*" (du Plessis 1997: 4) and what is akin to what Feinberg (1986: 190 et seq.) calls "compulsion". We find a description of this kind of coercion at the beginning of the third book of Aristotle's Nicomachean Ethics:

> Those things, then, are thought involuntary, which take place under compulsion (bia) or owing to ignorance (agnoia); and that is compulsory of which the moving principle (arche) is outside, being a principle in which nothing is contributed by the person who acts or is acted upon, e.g. if he were carried somewhere by a wind, or by men who had him in their power. (Aristotle 1985: 1009b 35 et seq.)

Since I am not interested in the voluntariness of the person coerced or paternalized in the first place, we can leave aside cases of ignorance – although we will encounter them again in the context of possible justifications for paternalistic actions. The essential feature of *vis absoluta* relevant for paternalism is that it simply happens to the person coerced. *Vis absoluta* is – in the situation at hand – in no sense mediated by the will of the person coerced. It is obvious that this kind of coercion can be found in a vast amount of instances of paternalism, namely whenever there is no opportunity of avoiding the good being forced upon oneself. When

---

[9] See Anderson (2009) for this sense.

someone is locked up in a psychiatric clinic and has no opportunity to get out– e.g. by accepting another evil like taking mandatory medicaments he otherwise would not take – he is exposed to *vis absoluta*. The same holds true for measures like mandatory contributions to social security that are subtracted from someone's income before it is disbursed.

Whereas *vis absoluta* in Roman Law often is defined by reference to a particular means like direct physical force, this restriction should be avoided when it comes to paternalism. Someone who is locked in a room or automatically has a certain share of his income subtracted for social security before he receives it may not be exposed to direct bodily force, but shall be understood as exposed to *vis absoluta*. I hope that this reading of *vis absoluta* also includes those cases that Gert and Culver invoke in order to show that coercion is not necessarily involved in paternalistic actions:

> Consider a case where a doctor lies to a mother on her deathbed when she asks about her son. The doctor tells her that her son is doing well, although he knows that the son has just been killed trying to escape from prison after having been indicted for multiple rape and murder. The doctor behaved paternalistically but did not attempt to control behavior, to apply coercion, or to interfere with liberty of action. (Gert and Culver 1976: 46)

The mother does not have any influence on getting to know the truth about her son. Therefore, *vis absoluta* should cover those kinds of interferences that do not leave the person coerced an option in influencing the run of events, i.e. cases in which his will does not play a mediating role.

3.3.2 Vis Compulsiva *(Coercion as Mediated by the Will)*

Whereas the standard view in Roman Law and up until Samuel von Pufendorf in the early enlightenment was that *vis absoluta* is the only real kind of coercion, i.e. the only kind of coercion that invalidates contracts and undermines voluntariness (Gutmann 2001: ch. 3.1.3), this view shifted to the opposite in the second half of the twentieth century. In his influential article that started a debate on the proper understanding of coercion (Pennock and Chapmann 1972), Robert Nozick claims:

> [S]ome writers (e.g. Bay) say that all infliction of violence constitutes coercion. But this is, I think, a mistake. If a drunken group comes upon a stranger and beats him up or even kills him, this need not be coercion. For there need have been no implicit threat of further violence if the person didn't comply with their wishes, and it would indeed be difficult for this to be the case if they just come upon him and kill him. (Nozick 1969: 444)

This observation only shows that violence is not sufficient for coercion. But Nozick discards the phenomena of *vis absoluta* completely and narrows the notion of coercion to what shall be called '*vis compulsiva*' or coercion in the narrow sense. This notion of coercion is tied to conditional threats that try to make a person perform a particular action by threatening a supposedly greater evil for not performing this action. In the passage following the one already quoted, Aristotle describes this kind of coercion as follows:

But with regard to the things that are done from fear of greater evils or for some noble object (e.g. if a tyrant were to order one to do something base, having one's parents and children in his power, and if one did the action they were to be saved, but otherwise would be put to death), it may be debated whether such actions are involuntary or voluntary. Something of this sort happens also with regard to the throwing goods overboard in a storm; for in the abstract no one throws goods away voluntary, but on condition of its securing the safety of himself and his crew, any sensible man does so. Such actions, then, are mixed, but are more like voluntary actions; for they are worthy of choice at the time when they are done, and the end of an action is relative to the occasion. (Aristotle 1985: 1110a 4 et seq.)

For the purpose of defining paternalism, we do not have to decide which action counts as voluntary and which does not. The essential difference between *vis absoluta* and this notion of coercion is the latter's being mediated by the will of the coerced person. The coercing agent (P) threatens the coerced person (Q) with a particular consequence if Q does not follow the course of conduct that P wants Q to follow. But different from *vis absoluta*, Q is left with the decision to do what P wants him to do or to bear the consequence P threatens. Depending on the coercive pressure and the character of the threatened consequence – issues I leave aside for the moment[10] – there may remain only one reasonable choice, which approximates instances of *vis compulsiva* to *vis absoluta*. But the principled difference is that Q is left with the choice.

Although there is no commonsensical definition of *vis compulsiva* or coercion in the narrow sense, I would suggest Nozick's (1969) definition as a helpful starting point in most contexts of paternalism. I think it has advantages over other definitions since it pays special attention to the intentions or motives of the coercing agent and is less concerned with the remaining responsibility of the person coerced. This makes Nozick's definition more suitable than the one Feinberg provides in Harm to Self (1986: 196), since Feinberg primarily asks whether the person coerced is still responsible for the action performed under coercive pressure, whereas I am interested in a definition of coercion and paternalism that is suitable for discussing the legitimacy of the action of the paternalizing agent, without deciding this beforehand. Although I cannot go into the details of Nozick's complex definition, some proposed modifications that follow from what has already been said shall be mentioned nevertheless. Scott Anderson highlights the following aspects as characteristic for Nozick's approach, namely:

[…] that (1) it associates coercion only with proposals (e.g. conditional threats) and excludes direct use of force or violence; (2) it insists that coercion takes place only when the coercee acquiesces to it; and (3) it makes coercion explicitly dependent on the coercee's choice to take or not take a specific action Z, and mandates that a judgment about coercion must refer to facts about the coercee's psychology, such as her assessment of the consequences Z-ing in light of the coercers proposal. (Anderson 2009; changed variables)

Regarding the first feature, I already mentioned that *vis compulsiva* should be supplemented with *vis absoluta*. For this reason, I treat coercion in Nozick's notion as only one of two kinds of coercion. Regarding the second feature, I would – in analogy to what has been said regarding a success condition in paternalistic

---

[10] For a helpful discussion of three different measurements of coercive pressure, see Gutmann (2001: ch. 3.2.7).

actions – avoid taking success as a necessary condition, since my focus is the coercer and not the person coerced. Whether or not one treats success of a coercive action as making a difference with regard to the moral quality of this action seems to be similar to the position one is inclined to take in the debate concerning moral luck. Regarding the third feature – which needs a more detailed discussion of coercive pressure – I will at this point only distinguish between cases in which the coercing agent has knowledge about the psychology and preferences of the person coerced and cases in which he does not. In the former cases, the coercer has precise knowledge about the strength of the coercive pressure he is applying. In the latter, it has to be relied on standards of what normally or by the average person would be perceived as strong, weak or no coercive pressure. Hence, for applying coercive pressure, it is not necessary to have knowledge about the coercee's psychology, some basic knowledge about the human condition is sufficient.

3.3.3   Threats and Offers

One last aspect from the debate on coercion (in the sense of *vis compulsiva*) shall be mentioned. There has been an intensive discussion on how to distinguish between threats and offers in the context of coercion (Gutmann 2001: 149 et seq.). This distinction bears some difficulties, since on the level of language both show some similarities, i.e. both are conditional proposals of the form:

If you do X, I will do Y & if you do not do X, I will not do Y.

In the case of a threat, the proposal is "If you refuse to do what I want, I will bring about a consequence that you (likely) want to avoid". In the case of an offer, the proposal is "If you do what I want, I will bring about a consequence that you (likely) see as a benefit." This now looks as if an offer is a consequence provided for the case of following the demanded course of conduct, whereas a threat is a consequence provided for not following the demanded course of conduct. But this distinction does not lead too far, since it is easy to phrase something that is obviously a threat in the form of an offer, e.g. "If you hand over your money to me, you can keep your life (which I would otherwise take)."

For that reason, it seems to be necessary to rely on an understanding of what the normal course of events would be. Since it is not normal to take someone's life, this clearly worsens the coercee's situation compared to the normal course of events. If the proposed consequence retains or improves the situation compared to the normal course of events, it is an offer. The problem now is the understanding of 'normal' in this context, because it oscillates between the somehow statistically normal course of events and the morally expected course of events. Nozick (1969: 450) discusses this with the example of a slave who gets beaten every morning and gets the opportunity for the next morning to get either beaten like every day or do something else the master would like him to do. Whether this is a threat or an offer depends on the kind of 'normality' that is chosen for drawing the line.

Fortunately, I do not think that we have to decide this question as long as we are concerned solely with the *concept* of paternalism, since I would see it as an advantage if the concept of paternalism allowed for *paternalistic offers* as well. A father who offers his offspring a nice car, if she decides to study law instead of philosophy might be labelled as paternalistic, as well as a health insurance system that provides incentives for certain healthy ways of living – insofar that this is a benevolent offer and does not (only) aim at reducing costs. Hence, we can – and should – avoid the exclusion of paternalistic offers on the conceptual level. Nonetheless, the problem of distinguishing between threats and offers may return when it comes to the justification of paternalistic practices, since there might be a difference regarding the moral value between threats and offers as means of promoting a person's good.

3.4 Conclusion

To sum up: When developing a concept of paternalism that does not itself embody a decision regarding the justification of paternalistic actions and is therefore morally neutral, it makes sense to stick to a roughly Dworkinian conception of paternalism. Such a definition treats three features of an action as necessary and jointly sufficient for rendering an action paternalistic. The first feature, coercion, should be understood in a broad sense in two ways. First, it should not only include coercion in the narrow sense, i.e. *vis compulsiva*, but also incorporate *vis absoluta*. Second, within *vis compulsiva*, it should allow not only for threats, but for offers as well. The second feature is the absence of consent, where consent should be understood in a narrow sense as including only actual consent. The third feature is the benevolence of the paternalizing agent towards the paternalized. Benevolence should make reference to the intentions of the paternalizing agent, not to the factual outcome of the paternalistic action.

References

Anderson, Scott. 2009. Coercion. In *Stanford encyclopedia of philosophy*, (Spring 2009 Edition), ed. Edward N. Zalta. URL: http://plato.stanford.edu/archives/spr2009/entries/coercion/.
Archard, David. 1990. Paternalism defined. *Analysis* 50(1): 36–42.
Aristotle. 1985. Nicomachean ethics. In *The complete works of Aristotle*, ed. Jonathan Barnes, 1729–1867. Princeton: Princeton University Press.
Beauchamp, Tom L. 1977. Paternalism and biobehavioral control. *The Monist* 60(1): 62–80.
Buckley, Francis H. 2009. *Fair governance. Paternalism and perfectionism*. Oxford: Oxford University Press.
Clarke, Simon Robert. 2002. A definition of paternalism. *Critical Review of International Social and Political Philosophy* 5: 81–91.
Clarke, Simon Robert. 2006. Debate: State paternalism, neutrality and perfectionism. *The Journal of Political Philosophy* 14(1): 111–121.

du Plessis, Jaques E. 1997. Compulsion in Roman law. Doctoral thesis [portion], University of Aberdeen. URL: http://aura.abdn.ac.uk/bitstream/2164/91/1/060410-002.pdf.

Dworkin, Gerald. 1972. Paternalism. *The Monist* 56: 64–84.

Dworkin, Gerald. 1983. Paternalism: Some second thoughts. In *Paternalism*, ed. Rolf Sartorius, 105–111. Minneapolis: University of Minnesota Press.

Dworkin, Gerald 2010. Paternalism. In *Stanford encyclopedia of philosophy*, (Summer 2010 Edition), ed. Edward N. Zalta. URL: http://plato.stanford.edu/archives/sum2010/entries/paternalism/.

Fateh-Moghadam, Bijan. 2010. Grenzen des weichen Paternalismus. Blinde Flecken der liberalen Paternalismuskritik. In *Grenzen des Paternalismus*, ed. Bijan Fateh-Moghadam, Stephan Sellmaier, and Wilhelm Vossenkuhl, 21–47. Stuttgart: Kohlhammer.

Feinberg, Joel. 1971. Legal paternalism. *Canadian Journal of Philosophy* 1(1): 105–124.

Feinberg, Joel. 1986. *The moral limits of criminal law*, Harm to self, vol. III. Oxford: Oxford University Press.

Gert, Bernard, and Charles Culver. 1976. Paternalistic behavior. *Philosophy and Public Affairs* 6: 45–57.

Grill, Kalle. 2007. The normative core of paternalism. *Res Publica* 13(4): 441–458.

Grill, Kalle. 2012. Paternalism. In *Encyclopedia of applied ethics*, vol. 3, ed. Ruth F. Chadwick, 359–369. Amsterdam: Academic.

Gutmann, Thomas. 2001. *Freiwilligkeit als Rechtsbegriff*. München: C. H. Beck.

Gutmann, Thomas, and Bijan Fateh-Moghadam. 2014. Governing [through] autonomy. The moral and legal limits of "soft paternalism". *Ethical Theory and Moral Practice* 17: 383–397.

Husak, Douglas N. 2003. Legal paternalism. In *The oxford handbook of practical ethics*, ed. Hugh LaFolette, 387–412. Oxford: Oxford University Press.

Kant, Immanuel. 1974. *On the old saw: That may be right in theory but it won't work in practice*. Philadelphia: University of Pennsylvania Press.

Kleinig, John. 1983. *Paternalism*. Manchester: Manchester University Press.

Mill, John Stuart. 1977. On liberty. In *The collected works of John Stuart Mill*, vol. XVIII, ed. John M. Robson, 213–310. Toronto: Toronto University Press.

Moggach, Douglas. 2009. Freedom and perfection. German debates on the state in the eighteenth century. *Canadian Journal of Political Science* 42(4): 1003–1023.

Nozick, Robert. 1969. Coercion. In *Philosophy, science, and method. Essays in honor of Ernest Nagel*, ed. Sydney Morgenbesser, Patrick Suppes, and Morton White, 440–472. New York: St. Martin's Press.

Nozick, Robert. 1974. *Anarchy, state, and utopia*. New York: Basic Books.

Quong, Jonathan. 2011. *Liberalism without perfection*. Oxford: Oxford University Press.

Pennock, J. Roland, and John W. Chapmann (eds.). 1972. *Nomos XIV: Coercion*. Chicago: Aldine-Atherton.

Raz, Joseph. 1986. *The morality of freedom*. Oxford: Clarendon.

Raz, Joseph. 1987. Autonomy, toleration, and the harm principle. In *Issues in contemporary legal philosophy. The influence of H. L. A. Hart*, ed. Ruth Gavison, 313–333. Oxford: Clarendon Press.

VanDeVeer, Donald. 1986. *Paternalistic intervention: The moral bounds on benevolence*. Princeton: Princeton University Press.

Chapter 4
Antipaternalism as a Filter on Reasons

Kalle Grill

4.1 Introduction

The charge of paternalism is a common objection to the actions of political and other authorities. Sometimes the charge is only that the authority has undervalued typical liberal values like freedom and autonomy relative to other values, such as physical, mental or financial wellbeing. Making this objection is consistent with accepting that in some cases, wellbeing outweighs freedom and autonomy and should be furthered at their expense. Other times, however, the charge of paternalism is more principled. The objection is not that wellbeing considerations are overstated, but that they are allowed to weigh in on the matter at all. This is the sort of antipaternalism that I will analyze in this article. My discussion and my proposals are meant to be helpful to the antipaternalist, and to anyone who wants to understand her. However, I should state at the outset that the antipaternalist position I describe and develop is not one I endorse.

I propose that principled antipaternalism entails that certain facts are prevented from playing the role of reasons they would otherwise play. Using an obvious metaphor, I call this the *filter approach* to antipaternalism: The potential reasons provided by some facts are filtered out and so do not play the role of reasons. Exactly which these facts are determines the precise normative content of antipaternalism. I take no stand on this issue. My thesis is a conceptual thesis on the structure of antipaternalism and not a normative position.

The main competitor to the filter approach is the *action-focused approach* according to which antipaternalism entails that certain actions or policies are paternalistic and therefore impermissible or otherwise morally problematic. The action-focused approach is the dominant approach in academic discussions of paternalism

K. Grill (✉)
Department of Historical, Philosophical and Religious Studies,
Umeå University, Umeå, Sweden
e-mail: kalle.grill@umu.se

© Springer International Publishing Switzerland 2015   47
T. Schramme (ed.), *New Perspectives on Paternalism and Health Care*, Library
of Ethics and Applied Philosophy 35, DOI 10.1007/978-3-319-17960-5_4

and antipaternalism. A significant problem with this approach is that it requires the sorting of actions into paternalistic and nonpaternalistic. It has proven quite difficult to define the criteria for such sorting. On the filter approach, actions need not be sorted in this way: Antipaternalism simply demands that some reasons are filtered out, without categorizing actions as either paternalistic or not. The filter approach therefore provides more straightforward normative implications, and, arguably, just the sort of normative implications antipaternalists typically intend.

The filter approach is independent of more fundamental questions concerning what is valuable and what we have reason to do. The filtering can be motivated in several different ways, including by practical concerns with the ability of certain agents (typically government agents) to properly consider certain reasons, by more normative concerns with the appropriateness of these agents considering these reasons, independently of effects, and by stronger normative principles according to which certain facts simply do not provide reasons in certain situations.

Both the action-focused and the filter approach are approaches to *antipaternalism*. However, the action-focused approach is dependent on a definition of paternalistic action, or in other words of *paternalism*. I will therefore also discuss some proposed definitions of this related concept. However, I defend no thesis on the best understanding of paternalism.

This article is outlined as follows: In Sect. 4.2, I specify the principled objection to paternalism that I am concerned with and contrast it with other objections. In Sect. 4.3, I present the filter approach in further detail. In Sect. 4.4, I contrast the filter approach with the dominant action-focused approach.

4.2   Antipaternalisms

I mentioned two variations on the paternalism charge in the first paragraph of the introduction, but in fact there are several. In this section, I will survey four and explain why my interest here is only with one of them. I will, for ease of presentation, throughout use "liberty" as a placeholder for various liberal values like freedom, autonomy, self-determination etc., and "limiting liberty" as a placeholder for the infringement, diminishing, disrespect etc. of some such value. I will also make a controversial assumption: That antipaternalism only applies to cases where the paternalist promotes the good of the very same person whose liberty she is limiting (or aims to do so). Seana Shiffrin rather famously rejects this assumption, arguing that it is paternalism (and so objectionable) to limit the liberty of some person also to make things better more generally, as long as the infringement concerns this person's sphere of legitimate control (2000, p. 216). I make my controversial assumption only for convenience. It will enable me to speak of the relevant reasons in a straightforward manner and thereby much simplify my presentation. However, my analysis could be reformulated to accommodate Shiffrin's wider take on paternalism (given that her position can be made sufficiently clear).

First of the four variations of the paternalism charge is the one that I will be exclusively concerned with after this present section: The principled objection that consideration of a person's good should not even count in favor of limiting her liberty. I propose that this objection is a plausible interpretation of Mill's liberty principle as well as Feinberg's antipaternalism (or "soft paternalism"). While Feinberg's account is perhaps unique in its detailed specification, the objection figures frequently, though more or less explicitly, both in philosophical accounts of paternalism and in applied ethical contexts and in the public debate. The objection is typically formulated as a rejection of the balancing or weighing of different considerations, in favor of the priority of liberty over other concerns.[1]

The second variation, mentioned in the introduction, is the common sense objection that considerations of a person's good are sometimes given too much weight in relation to respect for her liberty. In contrast to the first objection, this objection explicitly recognizes that liberty can be balanced against other good things for a person. The objection applies whenever some agent gets the balance wrong and so liberty is limited too lightly. Though determining the relative importance of liberty may be a very complex matter, the structure of the objection is straightforward – liberty is undervalued. The objection does not represent a normative principle or position distinct from the common-sense position that one should give each value its due.

Sometimes agents are liable to systematically undervalue liberty, for example because of overconfidence in their ability to force people to improve their own lives. There are arguably many different values or goods for individuals and so many ways to get the balance between them wrong. When such mistakes are systematic, however, there may be reason to make them more salient by giving them names, such as "paternalism". There may even be reason to give the opposition to such mistakes a name, such as "antipaternalism". However, this name should not then be taken to refer to anything beyond this opposition.

The third variation of the paternalism charge, not mentioned in the introduction, is the objection to one person taking action towards another based on the first person's view of what is good or best for the second person. Mill famously argued that each person best knows her own interests, which would seem to tell against acting on one's ideas of what is good for others, whether or not liberty is at stake (1859,

---

[1] Feinberg claims, for example, that "personal autonomy [...] is a moral trump card, not to be merely balanced with considerations of harm diminution in cases of conflict, but always and necessarily taking moral precedence over those considerations." (1986, p. 26) For a more recent example, Daniel Groll proposes that when it comes to benefitting a person, her will should be treated as "structurally decisive in determining what to do – it is meant to supplant the reason-giving force of other considerations not because it outweighs those other considerations but because it is meant to silence, or exclude, those other considerations from the practical deliberations of the subject of the demand" (2012, p. 701).

I should perhaps recognize that John Rawls is well known for his argument for the priority of liberty within the special context of distributive justice. Rawls, however, says very little about paternalism. He accepts paternalism since people's "capacity to act rationally for their good may fail, or be lacking altogether." (1999, p. 219) He does not discuss other, more problematic cases.

chapter IV, 4th paragraph). Immanuel Kant in a characteristically more principled manner stated that when it is our duty to promote others' happiness "[i]t is for them to decide what they count as belonging to their happiness" (1991/1797, 388). Dan Brock (1988) considers this objection central to the very concept of paternalism:

> [P]aternalistic interference involves the claim of one person to know better what is good for another person than that other person him- or herself does. It involves the substitution by the paternalistic interferer of his or her conception of what is good for another for that other's own conception of his or her good. (p. 559)

Gerald Dworkin (1983) too speaks of substitution in defining paternalism, but of "judgment" rather than of a conception of what is good. While it is clear in the context that Brock's "conception of what is good" refers to ideals about the good life, Dworkin's "judgment" seems to refer either to such judgments about ideals, or to more mundane judgments of how best to realize these ideals. These two possibilities are perhaps captured by his later distinction between strong and weak paternalism (2010).

When we substitute our judgment for someone else's, whether concerning what her good is or the best means of promoting it, this can lead us to limit their liberty. If it does, the principled objection applies and the common sense objection may apply depending on the balance of values. If our substitution of judgment does not lead us to limit someone's liberty, however, I do not see that there can be much of an objection. As Kant went on to say, "it is open to me to refuse them many things that *they* think will make them happy but that I do not, as long as they have no right to demand them from me" (1991/1797, 388). As Paul Guyer has argued, we should not think, nor take Kant to mean, that it is morally problematic that we make our own judgments about the best interests of other people (2014, 231–232).

For an illustration, assume that you judge that I am being overly friendly with you and should be less friendly, though I in no way impose on you. Perhaps you just believe that friendliness is a sign of weakness and therefore to be avoided. Suppose I know that this is your judgment but decide anyway to remain as friendly as ever, and that I do so partly or solely in order to get you to warm up and break your social isolation. This does not seem morally problematic and not the sort of thing liberals oppose. To the contrary, it seems a liberal ideal that I have a moral right to be as friendly as I choose, for whatever reason I find compelling, as long as I do not impose on others. Therefore, the third variation of the paternalism charge is not an independent objection. Only when a substitution of judgment leads to a limitation of liberty is it problematic, and so the problem lies with the limitation of liberty rather than with the substitution of judgment.

The fourth variation of the paternalism charge, also not mentioned in the introduction, is the objection to treating people in a condescending manner. This objection is often what people have in mind when they speak of "treating adults as if they were children" and the like (e.g. Szasz 1992, xiv, arguing against drug criminalization). Condescendence is an important part of Shiffrin's rich account of paternalism. Shiffrin claims that paternalism is always undertaken on the basis of a sort of disrespect toward the paternalized person's judgment or agency. This disrespectful attitude "is central to accounting for why paternalism delivers a special sort of insult to competent, autonomous agents" (p. 220).

It is not quite clear what Shiffrin means by an attitude in this context. Sometimes it seems that she means that paternalist behavior, in virtue of its interfering nature, essentially embodies a condescending or disrespectful attitude, regardless of the actual psychological state of the paternalist (who could for example reasonably believe that the target is incompetent). So for example she states that paternalistic behavior "manifests an attitude of disrespect" (ibid.). This line of thought, however, does not amount to an independent objection. Rather, it is an argument for the principled objection: We should not count a person's good as a reason for limiting her liberty, because doing so embodies a condescending attitude towards her.

Other times, however, it seems that Shiffrin means that an actual disrespectful attitude is essential to paternalist behavior. For example, she claims that "the paternalist's attitude shows significant disrespect" (ibid.), as if there could be an otherwise similar interferer whose attitude did not show disrespect and who therefore was not a paternalist. It is this interpretation that amounts to an independent variation of the paternalism charge – an objection to having, or acting on, improper attitudes.[2]

To consider this objection, note first that displaying a condescending or disrespectful attitude is not particular to paternalism. People can be condescending in all sorts of situations, many of which are clearly not paternalistic. For example, we may ridicule an unsuccessful competitor, or reject the advice of some well-meaning acquaintance as useless without hearing it. This shows that we do not need the concept of paternalism, or antipaternalism, to explain the possible moral problems involved in being condescending and disrespectful, i.e. treating people as if they were less competent than they are.

More to the point, many liberals feel that benevolent limitations of liberty are morally problematic whether or not they are accompanied with a condescending attitude. Benevolent *and condescending* limitations of liberty may be especially problematic, but so are benevolent *and unfair* limitations of liberty, and so on. Condescendence is a separate problem; what is particular to paternalism is the limitation of liberty.

To sum up, the charge of paternalism can, on closer inspection, amount to one of several different objections, or indeed to more than one. However, the objection from substitution of judgment and one interpretation of the objection from condescendence both presuppose an objection from the limitation of liberty. Another

---

[2] Other authors with a similar take on paternalism are no clearer than Shiffrin in this regard. For example, Jonathan Quong argues that the essence of paternalism is that action is "motivated by a negative judgment about the ability of others to run their own lives." (2010, p. 74). Paternalism is wrong, Quong claims, "because of the way it denies someone's moral status as a free and equal citizen." (Ibid.) It is not clear whether Quong thinks that being motivated by a negative judgment is a mental state distinct from simply judging oneself to have superior knowledge or ability in the particular case, or whether this (possibly true) judgment of superiority, regardless of other mental states, embodies a "negative judgment" that denies others' their moral status. Only the first understanding could be the basis of an independent objection.

interpretation of the objection from condescendence does not capture the heart of the matter but is a distinct concern with attitudes. Only the two objections from the limitation of liberty are independent and hit the right target. These two objections presuppose different views on the role of liberty. The common sense view assumes liberty to be one value among others, to be weighed against them. This view does not constitute a distinct normative view but is simply the view that each value should be given its due. In contrast, the principled view does constitute a distinct normative view, or rather a family of views with a shared structure. It is this structure that I will now go on to analyze.

4.3   Filtering Reasons

My thesis is that principled antipaternalism is best understood as a filter that prevents certain facts from playing the role of reasons (for certain actions). Given my designation of 'limiting liberty' as a placeholder for various allegedly illiberal impositions, and given my assumption that the relevant rationales have to do with the good of the person imposed upon, this can be put in somewhat more specific terms: The antipaternalist filter prevents the fact that the limitation of some person's liberty causes the promotion or protection of her good from playing the role of a reason for such limitation. With due care, this approach can also be applied to many-person cases, and so to public policy.[3]

It may seem an extreme position to hold that facts that play the role of reasons in other contexts are entirely blocked from doing so. A more moderate position would be to admit that they play the role of reasons, only with reduced strength. I believe the typical antipaternalist position is that reasons are filtered out entirely and so this is the view I will discuss. However, the filter approach is entirely consistent with the filter being only partial, perhaps reducing the strength of reasons according to some complicated formula.

What I provide here is a generic or structural account of antipaternalism. Because it is quite general, I call it an "approach" and reserve the term "account" for more specific characterizations. The approach is non-committal in several respects. For starters, it is an open question whether the filtered-out facts are prevented from playing the role of reasons because they are not reasons, or because they are reasons but nevertheless should not play that role in the particular context specified by the details of the doctrine. It is also, relatedly, an open question what role exactly reasons normally play and so what role the filtered-out reasons are prevented from

---

[3] The formula cannot be applied directly. Strictly speaking, it is not correct to say that the filter prevents the fact that the limitation of *several* persons' liberty causes the promotion or protection of *their* good from playing the role of a reason for such limitation. This is because an action that limits the liberty of several people may promote the good of each one only or partly by limiting the liberty of the others. Antipaternalism does not apply to such liberty-limitation. For a thorough treatment of many-people cases, see Grill (2007, pp. 453–455).

playing. It is precisely in order to keep the approach non-committal in these ways that I use the phrase "play the role of reasons".

Without going into any detail, here is a simple map of the roles reasons can play: Reasons can play a role either in reasoning processes or in objective determinations of normative status. The latter include determining the permissibility of actions, the legitimacy of policies and the standing and authority of agents. Some believe these normative statuses are independent of anyone's reasoning about them, yet are dependent on (objective) reasons. Reasoning processes, in turn, include both more volitional processes of forming intentions or motives, whether individually or collectively, and more evaluative processes that aim for knowledge concerning the normative statuses just described, including the permissibility of actions and policies, whether one's own or those of others, and whether future, present or past. The antipaternalist filter can block all of these roles, or some of them.[4]

To see that the basic notion of a filter on reasons is quite straightforward, note that having some reason to prefer some alternative does not imply that one should consider this reason in one's practical deliberation. There are often obvious pragmatic reasons for why one should not, including lack of time and ability, as well as social coordination resulting in the distribution of deliberative tasks.

If antipaternalism is specified towards practical deliberation and if its normative underpinning is pragmatic, the doctrine may be simply a useful guide to deliberation, a way of avoiding mistake, and wasted time and effort. A bit more interestingly, antipaternalism directed at some sort of reasoning can have a more thoroughly consequentialist underpinning, being based on a combination of pragmatic considerations regarding this kind of reasoning and a commitment to the furthering of some value or values. Even more interestingly, antipaternalism can be a deontological principle, directly preventing facts from playing the role of reasons, whether in reasoning or more objectively, for nonconsequentialist moral reasons. Whatever the normative underpinning, antipaternalism adds to (other) pragmatic guides to deliberation a further constraint – a recommendation or requirement that certain facts not play the role of reasons.

I noted that antipaternalism on the filter approach is non-committal regarding whether or not the facts that do not play the role of reasons are reasons at all. This means that the doctrine does not presuppose any particular view on what kinds of reasons there are. Nor, of course, does it presuppose any particular view on how reasons are related to values. This flexibility means that antipaternalism is neutral concerning what is the correct moral theory on this fundamental level. It is a module that can be included in various moral and political theories.

That the filter approach avoids integration with controversial views on what has value, what reasons there are, and what role reasons play, is a great advantage. Antipaternalist doctrines can be and are in fact supported on very different grounds,

---

[4] To the extent that a filter account prevents the consideration of certain facts, it is in a sense focused on action, namely on the mental action of considering facts. This does not, however, amount to an overlap with the action-focused approach, which is concerned with physical actions (and omissions), policies, laws etc.

and are usually intended not to depend on controversial positions on the nature of reasons, of justification, or of reasoning. When antipaternalism is separated out from other commitments, it can more easily be appraised.

The filter approach will hopefully strike many as quite intuitive, perhaps even obvious. It is, however, in opposition to almost all conceptual treatments of paternalism and antipaternalism currently on offer.

Joel Feinberg may at points seem to tend towards a filter approach, but in fact he never formulates it as an alternative. Feinberg's commitment to the action-focused approach is obscured by his ten liberty-limiting principles. These are formulated in terms of which reasons they sanction as good reasons, rather than in terms of which actions they sanction. Indeed, Feinberg defines his liberal position as the position that only two of these principles – the harm principle and the offence principle – are valid principles (e.g. 1984, pp. 14–15).[5] This directly implies a sort of filter on reasons, as Feinberg explicitly concludes: "Paternalistic and moralistic considerations, when introduced as support for penal legislation, have no weight at all."[6] (Ibid., p. 15) However, Feinberg does not stop at this point, but takes this conclusion to be but one step in his further argument. Rather than taking the liberty-limiting principles to operate directly on reasons, Feinberg takes them to determine which prohibitions are and which are not paternalistic. Later authors have followed Feinberg both in emphasizing the crucial role of reasons, and in nevertheless assuming that antipaternalism must target actions, policies or laws.

Douglas Husak has provided one of very few exceptions to the dominance of the action-focused approach.[7] Husak offers a filter account of antipaternalism in his proposal that

> a theory about the conditions under which paternalism is justified [...] might constrain the set of considerations to which legislators are allowed to appeal in their deliberations about whether to support or oppose a given piece of legislation.[8] (2003, pp. 391–392)

---

[5] For Feinberg, the harm principle is the principle that preventing harm to other persons than the actor is always a good reason for prohibition given that there are no better ways of preventing this harm.

[6] Towards the end of the four volumes, Feinberg retracts this claim and in fact states to the contrary that these considerations are "always relevant" (1990, p. 322). This could be understood as a complete abandonment of principled antipaternalism, along with principled antimoralism. However, Feinberg insists that the retraction is not that consequential. He still thinks that the considerations are "hardly ever" good reasons and "perhaps never" decisive (p. 323). Especially in the case of paternalism, Feinberg reaffirms his earlier stance that liberty, in the form of personal sovereignty, is a "trump" that "cannot be put on the interest-balancing scales at all" (p. 322). Tenable or not, his position, even here, is that there is a principled difference between different kinds of reasons, though perhaps a principle with some exceptions.

[7] One earlier example is C.L. Ten (1980), e.g. p. 40: 'There are certain reasons for intervention in the conduct of individuals which must always be ruled out as irrelevant'.

[8] Husak refers to Waldron's "Legislation and moral neutrality" (1989), where Waldron proposes a very similar interpretation of neutralism in politics. The debate on neutrality is in general more explicit and more consistent regarding the role of reasons than is the paternalism debate. Still, the filter approach may be useful for interpreting neutrality too. Rawls' idea of public reason, restricted in content, is a sort of filtering device (e.g. 1997, p. 776). Authors on neutrality such as Larmore (1987), De Marneffe (2010) and most explicitly Wall (1998) talk of neutrality as a constraint or restraint on some sort of reasons.

Husak goes on to more or less reject the action-focused approach, proposing that while reasons can be paternalistic, laws cannot. However, Husak then notes, correctly, that philosophers often talk of paternalistic laws. Apparently contradicting his own proposal, he claims that this talk is neither confused, nor a mistake (p. 390).[9]

I propose that it is indeed a mistake to suppose that the objection to paternalism can be fruitfully interpreted in terms of the rejection of certain laws or actions. I will argue this point at length in the following section.

I have myself previously (Grill 2007) argued that paternalism is essentially about action-reason compounds and so that neither actions nor laws can be paternalistic. I have also previously (Grill 2010) argued that antipaternalism is committed to what I then called the invalidation of reasons, which is a form of normatively based filtering. In relation to my earlier treatments, the filter approach is more general and more clearly positioned in relation to general moral theory and practical reasoning.

4.4 Filter vs. Action-Focus

Principled antipaternalism is the doctrine that a person's good should not count in favor of limiting her liberty. The standard interpretation of this doctrine is action-focused, taking it to be an objection to actions (including government or organizational actions) that limit some person's liberty and that are supported by reasons provided by the protection or promotion of this person's good.[10] Paternalism is typically defined as the performance of such actions. Sometimes it is further assumed or argued that antipaternalism only targets actions that are motivated by a view of some person's good that differs from her own view, or that it only targets actions performed with a condescending attitude. However, I argued against these understandings of antipaternalism above and now disregard them.

The normative debate on the limits of benevolent limitation of liberty is intertwined with a parallel debate on the concept of paternalism, which is partly independent of normative concerns. This wider conceptual paternalism debate is peculiar because, on the one hand, it largely relies on linguistic intuitions about what cases are properly called paternalism, while, on the other hand, it seems to engage authors because they are interested in normative issues to do with antipaternalism. This is peculiar because it seems obvious that the normatively most plausible version of antipaternalism need not target the linguistically most accurate characterization of paternalism.

---

[9] Husak recapitulates these points in a more recent contribution on penal paternalism (2013, p. 40–41).

[10] Important adherents to the action-focused approach include (Dworkin 1972; Gert and Culver 1976; Arneson 1980; Kleinig 1983; VanDeVeer 1986; Archard 1990; Shiffrin 2000; De Marneffe 2006; Dworkin 2010). All of these authors focus their conceptual concern on paternalism rather than on antipaternalism, but their contributions are motivated by the observation that paternalism is, or is allegedly, morally wrong.

Having acknowledged these reservations about the relevance of the wider conceptual debate, I will in the remainder of this section consider and criticize definitions of paternalism in terms of actions, first in general and then in the form of two accounts that are relatively recent and relatively plausible – those of Seana Shiffrin (2000) and Peter De Marneffe (2006). Since the action-focused approach to antipaternalism is dependent on the identification of paternalistic actions, a critique of the project of defining paternalism in terms of actions is also a critique of the action-focused approach to antipaternalism.

On the filter approach, antipaternalism is normatively straightforward. I observed above that reasons may either objectively determine normative status or may figure in reasoning processes. If facts are prevented from playing a role in objectively determining normative status, they simply do not weigh in on the matter, they do not affect the balance of reasons. If facts are instead prevented from playing a role in some reasoning process, antipaternalism is directly action-guiding in prohibiting certain well-defined (mental) actions. The exact scope of the doctrine determines to what status determinations or what reasoning processes it applies exactly, but whatever the scope it is quite clear what it means to abide by the doctrine when it does apply.

For example, if the doctrine applies to reasoning about public policy, then the fact that some policy furthers the good of a person by limiting her liberty does not play the role of a reason for that policy. It is either a practical mistake or a moral failure to be persuaded by such facts to, for example, vote for the policy or to enact it or to abstain from revoking it. If the doctrine applies to reasoning by physicians about the treatment of their patients, then the fact that some treatment furthers the good of a patient by limiting her liberty (e.g. because it is coercive or manipulative) does not play the role of a reason for that treatment. It is either a practical mistake or a moral failure to be persuaded by such facts to, for example, provide the treatment or urge colleagues to provide it.[11]

In contrast, the normative implications of a policy's being paternalistic on the action-focused approach are quite unclear. The action-focused antipaternalist must explain how we should respond to the fact that an action or policy is paternalistic. The most typical explanation is probably that paternalistic actions should not be performed, and policies not enacted. This, however, leaves many questions open,

---

[11] Like most other moral doctrines, antipaternalism, on the filter approach, has no obvious implications for how transgressions should be evaluated. It may be tempting to call "paternalism" failures to abide by antipaternalism directed at deliberation. However, such failures can be trivial in the sense that they do not affect the outcome of deliberation. It is not clear to me whether such trivial failures should be called "paternalism" or what would be gained by doing so.

It may seem extreme to hold that furthering people's good should not even be an operative reason to limit their liberty, in public policy and perhaps especially in the context of medical care. However, it should be noted that the *preferences* of citizens, residents or patients can still provide operative reasons. The principally antipaternalist physician would presumably treat patients in accordance with their preferences and not in accordance with their best interest (and if these are identical, her operative reasons would still be based on patient preference as preference, not as constituent of the good).

such as whether or not paternalistic actions and policies should be prevented by third parties, and if they should be revoked once enacted. To some extent, these problems are shared by any rule formulated in terms of actions. That we should not lie and not murder does not entail that we should prevent lies and murders (and it could plausibly be argued that we should prevent murders but not necessarily lies). However, that it is shared does not make the problem less acute.[12] Furthermore, the problem is arguably greater for antipaternalism than for many other moral prohibitions because there is no obvious net harm to anyone from paternalism. Indeed, there is a net benefit, or at least an intended net benefit. This makes it less obvious how to respond to the categorization of some action as paternalistic. Moreover, as I will now go on to discuss, the concept of paternalism is particularly tangled up in reasons, to the extent that it is less a type of action than the combination of a type of action with a type of rationale.

The dominant approach to defining paternalism is quite preoccupied with reasons. Actions and policies are deemed paternalistic in large part depending on what reasons there are for them.[13] There are two main obstacles to specifying this condition on paternalistic actions: (1) There are different sorts of reasons, including motivational and justificatory. (2) Actions most often have mixed or multiple rationales.[14] Both problems are particularly acute in the political realm, where rationales are more thoroughly considered and are often collective, or are aggregates of many individual rationales, and so are quite diverse. While both the filter and the action-focused approach must identify which *reasons* antipaternalism applies to, the action-focused approach must in addition specify for any mix of reasons for an action whether or not this mix makes the *action* paternalistic. It is the necessity of this further conceptual work that is the main weakness of the action-focused approach.

De Marneffe (2006) aptly captures some of the problems caused by 1 and 2 in his critique of Shiffrin's (2000) definition of paternalism. Shiffrin provides her definition in the context of a defense of the unconscionability doctrine in contract law. De Marneffe points out that Shiffrin's definition is explicitly based on motives rather

---

[12] Amartya Sen attempts to deal with this problem by emphasizing our imperfect duty to aid (e.g. Sen 2012, p. 96, but also in 2009, chapter 17). As noted by Frances Kamm, however, Sen's duty to aid "seems merely to ask us to think and act appropriately about important matters" (2011, p. 94). This is not very helpful.

[13] Some examples: VanDeVeer (1986, p. 22) says that behavior towards a person S is paternalistic only if it has "the primary or sole aim of promoting a benefit for S", Archard (1990, p. 36) says that behavior by P towards Q is paternalistic only if "P's belief that this behaviour promotes Q's good is the main reason for P's behaviour", De Marneffe (2006, pp. 73–74) says that a policy is paternalistic towards A only if "the government has this policy only because those in the relevant political process believe or once believed that this policy will benefit A in some way", and Dworkin (2010) says that X acts paternalistically towards Y by doing Z only if "X does so just because Z will improve the welfare of Y".

[14] Feinberg clearly identifies both of these problems (1986, pp. 16–23). So does Husak, see esp. pp. 390–391. Feinberg uses "multiple" and "mixed" interchangeably. I will for the most part stick with "multiple". It should not be assumed that multiple rationales are a set of rationales that are individually sufficient (to motivate, to justify, etc.) – rationales may be of any strength or weight.

than on justificatory reasons.[15] However, he notes, her aim and strategy is to defend policies from the charge of paternalism by providing alternative *justifications* for them. This defense is problematic because motives and justifications are different sorts of reasons. As De Marneffe goes on to note, policies can be motivated by certain reasons even though other reasons, which happen not to motivate, would provide an adequate justification. Conversely, policies need not be motivated by certain reasons even though these reasons provide the only adequate justification. Therefore, providing alternative justifications does not directly affect the status of a policy as paternalistic or nonpaternalistic on a motivational account of paternalism. (De Marneffe 2006, p. 71)

On the action-focused account, Shiffrin's project seems somewhat incoherent. She condemns benevolently motivated limitation of liberty as deeply insulting, yet she accepts just such limitation in the case of unconscionability since there are good justificatory reasons for this doctrine, unrelated to benevolence (these reasons are roughly to avoid being complicit to exploitation). She does not discuss how these proper justificatory reasons should be balanced against the deep insult that presumably remains as long as the benevolent motive remains. Furthermore, it is far from obvious how these justificatory reasons could remove or even mitigate the insult.

On the filter approach, in contrast, Shiffrin's different positions can be consistently accommodated. Translated to this approach, Shiffrin seems to hold, first, that in considering unconscionability, the fact that the weak party is benefitted does not play the role of a reason for the doctrine. She also seems to hold that certain other facts do provide good reasons for the doctrine. This is quite clear and consistent. The only remaining question is whether or not the fact that the promoter of unconscionability is motivated by the benefit to the weak party plays the role of a reason against such promotion. To hold that it does amounts to a sort of extreme antipaternalism that goes beyond the filtering out of reasons.

The problems that de Marneffe identifies in Shiffrin's account are related to the interplay of motivational and justificatory reasons. However, after his critique of Shiffrin, de Marneffe goes on to argue against a justificational account of paternalistic policy as well. The problem with these accounts is, de Marneffe notes, that it is hard to see how exactly justifications would sort actions into paternalistic and nonpaternalistic.

Before I look closer at this problem, I must briefly note a complication in de Marneffe's presentation. In describing the problem with deciding which actions are paternalist by looking at their justifications, and indeed throughout the article, de Marneffe talks about "paternalistic reasons" (and this is not a term he picks up from Shiffrin, who does not use it). He specifies this at one point as reasons "that cite some benefit to A that A does not want" (p. 72). This is unduly narrow, however, since it may presumably be paternalism to provide A with a benefit he does want, if this is done in a manner which limits his liberty (and that he does not want). More generally, it is not obvious, and is indeed unintuitive, that reasons can be paternalistic on the action-focused approach to paternalism. However, for ease of presentation

---

[15] Other actions-focused accounts that are in this way motivational include (Kleinig 1983; VanDeVeer 1986; Archard 1990; Husak 2003; Dworkin 2010).

and to avoid introducing new terms, I will use the term 'paternalistic reason' as a placeholder for whatever reasons are such that they make an action paternalistic by being reasons for that action. This may be de Marneffe's implicit intention as well (cf. p. 74, footnote 18).

De Marneffe's considers and rejects three justificational accounts of paternalism. The first is that a policy is paternalistic if and only if it can be justified only by counting paternalistic reasons in its favor. Since de Marneffe interprets 'justified only by' to imply 'justified by', all paternalistic policies are justified on this account, which is obviously not acceptable. The second account is that a policy is paternalistic if and only if it cannot be fully justified unless paternalistic reasons are counted in its favor. Given that 'unless' has the same truth conditions as 'or', all unjustified policies are paternalistic on this account, which is also unacceptable. The third account combines the first two to say that a policy is paternalistic if and only if it (i) cannot be fully justified without counting paternalistic reasons in its favor, and (ii) would be fully justified if paternalistic reasons counted in its favor. On this account, a policy for which there are only paternalistic reasons but which is unjustified (since these reason are not sufficient to outweigh the reasons against it) is not paternalistic. This is also unacceptable.

De Marneffe's response to the identified difficulties is to propose that a definition of paternalism, at least for Shiffrin's purposes of reconciliation, incorporate both a motive and a justification component. His preferred version of such a hybrid definition is:

> A government policy is paternalistic toward A if and only if (a) it limits A's choices by deterring A from choosing to perform an action or by making it more difficult for A to perform it; (b) A prefers A's own situation when A's choices are not limited in this way; (c) the government has this policy only because those in the relevant political process believe or once believed that this policy will benefit A in some way; and (d) this policy cannot be fully justified without counting its benefits to A in its favor. (pp. 73–74)

Among all action-focused definitions of paternalism that I am aware of, de Marneffe's best captures the reason aspect of paternalistic action. However, the definition still has peculiar implications. First, the definition inherits a problem from the three rejected accounts: If a policy targets a group of x persons and if it is motivated and justified by its effects on any set of x-1 members of the group, then it is not paternalistic towards anyone, by virtue of both (c) and (d). For no member of the group is it true that the government has the policy only because of the belief that the policy will benefit her, since the belief that it will benefit everyone else is sufficient to motivate the policy. For no member of the group is it true that the policy cannot be justified without counting its benefits to her in its favor, since the benefits to everyone else are sufficient to justify the policy. This implication is obviously undesirable, as it means that almost no policies are paternalistic. The definition can be reformulated to avoid this implication, but this would make it even more complex.[16]

---

[16] The problem cannot be avoided by simply tweaking de Marneffe's definition so that the relevant question is whether the policy can be expected to benefit each member of some group to which A belongs, as this would imply, implausibly, that a policy is not paternalistic toward A even if it limits her choices, she does not want it, and its benefit to her is a necessary part motive and part justification, as long as the policy also affects some other member, who is not (expected to be) benefitted.

Other undesirable implications of de Marneffe's definition may be even harder to avoid. The definition is arguably both too wide and too narrow. It is too wide in that it implies that a policy is paternalistic towards A even if it limits A's choices (in a way that A prefers they were not limited) merely as a side effect of saving A from unwanted harm. For example, a policy may regulate the use of explosives and though A may prefer, *ceteris paribus*, to be free to choose whether or not to use explosives himself, this may be a fairly trivial preference in relation to his strong preference for general regulation. Still, the regulation is paternalistic towards A, given that benefits to A are necessary for its motivation and justification (or some weaker condition of this sort reformulated in light of the problem pointed out in the previous paragraph). Antipaternalists would not, I take it, be principally opposed to this policy on A's behalf.

De Marneffe's definition is too narrow in that it implies that a policy is *not* paternalistic towards a group even if it limits their choices against their will, promotes their good, is generally endorsed for that reason, is in fact unjustified because of its oppressive character, and was enacted with nonpaternalistic motives (perhaps a long time ago). Antipaternalists, I take it, would be opposed to this policy and would want their doctrine to condemn it.

Recall that there are two problems with action-focused definitions of paternalism: There are different sorts of reasons and actions most often have multiple rationales. De Marneffe attempts to deal with the first problem by including both motives and justifications in his definition. However, as shown by the 'overly narrow' objection, there are other sorts of relevant reasons, such as the reasons for which others than policy-makers endorse a policy. De Marneffe attempts to deal with the second problem by making reasons of both kinds necessary reasons rather than for example the only or main reasons, as on other proposed definitions.[17] However, as show by the 'overly wide' objection, the ways in which a policy can limit someone's choice and benefit her are more diverse than the definition can handle.

One might respond to the identified problems by constructing an even more intricate definition of paternalism, and undoubtedly someone will. The project of defining paternalistic action invites creative counter-examples, further specification and modification, further counter-examples, and so on. However, it is not clear that this method, fruitful in other contexts, is helpful in this case, since the filter approach avoids the identified problems, without defining paternalistic action at all.

De Marneffe's explicit aim in the article is to refute antipaternalism, or what he calls "the general presumption against paternalism" (p. 69). In the latter two thirds of his article, he rather convincingly does so by arguing that neither paternalistic motives nor paternalistic justifications are inherently problematic. In the introduction, de Marneffe states that if "a general principle of antipaternalism is valid, then we should evaluate ... policies by evaluating whether or not there is sufficient

---

[17] For example John Gray (1983) claiming that paternalism is "to coerce an individual solely in his own interest" (p. 90) and Archard (1990) objecting to this condition and proposing instead that promoting the good of the person is the "main reason" (p. 38).

*non*paternalistic reason for them." (p. 69) This brief statement takes us very close to the filter approach – policies should be evaluated without regard to paternalistic reasons. Just as I have done, de Marneffe leaves it open to specification exactly which reasons are paternalistic and which are nonpaternalistic.

The characterization of antipaternalism that can be extracted from De Marneffe's brief statement is unnecessarily restricted to evaluation, since there are other kinds of reason-taking contexts. It is also needlessly restricted to the issue of whether or not there are *sufficient* reasons for a policy. We may also be interested in for example whether there are stronger nonpaternalistic reasons for some policy than for an alternative policy. Still, this brief characterization is perhaps sufficient for de Marneffe's normative investigation and I do not see how the following seven pages, where de Marneffe labors with a definition of paternalistic policy, does anything to advance either our understanding of antipaternalism or his case against this doctrine.

De Marneffe's and Shiffrin's contributions to the conceptual paternalism debate are, I believe, the state of the art in this area. However, they do not provide definitions of paternalism that enable us to employ this concept fruitfully in normative contexts. Their definitions cannot be used to capture that which antipaternalism is opposed to. De Marneffe's two-line characterization of antipaternalism is more helpful than his seven-page definition of paternalism. None of this proves that the action-focused approach to antipaternalism is deficient through and through, but it is a strong indication. Considering also the general problems faced by any action-focused definition – the variety of sorts of reasons and the multiple rationales for most actions – the prospects for the action-focused approach are slim. Especially so since there is a ready alternative – the filter approach.

4.5 Conclusion

Principled antipaternalism is the only objection to paternalism that has substantial independent normative thrust. It should be understood as demanding that certain facts do not play the role of reasons. This is the filter approach. Typically, the facts are of the form that some person will benefit from having her liberty limited or her autonomy infringed. The approach, however, is structural and conceptual, independent of which facts are targeted exactly. The approach is also independent of what reasons there are more generally and what role reasons in general play.

In the conceptual debate on paternalism, it is widely assumed that what needs defining is paternalistic action, conduct, behavior, policy or law. Relatedly, antipaternalism is typically understood in terms of resistance to these things. This dominant action-focused approach leads to intricate and unnecessary problems, as illustrated by Seana Shiffrin's and Peter De Marneffe's discussion of normative issues to do with paternalism. The failure of the action-focused approach to capture the proper role of reasons should lead us to favor the filter approach. There may be no escaping talk of paternalistic actions and paternalistic policies as shorthand in less stringent contexts, but these concepts are ill suited for careful normative investigations of the moral problems that allegedly surround paternalism.

The filter approach makes the most of the traditional liberal opposition to benevolent interference. The strongest antipaternalist position is reached by abstaining from sorting actions into paternalistic and nonpaternalistic, in favor of designing a plausible filter between facts and operative reasons.

References

Archard, David. 1990. Paternalism defined. *Analysis* 50(1): 36–42.
Arneson, Richard. 1980. Mill versus paternalism. *Ethics* 90: 470–489.
Brock, Dan. 1988. Paternalism and autonomy. *Ethics* 98(3): 550–565.
De Marneffe, Peter. 2006. Avoiding paternalism. *Philosophy and Public Affairs* 34(1): 68–94.
De Marneffe, Peter. 2010. *Liberalism and prostitution*. Oxford: Oxford University Press.
Dworkin, Gerald. 1972. Paternalism. *The Monist* 56: 64–84.
Dworkin, Gerald. 1983. Some second thoughts. In *Paternalism*, ed. Rolf Sartorius, 105–111. Minneapolis: University of Minnesota Press.
Dworkin, Gerald. 2010. Paternalism. In *The Stanford encyclopedia of philosophy*, (Summer 2010 Edition), ed. E.N. Zalta. URL = http://plato.stanford.edu/archives/sum2010/entries/paternalism/.
Feinberg, Joel. 1984. *Harm to others*. Oxford: Oxford University Press.
Feinberg, Joel. 1986. *Harm to self*. Oxford: Oxford University Press.
Feinberg, Joel. 1990. *Harmless wrongdoing*. Oxford: Oxford University Press.
Gert, Bernard, and Charles M. Culver. 1976. Paternalistic behavior. *Philosophy and Public Affairs* 6(1): 45–57.
Gray, John. 1983. *Mill on liberty: A defence*. London: Routledge & Kegan Paul.
Grill, Kalle. 2007. The normative core of paternalism. *Res Publica* 13: 441–458.
Grill, Kalle. 2010. Anti-paternalism and invalidation of reasons. *Public Reason* 2: 3–20.
Groll, Daniel. 2012. Paternalism, respect, and the will. *Ethics* 122(4): 692–720.
Guyer, Paul. 2014/2006. *Kant*, 2nd ed. New York: Routledge.
Husak, Douglas. 2003. Legal paternalism. In *Oxford handbook of practical ethics*, ed. Hugh LaFollette, 387–412. Oxford: Oxford University Press.
Husak, Douglas. 2013. Penal paternalism. In *Paternalism: Theory and practice*, ed. Christian Coons and Michael Weber, 39–55. Cambridge: Cambridge University Press.
Kamm, Frances. 2011. Sen on justice and rights: A review essay. *Philosophy and Public Affairs* 39(1): 82–104.
Kant, Immanuel. 1991/1797. *The metaphysics of morals*. Trans. Mary J. Gregor. Cambridge: Cambridge University Press.
Kleinig, John. 1983. *Paternalism*. Manchester: Manchester University Press.
Larmore, Charles. 1987. *Patterns of moral complexity*. Cambridge: Cambridge University Press.
Mill, J.S. 1859. *On liberty*, (Many editions).
Quong, Jonathan. 2010. *Liberalism without perfection*. Oxford: Oxford University Press.
Rawls, John. 1997. The idea of public reason revisited. *University of Chicago Law Review* 64: 765–807.
Rawls, John. 1999. *A theory of justice*, 2nd ed. Cambridge, MA: Harvard University Press.
Sen, Amartya. 2009. *The idea of justice*. Cambridge, MA: Harvard University Press.
Sen, Amartya. 2012. The global reach of human rights. *Journal of Applied Philosophy* 29(2): 91–100.

Shiffrin, S. 2000. Paternalism, unconscionability doctrine, and accommodation. *Philosophy and Public Affairs* 29(3): 205–250.

Szasz, Thomas. 1992. *Our right to drugs: The case for a free market.* Syracuse: Syracuse University Press.

Ten, C.L. 1980. *Mill on liberty.* Oxford: Clarendon.

VanDeVeer, Donald. 1986. *Paternalistic intervention.* Princeton: Princeton University Press.

Waldron, Jeremy. 1989. Legislation and moral neutrality. In *Liberal neutrality*, ed. R. Goodin and A. Reeve. London: Routledge.

Wall, Steven. 1998. *Liberalism, perfectionism and restraint.* Cambridge: Cambridge University Press.

Part II
Justifying and Rejecting Paternalism

Chapter 5
Paternalism and Consent

Douglas Husak

5.1 A Relatively Clear Case

I begin with an ordinary, everyday example from which I hope to generalize about the justifiability of paternalism and, to a lesser extent, about the difficulties of justifying paternalism in the criminal law. When permitted to eat anything he chooses, 4-year-old Billy skips his vegetables altogether and eats only his ice cream dessert. His father has tried to explain the reasons to eat a balanced diet, but Billy is unmoved, and has not changed his behavior. Suppose his father comes to you for advice about what to do at their next dinner. I stipulate that the father's only reason for seeking advice is to improve Billy's health and welfare by ensuring that he eats a more nutritious meal than if left to his own devices. It seems reasonable for you to recommend that Billy not be permitted to eat his ice cream unless and until he finishes his vegetables. Suppose his father decides to follow your advice. This example not only describes a situation in which Billy is treated paternalistically but also represents a relatively clear case in which the paternalistic treatment is justified.[1] In any event, I make these two assumptions about this case.

I stipulate that the father's only reason for withholding ice cream is to improve Billy's health and welfare because I construe paternalism to be a function of the *motives* for interfering in the liberty of another. Paternalism should not be defined in terms of its beneficial effects or consequences, but rather in terms of the reasons for

---

This article first appeared in Franklin G. Miller and Alan Wertheimer (eds.), *The Ethics of Consent: Theory and Practice*, Oxford University Press 2010), 107–130, and is reprinted by kind permission of Oxford University Press.

[1] I do not contend that this second assertion is beyond serious dispute. See infra p. 109. Fortunately, nothing of importance turns on any particular example; I need only to assume that *some* case of justified paternalism can be described, and that its justification depends on the criteria I provide.

D. Husak (✉)
Distinguished Professor of Philosophy, Rutgers University, New Brunswick, NJ, USA
e-mail: husak@rci.rutgers.edu

© Springer International Publishing Switzerland 2015
T. Schramme (ed.), *New Perspectives on Paternalism and Health Care*, Library of Ethics and Applied Philosophy 35, DOI 10.1007/978-3-319-17960-5_5

which it is imposed. His father acts paternalistically even if he unwittingly worsens Billy's health or welfare. Because of this feature in my understanding of paternalism, few rules or laws are unambiguously paternalistic—that is, *purely* paternalistic.[2] Most (and perhaps all) rules or laws are promulgated by authorities or legislators whose motives for enacting the rule or law are a mixture of paternalistic and nonpaternalistic motivations. Laws requiring the wearing of seat belts, for example, probably are designed both to minimize the severity of automobile accidents and to reduce the insurance costs to all drivers. The case I have described, however, is a good candidate for an example of pure paternalism. It is hard to see what other reason his father might have for withholding ice cream from Billy. In any event, I stipulate that his only motive is paternalistic.

Why might you offer the aforementioned advice? Five criteria conspire to make this example a relatively clear case of justified paternalism. *First*, the intrusion is a fairly minor interference in Billy's liberty—as minimally intrusive as can be imagined to accomplish its objective. Billy is not beaten or deprived of something of great significance to induce him to change his behavior. *Second*, the objective sought by his father is obviously valuable. No one contests the importance of health. *Third*, the means chosen are likely to promote this objective. If Billy's desire for ice cream is sufficiently strong, he is likely to alter his behavior and eat his vegetables. And any competent nutritionist agrees that vegetables are an essential part of a healthy diet—more essential than ice cream. *Fourth*, Billy himself is not in a favorable position to make the right decision. Children have notorious cognitive and volitional deficiencies relative to competent adults that prevent them from recognizing their best interests, or from acting appropriately even when they do. *Fifth*, his father stands in an ideal relationship to Billy to treat him paternalistically. Parents have special duties to protect and enhance the welfare of their children. I believe that my example satisfies each of these five criteria.

If I have misapplied any of these conditions, I would have to withdraw my claim that Billy's case represents a clear instance of justified paternalism. Since I have a few reservations, I describe this case as *relatively* clear. It is surprisingly difficult to find uncontroversial examples of justified paternalism. In particular, the application of the third criterion to my case might be contested. Among other difficulties, the father's plan may backfire. Arguably, the paternalistic treatment to which children like Billy are subjected may induce them to eat more poorly in the long run, when they no longer remain under parental supervision. Applying criteria of when paternalism is justified will always raise controversies, some of which involve disputes about matters of fact. My main focus, however, is on the criteria themselves. With only a bit of ingenuity, I believe that most and perhaps all questions about the justifiability of any paternalistic interference can be raised within the parameters of these five criteria.

---

[2] Literally, rules or laws are not the kinds of thing that *can* be paternalistic. To say that a rule or law is paternalistic is best interpreted to mean that it is adopted or enacted largely from a paternalistic motive. Generally, see Douglas N. Husak (2004).

Four comments about these criteria are worth making. First, there are potential difficulties with my strategy of beginning with a relatively easy case, identifying what is easy about it, and applying these criteria to other examples. In particular, each of my criteria may not need to be satisfied to justify an instance of paternalism. Why, for example, must the subject be less than fully competent? Doesn't this criterion automatically preclude what Joel Feinberg calls "hard paternalism"?[3] In order to avoid such questions, I do not insist that these criteria must be satisfied before an instance of paternalism is justified. Instead, each criterion merely contributes to the judgment that a case is easy. Whatever else may be said about instances of hard paternalism, they surely are more difficult to justify than cases of paternalism in which the subject is less than fully competent. I take no firm position on what we should ultimately say about a case in which it is dubious whether one or more of these conditions are satisfied. I hold only that it progressively becomes less clearly justified, and eventually is clearly unjustified.

Second, conditions one and three are the most important of several reasons why *criminal* paternalism is so difficult to justify. Consider the first condition. A paternalistic interference becomes harder to defend when the means required to attain its objective involve a greater hardship or deprivation of liberty. The criminal law, by definition, subjects persons to state *punishment*. If the state must punish someone to protect his interests and well-being, we have reason to suspect that the cure is worse than the disease. It may be bad for persons to use drugs, for example, but it may be even worse to punish them to try to get them to stop. When punishments are severe, their gains typically will not be worth their costs for the persons on whom they are inflicted. But when punishments are not severe, they rarely will create adequate incentives for compliance and thus will fail to improve the behavior of the persons coerced. An acceptable set of constraints to limit the imposition of the criminal sanction will require that criminal laws must be reasonably effective in attaining their objectives.[4] A criminal law motivated by a paternalistic end will fail to satisfy this condition if it does not alter conduct or actually makes the subject worse off, all things considered. I doubt that paternalistic reasons will justify state punishment in more than a handful of cases.

Criminal paternalism also is jeopardized by the third condition. To be justified qua paternalism, the interference must actually benefit the person coerced. Laws are general, however, and apply to a great many persons in a variety of circumstances. Statutes requiring persons to buckle their seat belts or activate their air bags, for example, protect the vast majority of drivers, but actually increase the risk of harm for a minority. Persons who plunge into water, for example, are more likely to drown if they are wearing seat belts. In addition, drivers who are unusually short are much more likely to be injured by air bags than persons whose height is close to average. In principle, of course, criminal laws can create exceptions for given kinds of circumstances, either by allowing a defense or by including an exceptive clause in the offense itself. In practice, however, it is nearly inevitable that rules will be

---

[3] See Joel Feinberg (1985).
[4] See Douglas Husak (2008).

overinclusive and persons will be criminally liable despite the fact that they act in circumstances in which compliance with the law would not have benefited them. In a one-on-one confrontation, such as that involving Billy and his father, we need be less worried that the generality of a rule motivated by a paternalistic objective will actually operate to the detriment of some of the persons coerced.

Third, most proposals to treat competent adults paternalistically are rendered problematic by the fourth criterion. A diet consisting solely of ice cream is probably no less unhealthy for middle-age individuals than for Billy, but sane adults rarely suffer from the deficiencies of typical 4-year-olds. Of course, age is simply a crude proxy for what is relevant: the state of cognitive and volitional capacities character-istic of sane adults. An adult who is cognitively and volitionally comparable to a child is an equally plausible candidate for paternalistic intervention. Unfortunately, some such adults exist. Thus, I see no reason to suppose that the paternalistic treat-ment of adults is never permissible.

Fourth, the final criterion is the most questionable in the set. Suppose that some-one who does not stand in a special relationship to Billy has an opportunity to treat him in exactly the same way for exactly the same reason as his father, withholding ice cream until he finishes his vegetables in order to enhance his health by improv-ing his diet. May he do so as well? We might disapprove of his tendency to meddle, but should we conclude that his interference would be unjustified? In a genuine emergency, I am sure that the fifth condition becomes totally irrelevant. If a child is playing in the road in the path of an oncoming bus, the identity of the person who snatches him away is immaterial. But what should we say about less extreme cases, like that of Billy? I am unsure how this question should be answered, and it provides the main basis for the misgiving I will express near the end of this chapter. In any event, the importance of the remaining criteria seems more secure. Suppose that the child is quite a bit older and more competent, the end that is sought is less clearly valuable than health, the interference is less likely to attain its objective, and/or the means employed involve a greater deprivation of liberty. For example, 13-year-old Jimmy might be prevented from playing with his friends until he finishes practicing the bassoon. Clearly, this instance of paternalism is far more difficult to justify. As these examples suggest, each of these criteria involves a matter of degree. As I have indicated, at some point on a continuum what is otherwise a clear case of justified paternalism becomes less clear, and eventually is not justified at all. Reasonable minds will differ about the precise point along this spectrum—or, indeed, along the several spectra—at which a particular instance of paternalism crosses this elusive threshold and becomes unjustified.

The foregoing is helpful in introducing my central thesis. Suppose we are given one additional piece of information about the ordinary, everyday case of justified paternalism with which I began. Imagine we are told that Billy does not consent to the treatment I have proposed. He strongly objects to what his father does, and pro-tests loudly when his ice cream is withheld until he finishes his vegetables. I trust that no one who agreed with my initial verdict about this case would change his opinion in light of this new information. In fact, it seems odd to describe this piece of information as *new*; most readers would have assumed it to be true in their initial reflections about the case. In any event, it would be remarkable to suppose that Billy's lack of consent to his treatment is material to whether the act of paternalism is justified. When one person *A* treats another person *B* paternalistically and is justified in so doing, *B's* lack of consent is irrelevant. Much of the point of the example is to show that his father is justified in treating Billy paternalistically, even though his son does not consent to being treated in this way.

In fact, Billy's consent almost certainly would entail that the case no longer qualifies as an example of paternalism at all, quite apart from whether it is justified.[5] Suppose his father threatens to withhold ice cream, and Billy, an exceptionally precocious child, replies that the threat is unnecessary to ensure his compliance. His past behavior notwithstanding, he now has come to understand the importance of health and the instrumental value of a good diet. He resolves not to eat his dessert before finishing his vegetables, and proceeds to act accordingly. In such an event, I would say that his father threatened to treat Billy paternalistically, but did not actually have to do so, since Billy complied without the need for interference—that is, without the need for his father to make good his threat.[6] Billy has been persuaded, not coerced. The clearest cases of paternalism involve *coercion*, or an *interference* with liberty.[7] If I am correct, persons are not treated paternalistically when they consent to their treatment.

But not all cases are clear, and philosophers have challenged my claim that paternalism involves an interference in liberty and that the absence of consent is irrelevant to its justification. Much of this paper is designed to respond to this challenge. So-called *libertarian paternalism* poses a possible complication for my claim that paternalism involves an interference in liberty.[8] Libertarian paternalism works primarily by designing default rules to correct for well-known cognitive biases and volitional lapses, thereby minimizing the likelihood that persons will make decisions that are contrary to their own interest. Consider the following two examples. Rather than explicitly choosing to participate in an efficient company health plan, employees might be enrolled automatically unless they opt out. Seat belts might be constructed to buckle immediately upon closing a car door, although occupants would be able to unbuckle them if they chose to do so.[9] Might consent be crucial to the justification of libertarian paternalism? Perhaps. But are these provisions really paternalistic? If persons can change the impact of these rules, it is doubtful we should say that an *interference* with choice has occurred. Notice that it might be true

---

[5] Perhaps this conclusion can be applied to all attempts to justify paternalism by reference to consent—even when consent is noncontemporaneous. See Thaddeus Mason Pope (2005).

[6] It is not clear how a parent can threaten to treat someone paternalistically when paternalism is justified. Typically, threats are distinguished from offers because they make their recipients worse off. If Billy is indeed better off when treated paternalistically, as I have stipulated, his father's proposal is difficult to categorize as a threat.

[7] Some philosophers contend that not all cases of paternalism involve interference. Presumably, a doctor may treat an unconscious patient paternalistically, although he could hardly interfere in a choice the patient is incapable of making. See Bernard Gert and George Culver (1976).

[8] See Cass R. Sunstein and Richard H. Thaler (2003).

[9] See J.D. Trout (2005).

that individuals "can" alter the default rule in two senses. First, persons who elect not to participate in the company health plan face no legal penalty. Second, opting out is not onerous, requiring a mere stroke of a pen or click of a switch. When these two conditions are satisfied, it seems more appropriate to construe these rules as designed merely to *influence* persons to pursue their self-interest.[10]

Admittedly, some provisions appear paternalistic even though they actually expand choice. The Federal Trade Commission, for example, mandates a 3-day cooling-off period for door-to-door sales. It seems facetious to characterize this rule as interfering with the options of a buyer—unless we suppose that the state has interfered with his choice to make a spontaneous purchase that is irrevocable.[11] Instead of construing these provisions as paternalistic, I believe they are better understood as assisting persons in satisfying their preferences rather than as interfering with their liberty. But I do not insist that any of these devices cannot be conceptualized as paternalistic; they embody what might be called the *spirit* of paternalism. When the effort required to change the operation of a default rule becomes overly burdensome—involving reams of paperwork, for example—we may be tempted to think that an interference with choice has taken place. I see no reason to suppose that there always must be a "right answer" to how paternalism should be defined, or how the definition should be applied to particular examples. Apart from my claim that the presence of consent would disqualify the case as an instance of paternalism, I make little further effort to offer a definition. At some point or another, theorists must resort to stipulation, and further quibbles about the exact nature of paternalism become fruitless. I hope my failure to provide a precise definition does not undermine any of the points I will defend. What is controversial is whether and how any or all of these devices can be justified, not whether they "really" qualify as instances of paternalism.

On the topic of paternalism and consent, I believe that not much more needs to be said. Although many difficult questions surround consent— whether it is a mental state or a performative, under what conditions it is voluntary, whether it should be a defense for serious inflictions of injury, and the like—none of these issues need concern the paternalist.[12] Hard cases notwithstanding, lack of consent on the part of the person treated paternalistically simply is not relevant to whether the interference is justified.[13] If all cases were as clear as my example of Billy and his father, the topic of paternalism and consent would be straightforward and uninteresting.

---

[10] Taxes designed to discourage people from engaging in activities that create risks of harm, such as using tobacco products, probably should be conceptualized similarly. Unless rates of taxation become prohibitive, they should be thought to influence rather than to interfere with choice. Generally, see the discussion of the "robustness principle" in Jim Leitzel (2008).

[11] See Colin Camerer et al. (2003). Complications arise if the price for the spontaneous and irrevocable purchase is lower than that for the revocable purchase.

[12] For a nice discussion, see Peter Westen (2004).

[13] Formulations of the consent defense in criminal law accord with this position. The Model Penal Code provides that consent is "ineffective" if "it is given by a person whose improvident consent is sought to be prevented by the law defining the offense." American Law Institute, *Model Penal Code* §2.11(3)(c) (1962).

Alas, matters are not so simple. Consent seemingly becomes controversial in justifying paternalism because many examples deviate from the ordinary case I have described. In the kinds of cases I will discuss, consent to a given treatment is *non-contemporaneous*; that is, consent is withheld at the moment the paternalistic treatment takes place, even though it is given at some other time. Despite the complexities about noncontemporaneous consent I will examine, however, I believe that my thesis remains basically correct: The absence of consent is irrelevant to whether a case of paternalism is justified. I will, however, express a misgiving about my thesis—a misgiving that leads me to describe my thesis as tentative. If consent is relevant to whether paternalism is justified, it is material to my fifth and final criterion: to the issue of *who* is entitled to treat another paternalistically. Ultimately, however, I am unsure whether this fifth criterion should be retained.

Apart from my reservation, it might be thought that consent is implicitly involved in the preceding case after all. I have simply assumed that his father is justified in treating Billy paternalistically. Even if my assumption is granted, we still may disagree about *why* his action is justified. According to Gerald Dworkin's pioneering article, consent plays a crucial role in answering this question. He alleges that what he calls "future-oriented consent" is the key to justifying paternalism. Dworkin writes: "Paternalism may be thought of as a wager by the parent on the child's subsequent recognition of the wisdom of the restrictions. There is an emphasis on what could be called future-oriented consent—on what the child will come to welcome rather than on what he does welcome."[14] Dworkin's proposal, as I construe it, is that the paternalistic intervention is justified if Billy subsequently comes to appreciate it, but is unjustified if he does not. If Dworkin is correct, my stipulation that the father is justified in withholding ice cream implies that Billy eventually will consent to the restriction.

Elsewhere, I have contended that this rationale fails for two related but distinct reasons.[15] First, criteria are needed to justify paternalism *ex ante*, when the parent must decide whether to impose it. We do not offer helpful advice to Billy's father if we inform him that no one can tell whether his proposed interference is justified until some future moment when Billy will decide whether or not to welcome what his father once did. And which of several possible future moments should we privilege? Billy may *vacillate*, changing his mind throughout his lifetime.[16] He might resist the interference for a short while, welcome it subsequently, only to resent it again later. As this possibility suggests, the fundamental problem with Dworkin's proposal is that Billy's ex post opinion is irrelevant to whether his father is justified—even if we could accurately predict Billy's *ex post* judgment ex ante. We should not conclude that his father is unjustified in treating Billy paternalistically simply because Billy never actually consents. Billy may fail to appreciate the wisdom of the restriction because he grows up to be stubborn or stupid, or—in the most extreme case—because he does not grow up at all. Suppose that Billy is hit by a bus

---

[14] Gerald Dworkin (1972).
[15] Douglas Husak (1980).
[16] See Tziporah Kassachkoff (1994).

and killed before he is old enough to assess his father's decision. Surely we should not conclude that his father's treatment was unjustified. The decision was justified *whatever* may happen to Billy at a later time.

A third difficulty is that Dworkin is not really talking about consent at all. It is unlikely that consent *can* be retrospective.[17] Even if consent can be retrospective in some unusual circumstances, I certainly do not consent to everything I subsequently come to welcome. Often I am in a better position to assess how events affect my welfare long after they occur, but this superior perspective should not be mistaken for consent if I later come to realize that the treatment I disliked at the time operated to my benefit. Suppose my wife runs off with another man and breaks my heart, and the details of how our property is to be divided depend on whether I consented to the separation. Suppose further that I find and marry a woman I adore even more, and come to believe that I never really loved my first wife at all. Someone would seemingly rewrite history if he claimed that I now consent to having been abandoned. I would agree that my first wife did me a favor by leaving me, even though I did not realize it at the time. But I would not say that I consented to her departure. Surely my first wife could not argue that I gave my future-oriented consent to the separation, so our property should be divided accordingly.

If consent ("future-oriented" or otherwise) does not justify his father's treatment of Billy, what does? In my view, paternalism is justified when it is reasonable, and the father must make a judgment of whether his restriction qualifies.[18] Obviously, no formula will govern determinations of reasonableness. But when each of the five criteria I have described is satisfied to a significant degree, I believe that paternalism will clearly be justified. In other words, paternalism is justified when it is reasonable, and the criteria I have provided will help us decide when this is so. Of course, some contractarians explicate reasonableness in terms of hypothetical consent. What is reasonable *is* what rational persons would agree to under appropriate conditions of choice. I need not try to dissuade these philosophers. Perhaps rational persons under appropriate conditions of choice would agree that paternalism is justified when each of my five criteria is satisfied to a significant degree. In any event, hypothetical consent simply is not *actual* consent, and my conclusion is that the latter, whenever conveyed, is irrelevant to the justifiability of paternalism.

5.2   Prior Consent: Self-Exclusion Programs

It would be hasty, however, to conclude that the absence of consent never is relevant to any determinations of whether paternalism is justified. In an interesting subset of cases, the justification of paternalism *seems* to originate in the actual consent of the

---

[17] For serious consideration of the possibility that consent can be retrospective, see Westen, *Logic*, 254–61.

[18] Elsewhere, I have suggested that paternalistic interferences are reasonable when they promote the conditions of personal autonomy. See Husak, "Legal Paternalism."

very subject treated paternalistically. Despite the consent of the person whose liberty is infringed, these cases still seem to qualify as genuine instances of paternalism. In the kinds of cases I have in mind, consent is real and given *ex ante*, not hypothetical or given *ex post*. Describing and assessing such cases will require a bit more effort than was involved in my previous example of Billy and his father.

Economists have come to appreciate that few of us are very proficient at maximizing our own happiness or utility.[19] This realization helps to justify a range of practices beyond the so-called libertarian paternalism I mentioned previously. Most of us recognize our own weaknesses and tendencies to perform acts that are bad for us and that we subsequently regret. If we are intelligent, we develop strategies to overcome these difficulties or to minimize the damage they cause. A number of prominent theorists, including Thomas Schelling,[20] Jon Elster,[21] George Ainslie,[22] and George Lowenstein,[23] have described several of these strategies in impressive detail. Suppose that painful experience leads Eric to understand his tendency to become intoxicated at parties. He may employ any number of *commitment strategies* to minimize the risk that he will suffer as a result of his behavior. For example, Eric may take a cab to the party so that he cannot drive home. These strategies involve what might be called *paternalism toward oneself*—a mode of paternalism that often is pure, not containing the mixture of paternalistic and nonpaternalistic motives so common for rules and laws imposed upon others. As far as I can discern, few interesting moral questions are presented when these commitment strategies do not enlist the assistance of others persons. These plans may be clever or dumb, effective or ineffective, but they rarely pose serious ethical issues. Moral difficulties arise, however, when a commitment strategy requires the cooperation of another party. These difficulties must be confronted because the second party may need to resort to coercion to ensure the success of the commitment strategy.

These moral issues are somewhat less acute (although not nonexistent) when a person specifically stipulates in advance how he wants to be treated when his contemporaneous consent cannot be given—because he will be unconscious, for example. Many individuals have executed "living wills" that specify their preferences if we are on life support and incapable of expressing our consent at the time a medical intervention is proposed. Moral problems are compounded, however, when we seek to provide in advance how we wish to be treated when we know that our contemporaneous consent can be given, but is likely to diverge from what we now believe will be in our best interest. Suppose that Eric drives to a party and entrusts his keys to his friend Jill,

---

[19] An enormous literature has grown around this topic. Generally, see Michael Bishop and J.D. Trout (2005).

[20] See Thomas C. Schelling (1984). Schelling lists several self-regulatory strategies, including relinquishing authority to someone else, disabling oneself, removing resources, submitting to surveillance techniques, incarcerating oneself, arranging rewards and penalties, rescheduling one's life, avoiding precursors, arranging delays, using teams, and setting bright line rules.

[21] Jon Elster (1983).

[22] George Ainslie (1992).

[23] George Lowenstein and Ted O'Donoghue (2006).

imploring her not to return them if he becomes drunk. Again, no difficulties are presented as long as he maintains his resolve. But moral problems must be confronted if Eric changes his mind and later decides that he no longer prefers to abide by the restrictions to which he had agreed. In this event, Jill must decide what she ought to do. Should she follow his earlier instructions and retain the keys, or comply with his present wishes and return them?

The first thing to notice about this kind of case is that it places Jill in an awkward position. On the one hand, Eric is likely to be angry with her today if she refuses to return his keys when he demands them. Jill will cite her earlier promise as her justification for noncompliance, but Eric (if he is sufficiently sober) will point out that promises ordinarily bind only as long as the promisee does not release the promisor from her promissory obligation. Both morality and law tend to privilege contemporaneous expressions of consent or nonconsent over prior conflicting preferences. Expressed in the simplest terms, persons generally are free to change their minds. On the other hand, Eric is likely to be angry with Jill tomorrow if she complies with his request to return his keys today. He will remind her that his sole reason for extracting her promise in the first place was to prevent him from changing his mind should this very contingency arise. Thus, he places Jill in a "lose-lose" predicament. One valuable lesson to be learned is that persons should be reluctant to make promises to cooperate with others who seek to attain paternalistic ends through a commitment strategy that enlists their assistance. Because we should be hesitant to place others in an uncomfortable moral position, we should make every effort to try to overcome our weaknesses without soliciting the help of others.

I propose to explore this sort of issue in the context of a fairly recent and fascinating phenomenon: *self-exclusion programs* that enable persons to voluntarily place themselves on a list to be barred from casinos. A majority of the 48 of 50 states that presently allow gambling have provided a device by which individuals can authorize casinos to eject them should they attempt to enter. The details of these programs vary enormously from one jurisdiction to another; generalizations are almost impossible to draw. New Jersey, for example, allows individuals to obtain forms by mail or over the Internet, but applicants must appear in person at a handful of designated locations to complete their enrollment.[24] Participants may request exclusion for a minimum of 1 year, for 5 years, or for life, and the exclusion is irrevocable throughout whatever period is elected. Casino personnel are instructed to refuse entry to persons on the list, or to prevent them from making wagers in the event they manage to gain admission. If participants in the program somehow gamble and win, their winnings are to be confiscated. If they lose, their losses are not to be returned. Participation in a self-exclusion program is an excellent example of a commitment strategy that requires the cooperation of another person. Individuals give their explicit consent to be excluded, but enlist the help of casino personnel to ensure that they maintain their resolve.

---

[24] Information and forms about this program are available at http://www.state.nj.us/casinos/forms/excludeform.pdf.

Like the previous examples I have discussed, no important ethical questions arise if the gambler conforms to his earlier position. No one need treat another paternalistically as long as the participant in the self-exclusion program does not attempt to gamble. In this event, these programs may be conceptualized as a helpful means to increase the probability that persons will attain objectives they recognize to be in their self-interest. Problems occur, of course, when the participant changes his mind. Suppose that Smith appears at a casino several years after having authorized a lifetime exclusion. He goes directly to the manager and explains that he has overcome the problems that led him to enroll in the program, and now wants to place a modest wager notwithstanding his prior request to be banned. The casino manager must decide whether to honor Smith's current preference or the preference he expressed in his distant past. In many respects, the manager's predicament resembles the uncomfortable position in which Eric placed his friend Jill when he sought her assistance in avoiding the consequences of his intoxication. The manager seeks advice from a moral philosopher. What advice should we offer?

The question I intend to raise might be construed somewhat differently. We want to know whether and under what circumstances a subject's prospective consent to a burden (which he undertakes for his own good) to which he subsequently objects remains *valid* or *effective* in morality—that is, whether his consent is sufficient in morality to permit the actor to impose the burden despite the subject's contemporaneous objection. Apart from the misgivings I describe later, my thesis is that consent does *not* make a difference to whether others are entitled to treat persons like Smith paternalistically. If it is permissible to treat him paternalistically, the ongoing validity of prior consent is not what does the justificatory work.

In assessing this thesis, notice how odd it would be to think that prior consent had any special significance when a given interference is motivated by a *non*paternalistic rationale. That is, the absence of consent gives us no reason to judge a deprivation to be impermissible when it is designed to prevent harm to others. Suppose Craig is painfully aware of his tendency to molest children, and requests city officials to escort him from a playground whenever he is found there. I stipulate that his sole reason for alerting the officials is to protect potential victims. Suppose that Craig appears at the playground, is asked to leave, and indicates that he withdraws his prior consent to depart. What should the official do? Whatever the answer to this question may be, I do not believe it differs from the answer the official should reach when confronted with Jason, whose tendency to molest children is known to be equally great but who has not issued an earlier request to be made to leave. Craig's prior consent is not effective in authorizing what would be impermissible in its absence. My tentative thesis about the irrelevance of consent entails that whatever is permissible to do to Craig is permissible to do to Jason. Later I will return to the issue of how the criteria to justify paternalistic interferences might be *un*like those that justify nonpaternalistic interferences. My present point is that these criteria do not appear to differ with respect to the relevance of prior consent.

Since paternalistic interferences are generally thought to be so much more difficult to justify than those grounded in a harm-to-others rationale, prior consent might appear far more significant in cases such as self-exclusion programs from casinos.

The crucial test of my thesis is as follows. Imagine Jones, a second gambler who is identical to Smith in all relevant respects except for the fact that he has not given his prior consent to be placed on the self-exclusion list. From a moral perspective, my thesis entails that the manager would be warranted in treating Jones similarly to Smith, since the criteria I have identified would be applied in exactly the same way to both persons. If Jones, who has not consented, should be treated exactly like Smith, who has consented, it follows that consent is irrelevant to whether paternalism is justified.[25]

My tentative thesis does not dictate how any of the persons in the examples I have presented *should* be treated. I am not confident how to answer the question of whether Smith or Jones should be admitted or excluded from the casino; I only conclude that they should be treated identically. More to the point, I contend that no general answer to this kind of question should be given. In other words, no one-size-fits-all solution is optimal for each of the Smiths and Joneses I have described thus far. Admittedly, the answer is relatively clear in some kinds of cases. One might think that the decisive factor in favor of honoring Eric's earlier preference rather than his later demand is that he was more competent at the time he formed it.[26] Eric is to be commended for anticipating his future impairment and for enlisting someone to protect him from the consequences of his subsequent behavior. If I am correct that consent is irrelevant to the justifiability of paternalism, however, one must appeal to factors other than his prior request to explain why this case is easy.[27] Indeed, Eric's case *is* easy, but differs from Smith's in several important respects— differences that make it hard to know whether to provide the same answer.

It may be true that Smith, like Eric, knew exactly what he was doing when he decided to place himself on the lifetime self-exclusion list. But why suppose that his original judgment must be respected for all time? Curiously, Feinberg seemingly believes not only that prior fully voluntary consent is relevant, but also that it is decisive. In fact, he would always privilege the earlier judgment. Feinberg claims

> when the earlier self in a fully voluntary way renounces his right to revoke in the future (or during some specified future interval), or explicitly instructs another, as in the Odyssean example, not to accept contrary instructions from the future self, then the earlier choice, being the genuine choice of a sovereign being, free to dispose of his own lot in the future, must continue to govern.[28]

---

[25] At least in this case. Admittedly, a factor that is irrelevant in one pair of cases need not be irrelevant in all such pairs. Generally, see the discussion of the "Principle of Contextual Interaction" in F.M. Kamm (2007).

[26] According to Joel Feinberg, we should rely on the subject's most "voluntary" decision in cases of conflict. See *Self*, 83.

[27] The time at which the person is more competent is not the only basis for privileging Eric's judgment, even if it is the most important. Suppose that Alan, who consented to cosmetic surgery in a sober moment, becomes terrified when the operation is about to be performed. Clearly, he may withdraw his consent at this later time, even though his judgment is likely to be impaired by his fear.

[28] Feinberg, *Self*, 83.

But this position pushes the idea of personal sovereignty too far. In addition, it is at odds with a wealth of empirical research. An abundance of data confirms that persons are notoriously poor in predicting what they will want at a later time under different circumstances. Young adults often proclaim that they would prefer to forego treatment and die rather than to live with a severe disability that would dramatically decrease the quality of their lives. When they actually suffer from the very condition they fear, however, they frequently cling to life. Why privilege their earlier judgment when they express a preference for a future contingency they can barely imagine?[29] Arguably, they are in a far better position to recognize their true preferences when they experience the very disability in question.

Someone may respond that gambling is different from an ordinary disability. Gambling is an addiction, all addictions compromise cognition or volition, and it is in the nature of addictions that no one can be cured.[30] This response, I think, involves more ideology than sound social science. Even if gambling qualifies as a genuine addiction, and addictions undermine voluntary choice, why suppose that someone who once was addicted will not be able to moderate his behavior in the future without relapsing into his prior addictive state?[31] As individuals mature, many learn to moderate their addictive behaviors. With hindsight, the decision to exclude oneself permanently from a casino seems a particularly rigid solution to an acknowledged gambling problem that might have been addressed more effectively by a commitment strategy that allows greater flexibility.

In addition, Smith need not have been an addict in the first place.[32] His earlier decision to enroll in the lifetime exclusion program may have been rash or the product of external pressure, reflecting less competence and cool deliberation than he now displays when requesting to be allowed to gamble. Perhaps his wife, morally opposed to gambling, threatened to leave him should he set foot in a casino, and Smith loved his wife more than he liked to gamble. Desperate to keep his wife, Smith may have enrolled in the selfexclusion program, even though he did not have a gambling problem at all. But imagine that his wife left him anyway, and Smith's second wife does not share her predecessor's moral aversion to gambling. The general point is that persons who oversee self-exclusion programs have no means to determine why applicants sought to exclude themselves; their own decisions in the matter are final and irrevocable. Moreover, unlike the case of Jill and Eric, the casino manager is not in an ideal position to observe whether Smith still is vulnerable to whatever compulsive tendencies he may have had. The manager cannot determine whether admission is likely to harm Smith—the third condition in my

---

[29] Questions of advance directives that allegedly bind demented patients raise problems of personal identity that are not clearly replicated in my example of self-exclusion programs. See, for example, Allen Buchanan (1988).

[30] See Constance Holden (2001).

[31] The debate about whether addictive behaviors can be moderated is waged most fiercely in the context of alcoholism. See Frederick Rotgers et al. (2002).

[32] Some commentators appear to assume that self-excluded gamblers must be addicts. See the otherwise informative contribution by Justin E. Bauer (2006).

criteria of when paternalistic interferences are justified. Although mistakes always are possible, Jill is better able to detect whether Eric is intoxicated and should not be given his keys. Thus, even if compulsive gambling is an addiction, and addictions are an incurable disease, there is no good reason to infer that Smith ever was afflicted with it, is less rational today than when he made his irrevocable commitment, or would actually be harmed were he allowed to change his mind.

But didn't Smith make more than a vow or a pledge not to gamble? Didn't he make a promise—perhaps even a contract—not to enter a casino? Of course, the whole point of a promise or contract is to prevent persons from changing their minds by requiring them to pay damages in the event they default. If we think of Smith as having made a promise or a contract with the casino to treat him paternalistically, we may feel somewhat more comfortable about excluding him. For two reasons, however, we should not conceptualize these self-exclusion agreements as creating contractual obligations between Smith and the casino. Perhaps my conclusions can be avoided by supposing that the promise is made to (or the contract is made with) a party other than the casino—say, to the state agency that establishes the self-exclusion program. The same problem would arise, however, if Smith asked an agent of the state to release him from his promise (or contract[33] First, nearly all contracts are reciprocal and involve a bargain, conferring what each of the parties regards as a benefit. In this case, however, it is unclear how the casino gains from the agreement. In short, the absence of consideration is likely to render this so-called contract unenforceable.[34] More important, a contract model fails to explain why the casino manager would lack the power to release Smith from any promise he has made. Both contract law and the moral conventions surrounding the institution of promises allow parties to amend their agreements by mutual consent. Some theoreticians have proposed ingenious devices to preclude parties from subsequently modifying their prior agreement, but none has proved especially effective in law or appealing in morality. If an automatic preference for honoring the earlier judgment were desirable, one might reasonably anticipate that mechanisms in law and principles in morality would be available to ensure this result.[35]

As Peter Westen indicates, "nonreciprocal irrevocable commitments are sufficiently rare that the paradigm for it comes not from law but [from fiction]: from Homer's account of Odysseus' encounter with the Sirens."[36] The fictional Odysseus, however, resembles Eric more than Smith; the Sirens drove sailors mad, making them less competent than when their songs could not be heard. Even here, prior consent does no substantive work. If Odysseus had not issued his prior command to

---

[33] Perhaps my conclusions can be avoided by supposing that the promise is made to (or the contract is made with) a party other than the casino—say, to the state agency that establishes the self-exclusion program. The same problem would arise, however, if Smith asked an agent of the state to release him from his promise (or contract).

[34] Arguably, this technical problem could be overcome if Smith paid consideration—say, a sum of $10—in exchange for the casino manager's promise to exclude him.

[35] See Kevin Davis (2006).

[36] Westen, *Logic*, 253.

remain tied to the mast, his crew would have been equally justified in ignoring his subsequent pleas. Why heed the commands of a madman who instructs his sailors to steer to their doom? By contrast, Smith's competence does not clearly vary from one time to another.

Thus, I assume that the manager should not automatically defer to Smith's prior request to be excluded from the casino for life. It is even easier to show that Smith's later demand to be admitted is not automatically entitled to deference. Morality should not contain an absolute bar against enlisting the assistance of others in devising a commitment strategy. Without cooperation, we sometimes cannot design an effective means to protect ourselves from our own weaknesses and tendencies to perform acts that we recognize to be bad for us. Few respondents believe that Eric's later demand for his car keys (or Odysseus's pleas to be untied) must be honored because contemporaneous preferences invariably trump those expressed at an earlier time.

If the casino manager should automatically defer neither to Smith's earlier preference nor to his current decision, what should he do? It is important not to misconstrue the nature of this question or to confuse it with three others that might be posed. First, I am not concerned with the self-interest of the casino manager. Even from this perspective, the answer is uncertain. On the one hand, it is evident that casinos make money by admitting patrons, not by excluding them. Persons who are barred by selfexclusion programs probably represent a significant loss of revenue for casinos.[37] On the other hand, compliance with these programs may generate favorable publicity for a beleaguered industry. Casinos might prosper more in the long run by maintaining a policy of refusing admission to persons who admit their gambling problem. Second, I am not concerned with the applicable law. Special statutory provisions govern self-exclusion programs in the several states, and the hands of a manager may be tied by a particular law to which he is subject. He may incur liability in the event he makes the wrong decision—whatever that decision may be. Perhaps Smith can recover damages from the casino if it culpably admits him.[38] Or perhaps the casino must pay a fine to the state or risk the loss of its license.[39] But suppose that no statutes clearly specify what the manager is legally obligated to do. In this instance, it is doubtful that courts should impose liability on a casino manager who does not make whatever decisions we believe to be correct. His predicament is sufficiently difficult that we may want to protect him from liability for *either* choice he makes in good faith, even if we regard one outcome as better than the other. Finally, I am not concerned with the empirical question of whether this commitment strategy is effective.[40] Excluded gamblers may simply be displaced

---

[37] According to one study, compulsive gamblers provide between 30 and 52 % of all casino revenues. See http://www.casinofreephila.org/research/gambling-revenues-compulsive-gamblers.

[38] For a negative answer, see *Merrill v. Trump Indiana, Inc.*, 320F.3d 729 (7th Cir. 2003).

[39] For an affirmative answer, see id.

[40] According to one study, 30 % of the participants completely stopped gambling once enrolled in this kind of program. See Robert Ladouceur et al. (2000).

to other venues such as racetracks or state lotteries, where the odds of winning are even more remote than in casinos. Interesting though these three perspectives may be, I put each of them aside.

Instead, I want to inquire what the casino manager ought to do from the *moral* point of view. My central (but tentative) thesis in this chapter is that the absence of consent is irrelevant to the justification of paternalism, even when it is given explicitly in the past. If this thesis is correct, the casino manager should proceed in exactly the same way as Billy's father or Eric's friend Jill: He must determine what is reasonable. I have identified five criteria that I think should guide this determination. I do not pretend that the application of these criteria is simple: It is not nearly as easy as in Billy's or Eric's case. The following difficult issues must be addressed to make a decision. At what time was Smith more competent to assess his own interests and to make the better judgment? As I have indicated, this question is especially important in cases in which reasonable minds differ about whether the interference is really worth the costs to the person coerced. Smith appears to be an unimpaired adult who does not suffer from any of the obvious deficiencies of Billy or Eric, and I see no reason to suppose that there always is a particular time—in the past or in the present—when persons who want to gamble are better able to assess their own interests. Second, how important is Smith's liberty interest, and how severe is the interference with it? Unfortunately, we lack a convenient metric to evaluate the value of the many liberties we recognize. Intuitively, exclusion from a casino is a larger infringement of liberty than the denial of ice cream, especially when the ice cream is withheld temporarily rather than permanently. Still, the ability to gamble is not ranked especially high on most scales of liberties. The two states that ban gambling altogether—Hawaii and Utah—are not typically thought to violate significant liberties. Third, how valuable is the objective to be achieved? Preventing gambling addicts from losing large amounts of money can be a significant achievement, but I have already expressed reservations about whether persons on the list are addicts. Fourth, what is the likelihood that exclusion will be effective in preventing Smith from losing money? Empirical research is needed to shed light on this matter. Finally, is the casino manager in the appropriate position to treat Smith paternalistically? I will have more to say about this final condition in a moment. At the present time, I repeat my confidence about how these five factors should be balanced in Billy's or Eric's case, and my lack of certainty about how they should be balanced in Smith's case. We need far more information before we should be clear about our answer, and are likely to remain ambivalent even when all of the facts are known. My more modest goal, however, is not to resolve this difficult issue, but to examine the role consent plays within the framework in which the question should be addressed.

My tentative thesis is that consent does not enter into this moral framework at any point in the analysis. The fact that Smith gave his prior consent is not material to whether the manager should ban him for his own good.

Admittedly, this position seems somewhat counterintuitive—even to me. My own intuitions on this topic are frail and unstable. Can it really be true that prior consent plays no role whatever in the face of contemporaneous nonconsent? If so, why are so many philosophers inclined to believe otherwise? Three answers seem promising. First, consent may alter the burden of proof in determining whether or not paternalism is justified. It is almost never clear whether a particular instance of paternalism satisfies my test. Perhaps the burden of showing these criteria are *not* satisfied should be allocated to the person to be treated paternalistically when he has given his prior consent to the interference. A second point is closely related. We are entitled to try especially hard to persuade someone to act in his own interest when he has requested that we do so. Suppose, for example, that your friend urges you in the morning not to let him succumb to laziness if he fails to keep his promise to meet you in the gym later in the day. When he changes his mind and proposes to stay home, you are permitted to remind him forcefully of his previous request. If he continues to decline, however, I think we must respect his contemporaneous rather than his prior choice. Finally, and most obviously, consent appears to be important because it serves as *evidence* that some of my criteria are satisfied. In particular, it provides a reason to believe that Smith has a gambling problem he once thought to be sufficiently serious to warrant his permanent exclusion. In the absence of his earlier consent, the casino manager almost certainly will have more reason to believe that the ban protects Smith's interests more than those of Jones, the patron with the identical gambling problem. But I propose to put such epistemological considerations to one side. Suppose for the sake of argument that the casino manager happens to know just as much about Jones as he knows about Smith. As a matter of principle, I do not understand how consent should be a factor in our advice about whether either or both may be excluded. If I am correct, both Smith and Jones should be treated similarly, and the absence of consent is irrelevant to the question of whether their paternalistic treatment is justified.[41]

To bolster my thesis, we should notice that consent is equally irrelevant in deciding how Eric, the intoxicated but prudent guest, should be treated. Imagine that Jill finds the keys that Patricia, another guest, has misplaced at her party. Patricia is now as drunk as Eric, and demands that her keys be returned so she can drive home. Unlike Eric, Patricia has not voluntarily entrusted her keys to Jill should this very contingency arise. But if their circumstances are identical otherwise, it is hard to see why Jill should return Patricia's keys but withhold those of Eric. With the following caveat, each of my five criteria applies equally to both persons.

I confess to misgivings about denying an important (nonevidentiary) role to consent in the cases of Smith or Eric. Because of these misgivings, I have persistently qualified as tentative my thesis about the irrelevance of consent to the justifiability of paternalism. Arguably, Smith's prior consent has normative significance because it is material to the fifth criterion in my test of whether paternalistic interferences are reasonable and thus justifiable. Recall that parents stand in an ideal (or special) relationship to their children to treat them paternalistically. Biology and the duties

---

[41] This position resembles the controversial view Joseph Raz has defended in the context of analyzing political authority. According to Raz, "consent is a source of obligation only when some considerations, themselves independent of consent, vindicate its being such a source." Joseph Raz (2006).

conventionally attached to parents are not, however, the only source of special relationships. Smith's prior consent may create the special relationship between himself and the casino that entitles the manager to treat him paternalistically. Even though "special relationships" ordinarily are posited to justify the creation of *duties*, they also are capable of justifying the creation of privileges or permissions. In any event, no such relationship exists between Jones and the casino, or between Patricia and Jill. Is the existence of a special relationship needed before paternalism is justified? I am agnostic; my intuitions tug me in different directions.

But if my misgivings are sound, and the identity of the person who interferes is relevant to whether that interference is permissible, we have a possible basis for contrasting the justifiability of paternalism from that of nonpaternalism. Earlier, I suggested that Craig and Jason should be treated similarly if they have comparable tendencies to molest children. But it is hard to see why anyone would think that the identity of the individual who proposes to evict either Craig or Jason from a public playground should be a factor in determining whether the eviction is permissible. This fifth and final criterion in our test of when paternalism is reasonable has no clear analogue in cases in which the interference is motivated by nonpaternalistic considerations.

Suppose my misgivings are correct, and Smith's actual, prior consent is crucial to whether his paternalistic treatment is justified because it creates a special relationship with the casino manager. If so, we are left with an interesting result. Jones is (otherwise) identical to Smith. With respect to Jones, however, we would have a case of (otherwise) justifiable paternalism, with no one in an appropriate position to impose it. We could try to surmount this hurdle by multiplying the number of relationships we hold to be special. We might allege a relationship is special whenever one person is in a position to treat another paternalistically. Perhaps Jones's mere appearance in a casino creates a special relationship that would satisfy the fifth condition in my criteria. Maybe the act of hosting a party and finding Patricia's keys creates a special relationship that warrants paternalistic intervention. But this solution, though sensible in some contexts, has limits, and threatens to render my fifth criterion all but vacuous. Special relationships are *special*, after all. Unless the number of special relationships is multiplied beyond recognition, a plausible objection to a great deal of (otherwise) justifiable paternalism is that no one stands in a suitable relation to impose it on the person to be treated paternalistically.

If we hold the fifth criterion in my test of reasonableness to be important, we may have an additional reason to be skeptical of *criminal* paternalism—of laws that subject persons to punishment for their own good. Arguably, the state lacks an appropriate (or special) relation to its citizens to be eligible to treat them paternalistically. On some minimalist conceptions of the state, its only function is to prevent persons from harming others. Of course, a defense of this liberal (or libertarian), nonperfectionist political view requires nothing less than a theory of the state and a corresponding theory of criminalization— tasks well beyond the scope of this chapter.[42] Here I offer a single observation about why we should be reluctant to elevate my

---

[42] For further thoughts, see Husak, *Overcriminalization*.

misgivings into a general opposition to all legal paternalism. Political philosophers who resist a perfectionist theory of the state will be hard-pressed to defend the probable implications of their views for the justifiability of so-called libertarian paternalism. If the state does not stand in a proper relation to its citizens to treat them paternalistically, it is unclear why it has good reason to design default rules to protect persons from the consequences of their own weaknesses. This conclusion strikes me as counterintuitive, even if we are skeptical of paternalism in the criminal domain. After all, the state must provide *some* content to default rules. On what other basis should they be formulated? *Ceteris paribus*, why should the state be precluded from designing default rules to influence citizens to pursue their own good? No abstract argument against perfectionism and in favor of a liberal (or libertarian) theory of the state is likely to provide a satisfactory answer to this question. Generally, we should find it easier to resist criminal paternalism than state actions in (what I have loosely called) the spirit of paternalism pursued through noncriminal means.

Earlier, I suggested that the final criterion in my fivefold test of reasonableness is the most questionable. I conclude that insofar as we regard this fifth criterion as unimportant, we should not believe that Smith's previous decision to seek exclusion is relevant to how the casino manager should proceed. In this event, the case of Smith and Jones, as well as that of Eric and Patricia, stand or fall together. Moreover, their cases resemble that of Craig and Jason, whose liberty is deprived not for paternalistic reasons, but to prevent harm to others. Unless the final criterion in my test is retained, and the justifiability of paternalism depends partly on the identity of the person who imposes it, my thesis is that consent makes no difference to the criteria we should apply in deciding whether we are permitted to treat someone paternalistically.

Acknowledgement I would like to thank participants at the Criminal Theory Workshop at the International Congress of Political and Legal Philosophy at Krakow, Poland. I also received valuable help from members of the Department of Philosophy at Virginia Commonwealth University as well as from members of an NIH seminar at Georgetown University. Special thanks to Youngjae Lee, Frank Miller, Alec Walen, Peter Westen, and Alan Wertheimer, each of whom provided detailed written assistance on earlier drafts.

References

Ainslie, George. 1992. *Picoeconomics: The strategic interaction of successive motivational states within the person*. Cambridge: Cambridge University Press.

Bauer, Justin E. 2006. Self-exclusion and the compulsive gambler: The house shouldn't always win. *Northern Illinois Law Review* 27: 63.

Buchanan, Allen. 1988. Advance directives and the personal identity problem. *Philosophy and Public Affairs* 17: 277.

Camerer, Colin, et al. 2003. Regulation for conservatives: Behavioral economics and the case for 'asymmetric paternalism'. *University of Pennsylvania Law Review* 151: 1211.

Davis, Kevin. 2006. The demand for immutable contracts: Another look at the law and economics of contract modifications. *New York University Law Review* 81: 487.

Dworkin, Gerald. 1972. Paternalism. *The Monist* 56: 64.

Elster, Jon. 1983. *Sour grapes: Studies in the subversion of rationality*. Cambridge: Cambridge University Press.

Feinberg, Joel. 1985. *Harm to self*. New York: Oxford University Press.

Gert, Bernard, and George Culver. 1976. Paternalistic behaviors. *Philosophy and Public Affairs* 6: 46.

Holden, Constance. 2001. 'Behavioral' addictions: Do they exist? *Science* 294: 980.

Husak, Douglas. 1980. Paternalism and autonomy. *Philosophy and Public Affairs* 10: 27.

Husak, Douglas N. 2004. Legal paternalism. In *Oxford handbook of practical ethics*, ed. Hugh LaFollette, 387. New York: Oxford University Press.

Husak, Douglas. 2008. *Overcriminalization*. New York: Oxford University Press.

Kamm, F.M. 2007. *Intricate ethics*, 17. New York: Oxford University Press.

Kassachkoff, Tziporah. 1994. Paternalism: Does gratitude make it okay? *Social Theory and Practice* 20: 1.

Ladouceur, Robert, et al. 2000. Brief communications analysis of a Casino's self-exclusion program. *Journal of Gambling Studies* 16: 453.

Leitzel, Jim. 2008. *Regulating vice: Misguided prohibitions and realistic controls*, esp. 72–esp. 92. Cambridge: Cambridge University Press.

Lowenstein, George, and Ted O'Donoghue. 2006. 'We can do this the easy way or the hard way': Negative emotions, self-regulation, and the law. *University of Chicago Law Review* 73: 183.

Michael, Bishop, and J.D. Trout. 2005. *Epistemology and the psychology of human judgment*. New York: Oxford University Press.

Raz, Joseph. 2006. The problem of authority: Revisiting the service conception. *Minnesota Law Review* 90: 1003–1038.

Rotgers, Frederick, Mark F. Kern, and Rudy Hoeltzel. 2002. *Responsible drinking: A moderation management approach for problem drinkers*. Oakland: New Harbinger Publications, Inc.

Schelling, Thomas C. 1984. self-command in practice, in policy, and in a theory of rational choice. *American Economics Review* 74: 1.

Sunstein, Cass R., and Richard H. Thaler. 2003. Libertarian paternalism is not an oxymoron. *University of Chicago Law Review* 70: 1159.

Thaddeus Mason, Pope. 2005. Monstrous impersonation: A critique of consent-based justifications for hard paternalism. *University of Missouri-Kansas City Law Review* 73: 861.

Trout, J.D. 2005. Paternalism and cognitive bias. *Law and Philosophy* 24: 393.

Westen, Peter. 2004. *The logic of consent*. Burlington: Ashgate Publishing Co.

Chapter 6
Paternalism and Equality

Kristin Voigt

6.1 Introduction

Paternalistic interventions restrict individuals' liberty or autonomy so as to guide their decisions towards options that are more beneficial for them than the ones they might choose in the absence of such interventions. Although some philosophers have emphasised that there is a case for justifiable paternalism in certain circumstances (e.g., De Marneffe 2005; Wilson 2011), much of contemporary moral and political philosophy works from a strong presumption against paternalistic interventions: because paternalistic interventions restrict individuals' liberties and treat them as less than fully capable of making decisions that are in their own best interest, they are generally considered impermissible, barring very exceptional circumstances.

Richard Arneson has argued that there are egalitarian reasons that support the case for paternalism: paternalistic interventions can protect poor decision-makers from making 'bad' choices, thus preventing inequalities between them and those with better decision-making skills. This line of argument can be applied to a range of contexts. For example, paternalistic restrictions on participation in biomedical research have been supported by concerns about equality: differences in individuals' decision-making capacities result in an unfair distribution of the costs and risks associated with participation in such research.

This work was first presented at the 'New perspectives on medical paternalism' workshop at the University of Hamburg in March 2012 and benefited greatly from the comments received. I would also like to thank Kalle Grill and Thomas Schramme for their helpful comments on an earlier draft.

K. Voigt (✉)
Department of Philosophy and Institute for Health and Social Policy,
McGill University, Montreal, QC, Canada
e-mail: kristin.voigt@mcgill.ca

© Springer International Publishing Switzerland 2015
T. Schramme (ed.), *New Perspectives on Paternalism and Health Care*, Library
of Ethics and Applied Philosophy 35, DOI 10.1007/978-3-319-17960-5_6

This paper aims to clarify and advance our understanding of the egalitarian argument for paternalism. Arneson's argument adds an important and often neglected dimension to the debate about paternalism. However, the argument also raises a number of questions about equality, paternalism and the relationship between the two.

I begin by restating Arneson's argument (Sect. 6.2) before highlighting a number of complexities surrounding it (Sect. 6.3). First, with respect to what kinds of choices does Arneson's argument hold? Second, what kinds of outcomes should we be concerned with when assessing whether or not a particular intervention is really in the interest of the person interfered with? Third, what types of paternalistic interventions lend themselves to Arneson's argument? Section 6.4 reconsiders the concern that paternalistic interventions treat as less than equal those whose liberties they restrict. Arneson's argument, with its focus on 'good' and 'bad' choosers seems particularly susceptible to this kind of concern. Any gains in distributive equality that can be garnered from paternalistic interventions must be weighed against possible negative effects on these 'relational' aspects of equality. Section 6.5 concludes.

6.2   Paternalism and Its 'Distributive Dimension'

Like many discussions of paternalism, I will start from the definition proposed by Gerald Dworkin. According to this definition,

> X acts paternalistically towards Y by doing (omitting) Z:
> Z (or its omission) interferes with the liberty or autonomy of Y.
> X does so without the consent of Y.
> X does so just because Z will improve the welfare of Y (where this includes preventing his welfare from diminishing), or in some way promote the interests, values, or good of Y. (Dworkin 2011)

In two papers, Arneson (1989b, 2005) connects paternalism with considerations of distributive equality, emphasising that we should recognise what he calls the 'distributive dimension' of paternalism. He starts from the observation that there are significant variations in individuals' decision-making skills: 'people differ widely in their native capacities for deliberation about plans and skilful execution of them' (Arneson 1989b: 412). Starting from a baseline of equality, if 'bad' options are available for people to choose, then these systematic differences in people's decision-making skills will lead to inequalities between 'good' and 'bad' choosers, as the 'bad' options can be avoided by the good but not by the bad decision-makers:

> A ban on paternalism in effect gives to the haves and takes from the have-nots. Left unrestrained in self-regarding matters, more able agents are more likely to do better for themselves choosing among an unrestricted range of options, whereas less able agents are more likely to opt for a bad option that paternalism would have removed from the choice set. (Arneson 1989b: 412)

Thus, paternalistic interventions that remove particularly bad options from people's option sets can prevent inequalities. Anti-paternalism, Arneson concludes, 'looks to be an ideology of the good choosers, a doctrine that would operate to the advantage of the already better-off at the expense of the worse-off, the needy and vulnerable' (Arneson 2005: 276).

How might this argument work in practice? One application, which I will keep coming back to in this paper, is suggested by Jansen and Wall (2009), who focus on individuals' participation in clinical research trials. Jansen and Wall defend hard paternalism in the case of trials that have an unfavourable balance of risk and benefit (in the case of therapeutic trials) or that impose more than minimal risk on participants (in the case of non-therapeutic trials). With respect to such trials, they argue, we must not simply rely on individuals' consent when it comes to their participation as research subjects; a hard paternalistic approach to trial participation is justified because of considerations of fairness.

For the most part, their argument follows Arneson's (though I discuss below one issue on which they deviate from Arneson). Like Arneson, Jansen and Wall emphasise that people differ in their decision-making skills and capacities:

> Some are wise, some are foolish; some are careful in their deliberations, some are rash. Some are subject to cognitive and emotional deficiencies of one type or another, some are relatively free from such deficiencies. Some are too trusting of authority, some are too independent minded. And some are better able than others to take in and process the information relevant to the decision to participate in a given trial. (Jansen and Wall 2009: 176)

Thus, different approaches to restrictions on clinical research differ in their impact on potential research participants: stringent restrictions will benefit those with lesser decision-making skills, while loose restrictions will advantage those whose decision-making capacities are greater. Because those with lesser decision-making skills are likely to be (or become) worse off overall, considerations of distributive equality are relevant. A concern with distributive fairness, Jansen and Wall argue, 'will require us to compare the likely distributive outcomes (in terms of the welfare impact on the population of potential research subjects) of different regulatory policies' (Jansen and Wall 2009: 175).

The fact that paternalism can, as this line of argument suggests, be equality-promoting gives us *one* reason to support paternalistic interventions. At the same time, of course, equality is not the only consideration at stake and any distributive advantage of particular paternalistic interventions may well be outweighed by other relevant considerations. The objections to paternalism – that it involves undue interference with individuals' liberties and that it interferes with individuals' authority over important aspects of their lives – still stand and are not necessarily outweighed by distributive concerns. However, Arneson's argument emphasises the distributive implications of paternalistic interventions as an important – and often neglected – consideration that must be weighed against other, possibly competing concerns when evaluating the case for particular paternalistic interventions.

6.3   The Scope of Arneson's Argument: Choices, Outcomes and Interventions

In this section, I consider three questions that arise in connection with Arneson's argument about the 'distributive dimension' of paternalism. First, we need an argument as to why the inequalities prevented by a particular paternalistic intervention would have been *unfair* inequalities; the fact that the inequality would have been the result of the agent's choice could lead us to think that the inequality would have been fair. Second, the exact implications of the argument will depend on the metric we use to determine whether and to what extent someone is actually made 'better off' by a paternalistic intervention. Finally, while Arneson's argument focuses on paternalistic interventions in which particular options are removed entirely, many paternalistic interventions rely on different mechanisms to shape individuals' choices. These kinds of interventions can raise distributive concerns not addressed in Arneson's argument.

6.3.1   Choice

Central to Arneson's argument is that, under certain conditions, paternalistic interventions are beneficial from the perspective of distributive fairness because they can prevent unfair inequalities. Importantly, Arneson's argument applies only if the inequalities that the paternalistic intervention prevents would in fact have been *unfair*: distributive fairness is not improved when we prevent inequalities that would have been fair. However, paternalistic interventions work by interfering with individuals' choices, and choices are often considered to be a source of *fair* rather than unfair inequalities. How does this affect Arneson's argument?

Arneson addresses this issue when he argues that his argument supports (at least some) *hard* paternalistic interventions. The distinction between hard and soft paternalism is frequently referred to in the paternalism debate. This distinction captures the difference between interventions that restrict *voluntary* choices (hard paternalism) and those that restrict *non-voluntary* choices (soft paternalism) (see Feinberg 1986). For example, if I hide my friend's cigarettes after she has repeatedly expressed frustration about her nicotine addiction and inability to quit, this may be considered an instance of soft paternalism: to the extent that my friend's decision to smoke is driven by addiction and not endorsed by her, the intervention is interfering with a non-voluntary choice. If, on the other hand, her choice to smoke is fully voluntary, my hiding her cigarettes should be described as an instance of hard paternalism.

In the literature on distributive equality, the criterion of voluntariness is also invoked, including perhaps most prominently in Arneson's contributions to that debate. Arneson is one of the original proponents of (a particular version of) luck egalitarianism (Arneson 1989a, 1991), which he has since abandoned in favour of what he calls responsibility-catering prioritiarianism (Arneson 2000). However, both his egalitarian and prioritarian positions are meant to be responsibility-sensitive, which Anderson understands as the requirement that 'it is morally wrong if some people are worse off than others through no fault or voluntary choice of their own' (Arneson 1990: 177). Accordingly,

> distributive justice does not recommend any intervention by society to correct inequalities that arise through the voluntary choice or fault of those who end up with less, so long as it is proper to hold the individuals responsible for the voluntary choice or faulty behavior that gives rise to the inequalities. (Arneson 1990: 176)

At first sight, Arneson's endorsement of voluntary choice as a source of *fair* inequality sits uneasily with his suggestion that there is an egalitarian argument for hard paternalism: why would paternalistic interventions that prevent voluntary choices be advantageous from the perspective of distributive equality? In the remainder of this section, I discuss two possible strategies for making the case that the inequalities that are prevented by the interventions Arneson has in mind would have been unfair, even though they are the result of individuals' voluntary choices: one provided by Jansen and Wall, the other by Arneson. Arneson, I suggest, provides a more appropriate response to this problem than Jansen and Wall.

Jansen and Wall argue that, even if individuals are responsible for the choices they make, it is society as a whole that is responsible for creating or permitting the opportunities in which particular choices become possible. Take gambling as an example:

> a political society might decide to permit gambling houses in its territory in full knowledge that gambling will spell the financial ruin of many its members. In doing so, it would be vulnerable to the fairness objection we have been discussing. This remains true, even if it is also true that those who recklessly gamble are responsible, or at least partly responsible, for their fate. (Jansen and Wall 2009: 178)

On their account, then, '[i]t can be unfair to implement a regulatory scheme that is costly for bad decision-makers over one that is less costly to them, even if the bad decision-makers are responsible, or partly responsible, for their bad decisions' (Jansen and Wall 2009: 178).

However, this argument does not, in fact, address the problem at hand. If – as Jansen and Wall assume – the choice to gamble is a choice that individuals are responsible for in the sense required by distributive fairness, then any inequalities resulting from those choices are fair. If we make this assumption, then opportunities to gamble are simply opportunities for *fair* inequalities. However, the move from an equal, fair distribution to one that contains inequalities where all of these inequalities are fair, is a move that – from the perspective of distributive equality – we must be indifferent towards. More generally, as far as equality is concerned, responsibility-catering versions of distributive equality cannot distinguish between, on the one hand, a distribution that is equal and fair and, on the other hand, a distribution that contains some inequalities, as long as these inequalities are fair.

Arneson addresses this problem in a different way. He emphasises that even if we accept that it is fair for individuals to be better or worse off than others to the extent that they are responsible for such (dis)advantages, many important differences in people's decision-making capacities are due to luck and individuals therefore should not be held responsible for them:

> Whatever conception of fault one adopts, inequalities of welfare that arise through the individual's own fault as judged by that conception will neither violate the principle of equality nor count as unfair. But on anybody's conception of fault the prudential disabilities that separate more and less able agents are surely in very considerable part due to accidents of genetic endowment and variously favourable early childhood circumstances that do not lie within the agent's control and for which he cannot be either praiseworthy or blameworthy. So even if we accept that it is sensible to attribute some prudential failings of individuals to personal fault, these attributions cannot reconcile us to regarding as fair the great bulk of inequalities of welfare that separate more and less able agents. Paternalism remains in the running as one morally appropriate response to some of these pervasive and disquieting inequalities. (Arneson 1989b: 422–3)

Importantly, even if choices are 'voluntary enough' so that interference with them would be a case of hard rather than soft paternalism, that does not automatically make them the kinds of choices for which it would be fair to hold them responsible, for the purposes of distributive equality. Even though responsibility-catering egalitarians (including Arneson) often talk about 'voluntary choices' as the kind of choices that lead to fair inequalities, the conception of voluntariness they rely on is often very different from the notion of voluntariness that is used to draw the line between soft and hard paternalism. In fact, on more stringent interpretations of responsibility-sensitive egalitarianism, there will be few, if any, choices in the real world that would meet the requirements responsibility-sensitive egalitarians stipulate. Even 'voluntary' choices can be shaped by brute luck in such a way that any inequalities resulting from those choices would be considered unfair (Voigt 2007).

What does this imply for the egalitarian argument for paternalism that Arneson advances? Arneson's argument only applies when we are talking about paternalistic interventions that prevent *unfair* inequalities. Paternalistic interventions that restrict choices that would have led to *fair* inequalities do *not* improve distributive equality. The scope of Arneson's argument therefore depends on our ability to make the case that the choices restricted by particular paternalistic interventions are not the kinds of choices that would lead to fair inequalities – choices, that is, for which it would be fair to hold them responsible.

6.3.2 Outcomes

Another question we have to ask concerns the 'metric' we use to determine whether an individual is indeed made 'better off' by a particular intervention. Dworkin defines this aspect of paternalistic interventions broadly, suggesting that these interventions are to 'improve the welfare of Y … or in some way promote the interests, values, or good of Y' (Dworkin 2011). This idea can, of course, be fleshed out in very different ways. We may, for example, rely on the agent's own judgements and preferences to determine whether or not an intervention improves his or her well-being: paternalistic interventions that are based on the preferences of the agents interfered with are arguably much less problematic than interventions that do not defer to agents' preferences in this way (Goodin 1991).

Specifying the outcomes we are concerned with when we are intervening paternalistically can also help us respond to a line of criticism that has been put forward in this debate. In response to Jansen and Wall's argument about participation in research trials and the idea that prohibitions on particularly unfavourable risk-benefit ratios will protect 'poor' decision-makers, Edwards and Wilson have emphasised that, contrary to what Jansen and Wall assume, people may participate in risky research trials not because they are poor decision-makers and unaware of the risks or incapable of accurately evaluating them but because they have altruistic motivations and are happy to accept risks that will benefit others (Edwards and Wilson 2012).

Jansen and Wall anticipate this argument. They emphasise that if we are thinking about the issue from the perspective of regulation, practical limitations will prevent us from designing policies that distinguish between, and treat differently, altruists and poor decision-makers: whatever regulation we come up with will affect both of these groups. More importantly, they argue, some risks would simply be wrong for individuals to assume, *even for altruistic reasons*: 'Each person has a duty to respect herself' and we fail to meet our 'self-regarding duties' if we accept such risks (Jansen and Wall 2009: 179). We can then think of legal requirements that clinical trials not subject research participants to excessive risks as giving effect to this concern.

Jansen and Wall do not provide a full defence of this (clearly controversial) argument. What matters for the present argument is that our understanding of, and response to, the altruism challenge also depends on how we define what the relevant outcomes are. We could, for example, define individual well-being in such a way that acting on an altruistic motivation makes the agent better off in some respects, for example because the agent gets pleasure from thinking about the beneficial effects of her actions on others, or because there is objective good in acting from altruistic motivation. If there are such positive effects on well-being, then these may outweigh, at least in some cases, the costs or risks that altruistic agents accept. We can also allow for the possibility that the influence of altruism on people's choices can indeed be the kind of disadvantage that they should be compensated for. From the perspective of distributive justice, we could argue that altruistic choices can be problematic, even if they are 'voluntary' and reflect individuals' preferences: if, for reasons of brute luck, some people are more altruistic than others and therefore more likely to forego benefits for themselves so as to aid others, it is certainly not obvious that the sacrifices they make as a result should *not* be considered unfair disadvantages. What this highlights is that it is important to specify what outcomes a paternalistic intervention is meant to improve; our answer to that question will also influence how we think about costs that people bear as a result of altruistically motivated choices.

This hints at a broader problem for the translation of Arneson's argument into specific policy solutions: we may not be able to identify any options that are unequivocally bad for everyone who is affected by a particular paternalistic intervention, irrespective of their specific interests and preferences. Even in the context of health, which is often considered the kind of good that everyone needs, it is not the case that we can identify specific risks that would be 'bad' options for everyone. In many cases, people make trade-offs between health and other goods that may be important to them. For example, some women choose to continue with high-risk pregnancies because they place a very high value on having a child that is biologically related to them. Importantly, however, while this may limit the extent to which we can translate Arneson's argument into policy prescriptions, it is not a problem with the theoretical underpinnings of his argument.

6.3.3   Interventions

Arneson implicitly assumes that we are working with paternalistic interventions that fully remove welfare-reducing options from people's option sets. Is Arneson's egalitarian argument relevant when we are looking at paternalistic interventions that rely on different mechanisms to shape individuals' choices? In this section, I discuss two types of paternalistic interventions that rely on different mechanisms: those where information is withheld from individuals and those that change the relative cost associated with particular options rather than removing them entirely.

While most discussions of paternalism focus on interventions where particular options are blocked entirely, this is not the only – and perhaps not even the most common – way to interfere paternalistically with someone's decisions. One type of intervention that could be guided by paternalistic motivations is that of withholding information. With this type of intervention, the concern is that the provision of information has a negative impact on individuals, for example by causing distress or leading individuals to make decisions that are likely to have negative consequences for them.

An interesting example of this kind of intervention is physicians' decision not to disclose information to patients about unsubsidised medication. In some countries, medications that are not funded or subsidised through the health care system may nonetheless be available for patients to purchase at their own expense, often at very high prices. Sometimes, such drugs are not subsidised because they have only just become available and have not yet been approved for subsidised provision through the health care system. In other cases, drugs may not be subsidised because they are not considered good 'value for money': the drug may be perfectly safe but the expected patient benefit does not seem significant enough to warrant public provision at the price attached to it. When such drugs are available and suitable for particular patients, doctors will have to make decisions about whether or not to inform these patients. Some doctors appear to be reluctant to do so (Jefford et al. 2005).

Of course, a doctor's decision not to let a patient know about such a drug is not necessarily paternalistic. For example, doctors may be withholding this information because they are really committed egalitarians: they oppose the idea that wealthy patients can gain access to medications that poorer patients cannot afford. Or, more realistically, they may simply not have the time to inform patients about the existence of such drugs. In such cases, the decision not to inform patients about an unsubsidised drug would not count as paternalistic. For the purposes of the argument, I am interested in scenarios where the decision to withhold information about unsubsidised treatments is paternalistic: doctors choose to withhold such information because they believe that this information would be detrimental to the patients' well-being, for example because it would cause them distress.[1]

One interesting complication in this particular example that does not arise for the scenarios Arneson discusses is that here people's decisions – and the benefit they derive from the paternalistic intervention – depends at least in part on their financial situation. For wealthy patients, the cost of the drug may not make much difference, whereas poorer patients may find that they have to make significant sacrifices in order to purchase the drug. It may be the case that for these poorer patients, the case for paternalistic interference is greater simply because *these patients have more to lose*. If we are concerned with the distributive dimension of paternalistic interventions, this is certainly a relevant consideration – and perhaps more significant in its effects than the concerns about differences in decision-making capacities that are central to Arneson's argument.

Similar concerns arise in connection with paternalistic interventions that, instead of removing 'bad' options entirely, make such options more expensive relative to their alternatives. We can, for example, attach a financial penalty to welfare-reducing options, or make welfare-enhancing options more attractive by using incentives. Depending on how strictly we define what it means to 'remove' an option, many paternalistic policies will in fact be based on this approach. Few interventions are able to block people's access to particular options *entirely*; even legislation that enforces the use of seatbelts and safety helmets is effectively an intervention that changes the relative cost of the options involved: the option of not wearing a seatbelt or a helmet is not removed entirely but fines and penalties make it significantly more expensive.

In the health context, taxation is often used to make unhealthy products more expensive. Tobacco is perhaps the most prominent example here but taxes have also been introduced or considered to lower the consumption of alcohol as well as fatty and sugary foods. Positive incentives are also increasingly used. In the US, for example, many employers will lower health insurance contributions for employees who are non-smokers and whose weight is within the 'normal' range (Schmidt et al. 2010). As a matter of fact, of course, at least some of these policies are more likely

---

[1] In Jefford et al.'s (2005) study with Australian oncologists, the most commonly voiced concerns about giving patients information about unsubsidised drugs were about causing the patient and their family distress and mentioning a drug to patients even though they probably wouldn't be able to afford it.

to be motivated by considerations of cost rather than concerns about individuals' well-being. However, these interventions *could* be paternalistically motivated: they could have been put in place so as to steer people towards choices that are 'better' for them.

A related type of intervention, which is also focused on the relative cost of particular options, is the regulation of financial inducements. This is a concern that is often raised in health contexts. For example, there are restrictions on how much people can be paid to participate in research trials, to act as pregnancy surrogates or to donate blood or organs.[2] Again, these policies do not necessarily have a paternalistic motivation but certainly paternalistic arguments could be made to support them.

As in the earlier example of the doctor deciding whether to disclose information about an unsubsidised drug, individuals' financial situation is likely to influence what decisions they make and whether or not they will benefit from paternalistic interventions. With financial inducements, the concern is that they will have more of an effect on someone who is poor than on someone who is wealthy. In the US, for example, critics have noted that, in the context of phase 1 trials, the payments for participation have become 'high enough to make participating in trials more lucrative than holding a minimum-wage job' and as a result many poor people are relying on trial participation as a source of income (Elliott and Abadie 2008: 2317). From the perspective of equality, the worry is that because people are making choices against such unequal background conditions, people on low incomes would be more willing to take risks than people on higher incomes: 'a sum of money that the wealthy can easily resist may be very tempting for poorer people' (Elliott and Abadie 2008: 2316).

With incentives for healthy behaviour (and taxation on unhealthy products), one claim that is often made is that such interventions will have a greater effect on low-income groups, therefore helping to improve health outcomes among individuals who tend to be of poorer health. This mechanism would make incentives beneficial from the perspective of health equality. At the same time, we cannot be certain that these are the effects that incentives are in fact going to have: wealthier people are often in a better position to take advantage of available opportunities to adopt healthier behaviours, whereas for poor people, the existence of other constraints means that even with additional incentives in place, they may not be able to adopt healthier behaviours (Voigt 2012). If it is this second mechanism that becomes dominant, then incentives effectively create opportunities for inequalities that did not exist previously: for example, poor smokers may face the additional financial burden of higher taxes without reaping any of the health benefits that policy-makers sought to achieve. To the extent that such effects are likely to occur, it may actually be better – from the perspective of distributive equality – to completely *remove* options rather than change the cost attached to them. For example, from the perspective of equality, it may be better to completely ban tobacco than to increase

---

[2] Note that strictly speaking we are dealing with impure paternalism here, where third parties are restricted in their liberties so as to protect other agents from making 'bad' choices.

tobacco taxes (see also Voigt 2010). Thus, even though approaches that merely change the relative cost of options are often considered more appealing because they are less restrictive, they may have distinct disadvantages as far as distributive equality is concerned (Voigt 2012).

6.4 Paternalism: Distributive Versus Relational Equality

The previous section highlighted some of the complexities surrounding Arneson's argument and the importance of specifying how exactly specific components of the argument should be defined or interpreted. Arneson's argument focuses on the *distributive* implications of paternalistic interventions: how do paternalistic policies increase – or decrease – equality in distributive outcomes? However, critics have pointed out that distributive concerns should not be regarded as central to egalitarian justice; instead, equality should be conceived of as concerned primarily with the nature and quality of relationships between individuals (Anderson 1999; Scheffler 2003).[3] Distributive considerations may still be relevant – because distributive inequality is likely to undermine relational equality – but distributive equality is only instrumentally, not intrinsically, important.

Similar 'relational' concerns are raised about paternalism. That paternalism may communicate disrespect towards the individuals interfered with has been a concern in the debate, even if the link to the relational equality literature is not made. For example, Seana Shiffrin underlines the way in which paternalistic interventions implicitly stipulate an asymmetry of knowledge and competency between the two agents involved. This makes the expression of disrespect a central feature of paternalistic interventions:

> The essential motive behind a paternalistic act evinces a failure to respect either the capacity of the agent to judge, the capacity of the agent to act, or the propriety of the agent's exerting control over a sphere that is legitimately her domain…. Paternalistic behaviour is special because it represents a positive… effort by another to insert her will and have it exert control merely because of its (perhaps only alleged) superiority. As such, it directly expresses insufficient respect for the underlying valuable capacities, powers, and entitlements of the autonomous agent. Those who value equality and autonomy have special reason to resist paternalism toward competent adults. (Shiffrin 2000: 220)

What does this mean for Arneson's argument? If Arneson is right, then paternalistic interventions can have important benefits for distributive aspects of equality. At the same time, however, paternalistic interventions can be seen as problematic from the perspective of *relational* equality. Moreover, it is arguably one of its central assumptions – that some people are simply better or more competent decision-makers than others – that makes Arneson's argument particularly susceptible to this

---

[3] On the relationship between the distributive and relational views, see also Schemmel (2012). The possible implications of the relational approach for questions surrounding health are considered in Voigt and Wester (forthcoming).

line of criticism. Arneson's argument is not unique in focusing on problems with individual decision-making. Richard Thaler and Cass Sunstein's argument for libertarian paternalism, for example, is similarly based on concerns about choice heuristics and cognitive biases that affect individuals' choices (Thaler and Sunstein 2008). However, while such biases will likely affect all of us to some extent in different situations, it is a central aspect of Arneson's argument that some people are *systematically* worse than others when it comes to making choices. When an argument relies on drawing a clear line between 'good' and 'bad' choosers, concerns about disrespect cannot easily be dismissed.

Can we defend Arneson's argument against this challenge? One possible response is that concerns about disrespect can – at least to some extent – be accommodated within the argument. As I suggested in Sect. 6.3 above, we need to specify in terms of what outcomes we decide that individuals' welfare or interests are indeed served by a particular paternalistic intervention. Some metrics, such as welfare, may well capture some of the negative effects of individuals feeling that certain restrictions express disrespect towards them. Depending on how significant these negative effects are, they may outweigh whatever benefits we expect from the paternalistic interference. However, this response does not address the core concern about relational equality, which is a concern about how individuals treat and relate to each other *independently* of any effects such treatment may have on anyone's well-being.

A second response is proposed by Jansen and Wall, who explicitly address this concern. They argue that paternalistic policies need not imply disrespect to anyone. They suggest that as long as these policies are sufficiently broad, they can avoid the kinds of effects anti-paternalists like Shiffrin are worried about. With respect to their own argument about participation in research trials, they emphasise that 'fairness-based opposition to anti-paternalism has an impersonal dimension' (Jansen and Wall 2009: 181). They conclude,

> the paternalistic… restrictions that the fairness argument would justify must be formulated in general terms that apply broadly to the entire set of potential research subjects. They do not target specific individuals but rather groups of people. No person should conclude therefore that the paternalistic restrictions express the message that he or she lacks good judgement or good decision-making abilities. At most, the restrictions express the message that some (unspecified) members of the population of potential research subjects lack good judgement and decision-making abilities. And this message need not be insulting to any person in particular – indeed, it may be a message that nearly all would assent to. (Jansen and Wall 2009: 181)

This response is not entirely satisfactory. While it is possible to design policies that do not make reference to particular individuals and their decision-making capacities, whenever individuals find that particular choices are blocked for paternalism-for-equality type reasons, this will indicate to them that the options they would have chosen were considered to reflect poor judgement. Some people will find that their liberties are restricted by a paternalistic policy whereas for others, the restrictions do not interfere with the decisions they would like to make. If these restrictions are supported by an argument such as Arneson's, the restrictions may well communicate disrespect to the former group; this is not the case for those whose choices remain unaffected.

Given the assumption of unequal decision-making capacities underlying his argument, it seems that Arneson's position is susceptible to concerns about relational inequality. Of course, this does not invalidate Arneson's argument; rather, it draws attention to a facet of equality that remains unexplored in his approach. Paternalistic interventions require careful weighing of the different considerations at stake. Distributive and relational concerns seem to pull in different directions in this case and it is far from clear how conflicts between relational and distributive equality should be resolved.

6.5 Conclusion

By drawing attention to the 'distributive dimension' of paternalism, Arneson highlights an important but often underappreciated aspect of paternalism. Distributive equality is an important goal and should be one of the considerations we take into account when considering whether or not particular paternalistic interventions are acceptable. Arneson's argument does not (and is not meant to) provide an all-things-considered, knock-down argument to defeat anti-paternalism. Rather, it adds an important nuance to the debate; it brings another consideration to the table that we have to take into account as we evaluate the costs and benefits of particular paternalistic policies or interventions.

This paper considered Arneson's argument in more detail so as to get a better sense for its scope, possible implications and the complexities it raises. Perhaps most importantly, Arneson's argument only works to the extent that the choices curtailed by particular paternalistic interventions would have led to *unfair* inequalities. Since choice is involved, we may think that the resulting inequalities would have been unproblematic. The scope of the argument depends on our ability to make the case that differences in choice-making capacities are matters of brute luck whose influence on distributions should be considered unfair. It is also far from straightforward to determine when individuals are indeed made 'better off' by a particular intervention and to translate Arneson's argument into policy proposals that would reliably make better off (in the required sense) those affected by the policies. Further, while Arneson focuses on paternalistic interventions that remove particular options, other – probably more common – types of paternalistic interventions (such as withholding information or changing the relative 'cost' of particular options) raise further issues of distributive equality not addressed by Arneson's argument. Finally, whatever contribution paternalistic interventions can make to distributive equality, there arguably is more to equality than outcomes. Paternalistic interventions, especially when they rely on distinction between 'good' and 'bad' choosers, may express disrespect towards those whose interests they are meant to protect. Such relational inequalities must be weighed against whatever improvements in distributive equality we expect paternalistic interventions to achieve.

Bibliography

Anderson, Elizabeth. 1999. What is the point of equality? *Ethics* 109: 287–337.

Arneson, Richard J. 1989a. Equality and equal opportunity for welfare. *Philosophical Studies* 56: 77–93.

Arneson, Richard J. 1989b. Paternalism, utility, and fairness. *Revue Internationale de Philosophie* 43: 409–437.

Arneson, Richard J. 1990. Liberalism, distributive subjectivism and equal opportunity for welfare. *Philosophy and Public Affairs* 38: 158–194.

Arneson, Richard J. 1991. A defense of equal opportunity for welfare. *Philosophical Studies* 62: 187–195.

Arneson, Richard J. 2000. Luck egalitarianism and prioritarianism. *Ethics* 110: 339–349.

Arneson, Richard J. 2005. Joel Feinberg and the justification of hard paternalism. *Legal Theory* 11: 259–284.

De Marneffe, P. 2005. Avoiding paternalism. *Philosophy and Public Affairs* 34: 68–94.

Dworkin, G. 2011. Paternalism. In *Stanford encyclopedia of philosophy*, ed. Edward N. Zalta. http://plato.stanford.edu/entries/paternalism/.

Edwards, Sarah, and James Wilson. 2012. Hard paternalism, fairness and clinical research: Why not? *Bioethics* 26: 68–75.

Elliott, Carl, and Roberto Abadie. 2008. Exploiting a research underclass in phase 1 clinical trials. *The New England Journal of Medicine* 358: 2316–2317.

Feinberg, Joel. 1986. *Harm to self.* Oxford: Oxford University Press.

Goodin, Robert. 1991. Permissible paternalism: In defense of the nanny state. *The Responsive Community* 1: 42–51.

Jansen, Lynn, and Steven Wall. 2009. Paternalism and fairness in clinical research. *Bioethics* 23: 172–182.

Jefford, M., J. Savulescu, J. Thomson, P. Schofield, L. Mileshkin, et al. 2005. Medical paternalism and expensive unsubsidised drugs. *British Medical Journal* 331: 1075–1077.

Scheffler, Samuel. 2003. What is egalitarianism? *Philosophy and PublicAffairs* 31: 5–39.

Schemmel, Christian. 2012. Distributive and relational equality. *Politics, Philosophy and Economics* 11: 123–148.

Schmidt, Harald, Kristin Voigt, and Daniel Wikler. 2010. Carrots, sticks, and health care reform – Problems with wellness incentives. *The New England Journal of Medicine* 362: e3.

Shiffrin, S. 2000. Paternalism, unconscionability doctrine, and accommodation. *Philosophy and Public Affairs* 29: 205–250.

Thaler, Richard, and Cass Sunstein. 2008. *Nudge: Improving decisions about health, wealth, and happiness.* New Haven/London: Yale University Press.

Voigt, Kristin. 2007. The harshness objection: Is luck egalitarianism too harsh on the victims of option luck? *Ethical Theory and Moral Practice* 10: 389–407.

Voigt, Kristin. 2010. Smoking and social justice. *Public Health Ethics* 3: 91–106.

Voigt, Kristin. 2012. Incentives, health promotion and equality. *Health Economics, Policy, and Law* 7: 263–283.

Voigt, Kristin, and Wester, Gry. Forthcoming. Relational equality and health. *Social Philosophy and Policy.*

Wilson, James. 2011. Why it's time to stop worrying about paternalism in health policy. *Public Health Ethics* 4: 269–279.

Chapter 7
Contested Services, Indirect Paternalism and Autonomy as Real Liberty

Thomas Schramme

In many countries medicine has become a service institution. People ask for all kinds of medical interventions that do not constitute treatments of diseases. Normally, for the very fact that these interventions are non-therapeutic, they have to be paid for by the person who seeks it. Many medical services are allowed, but there are limits. Not everything that is medically possible, required and would happily be privately paid for, is also allowed. In this paper, I am interested in the normative status of services that are either already banned or regarded as morally problematic, especially for paternalistic reasons. Examples include voluntary active euthanasia, surrogate motherhood, body modifications such as tongue splitting or stapling, and cognitive enhancements. Note that not all these services need to be offered by members of the medical profession to be deemed problematic, neither are they in reality brought about only by doctors. But an aspect of one potential justification of a ban has to do with the goals of medicine – hence my focus on the medical profession.

I will first introduce the notion of indirect paternalism. Indirect paternalism involves not just a paternalistic intervener and a person interfered with, but also another party, whom I call assistant. Indirect paternalism interferes with an assistant in order to prevent supposed harm to another person. This sounds like established cases of preventing harm to others, but an important aspect of indirect paternalism is the fact that the assistance is sought by the person whose good is supposed to be secured by intervention. In the second section I will introduce several strategies that paternalists can pursue to justify indirect paternalism. Some are geared toward indirect paternalism, but can be applied to other forms of paternalism as well. One strategy queries the voluntariness of choices to harm oneself, another one focuses on common affective reactions of people towards certain practices, such as disgust. A more effective strategy, to my mind, is then discussed in more detail in the third section of the paper. It specifically targets an element of

T. Schramme (✉)
Department of Philosophy, Hamburg University, Hamburg, Germany
e-mail: Thomas.Schramme@uni-hamburg.de

© Springer International Publishing Switzerland 2015
T. Schramme (ed.), *New Perspectives on Paternalism and Health Care*, Library
of Ethics and Applied Philosophy 35, DOI 10.1007/978-3-319-17960-5_7

assistance cases, namely the fact that people do not necessarily have a justified claim or entitlement to demand such assistance. To prevent people from providing assistance seems normatively different from preventing a person to do something to herself by her own means. Before I conclude I scrutinize this particular argument supporting indirect paternalism in the step in my argument in sections 4 and 5. I try to undermine the rationale of indirect paternalism by showing that there are at least two situations where it does not work. One such situation that undermines the justification of indirect paternalism is given when the offered service is itself harmless; another pertinent situation consists of a person necessarily requiring assistance to be really free.

7.1   Indirect Paternalism[1]

Cases where a person A requires the support or service of another person B to achieve a particular outcome or to perform an action can be called assistance cases. If the required assistance is forbidden, or by other means hindered or made impossible, for reasons of securing the good of person A, then we can deem these interventions instances of indirect paternalism. Indirect paternalism is therefore a form of multiple-party paternalism.[2] It might involve more than two parties, for instance in the case of surrogate motherhood where usually at least an in-vitro fertilisation (IVF) specialist, a surrogate mother and an egg or sperm donor are involved. Also, the paternalistic intent of a person or institution considering a ban on a particular service might not aim at the person who requests a service but at the good of a potential assistant, for example when active euthanasia is prohibited for reasons of preventing psychological harm to the person who kills another person on request. This would transfer the same case, which can be discussed under the rubric of indirect paternalism, into a common case of paternalism where the intervenee, i.e. the person interfered with, and the supposed beneficiary are the same person.

A central feature of many assistance cases is that a particular type of action, which is deemed an example of grave harm, changes its normative status by a seemingly tiny bit of addition: the voluntary consent of a person. An action by B done towards A – say, to cut flesh from his body – would normally be a crime, but is a body modification ("scarification") if requested. This ties in well with the legal principle *volenti non fit injuria*, which (roughly) translates "no one is wronged willingly". For anti-paternalists in the tradition of John Stuart Mill's harm principle,

---

[1] Some passages in this essay draw on an earlier version of my argument (Schramme 2013).

[2] Though at least theoretically there might be multiple-party cases of paternalism that are not forms of indirect paternalism, see Feinberg (1986, p. 9). Feinberg used the term "two-party cases", but this might be confusing as there are more than two parties involved in the practice of indirect paternalism. He obviously meant that two parties are the target of a paternalistic interference, where one party is interfered with and the other benefits.

only harm done to others (against their wishes) should be prevented, whereas "[o]ver himself, over his own body and mind, the individual is sovereign".[3]

Obviously, Mill's anti-paternalism has raised many doubts. Objections to the liberal harm principle might work in the following way: If harm is bad for a person, then it is always bad, whether it is wanted or chosen by the person herself or not. Indeed, harm, such as physical injury that involves pain, is intrinsically bad, so why should it matter for normative purposes whether the affected person desires it? There is, however, a convincing response to this objection. It stems from Joel Feinberg's interpretation of the harm principle.[4] Feinberg reads the harm principle as to require the prevention of wrongfully inflicted harms. Indeed, he offers two interpretations of the notion "harm", firstly to injure or damage, and secondly to set back interests. We might want to call the first conceptualisation "impersonal harm", because it does not necessarily involve a point of view of a person; it is simply something undergone, for instance an alteration of bodily structure. The second reading of the notion of harm might be called "personal harm", as it involves the standpoint of a person. Only things that happen to a person, which are deemed a setback of her interests, are instances of personal harm. For Feinberg this second reading leads to a proper understanding of the harm principle: It requires prevention of any wrongfully inflicted setback of interests. This principle obviously does not prohibit voluntarily chosen injuries, disadvantages or other detriments; indeed, these are not even considered harms, or personal harms in my own terminology. In short, according to Feinberg's account, we may stick to the general anti-paternalism implied by the harm principle and endorse the *volenti* maxim.[5]

7.2 Paternalistic Strategies

There are still plenty of strategies to defend paternalistic intervention: First of all, doubts might be raised regarding the voluntariness of particular choices. In cases of body modification willing interveners might want to quarrel with the reasonableness or, indeed, sanity of a desire to severely alter features of one's body (cf. Schramme 2008, p. 10 f.); as regards surrogate motherhood one might want to raise doubts regarding the voluntariness of choice by pointing out strong cultural influences in many countries on the service seekers' desire to have children, and, of course, the exploitative circumstances in which surrogate mothers normally find

---

[3] Because B seems to harm A one might think that these cases were already banned by the Millian harm principle. Yet it should be obvious that the voluntary consent changes the normative status of the same action here and, as we will see, it is even slightly misleading to say that B *harms* A.

[4] Feinberg (1986, p. 10 ff).

[5] Note that it is even possible to accept that impersonal harm is intrinsically bad, and still allow for other considerations, which have to do with personal interests, to outweigh this kind of harm and to conclude that there is no personal harm present where a person has an interest in an impersonal harm. A person may reasonably choose what is intrinsically bad, as long at it is not only intrinsically bad.

themselves. If these kinds of arguments would be successful, they would deem paternalistic intervention into these choices an instance of soft paternalism, as it would only account for a prevention of non-voluntary choices – something that is normatively less problematic than hard paternalism.[6]

I disagree with this strategy, mainly because I see voluntariness as a procedural feature that has to do with the way a choice has been reached. If no coercion or similar influences of will-formation are involved, a choice is voluntary.[7] We cannot identify involuntariness by the content of a choice, for instance by claiming that nobody would voluntarily choose to have his tongue split. Voluntariness and reasonableness are simply not the same – on whatever account of the reasonable we might come up with (Feinberg 1986, p. 104 ff.; cf. Möller 2005, p. 164 ff.). Although it is true that cultural and similar influences on choices can be strong, this is not by itself sufficient warrant for deeming certain choices involuntary. Indeed one might wonder how we would otherwise be at all able to draw a distinction between voluntary and involuntary choices, as every choice is strongly influenced by our circumstances, our upbringing, our friends etc. Roughly speaking, a choice is voluntary when a person is under no coercive influence and endorses, or identifies with, her choice.[8] There is no general argument that would exclude choices involving impersonal harm from the realm of voluntary choices.[9]

A second strategy of the paternalist is not really paternalistic in the narrow sense of the word. However, it is a common strategy used by people who would like to ban the kind of services we are considering, and it is often discussed in relation to paternalism, especially so-called moral paternalism. This approach refers to strong negative feelings or beliefs, such as disgust or strong contempt of common people towards a sought service; hence it seems to be an attempt to place the rationale for intervention in the traditional harm principle. Disgust or repugnance might be considered kinds of personal harms, hence as a ground for preventing the proposed service. A well-known example of such an argument can be found in Lord Devlin's *The Enforcement of Morals*, which sparked a famous debate with Herbert Hart (Devlin 1965; Hart 1963). Devlin's examples are homosexuality and prostitution, but this should not be our concern here. More importantly for our purposes, he justifies state or legal intervention by reference to common sense. The first step in his

---

[6] It is indeed arguable whether soft paternalism should be called "paternalism" at all (see Feinberg 1986, p. 12). But even if intervention into non-voluntary choices were not paternalism after all, this would of course still allow to regard the cases we now refer to under the umbrella term "soft paternalism" as unjustified.

[7] Obviously it is an important issue what kind of undue influences there might be, which consequently undermine consent to (impersonal) harm. I cannot discuss this question here, but see, for instance, Kleinig (2010, 13 ff).

[8] In these cases we might also want to use the notion "autonomous choice". I disregard the relation between autonomy and voluntariness for the purposes of this essay.

[9] There is an important debate regarding the possible coerciveness of inducements that I will ignore for the purposes of this paper (see, for instance Radcliffe Richards 2010).

argument is to claim that potentially every immoral action could be prohibited.[10] The realm of immorality itself is determined by "public morality", i.e. by what is considered immoral by the "right-minded person". Devlin is adamant that the latter is an idealization; it is not a reference to what happens to be regarded as immoral by the majority.[11] The second step consists in stating a threshold of the tolerable. Devlin does not want to allow bans on everything that is regarded immoral; he insists that there should be scope for individual liberty to perform even immoral acts. But there are limits. These limits are drawn by society, namely by common people's indignation or disgust towards a practice. In short, if there are strong feelings of rejection, we have sufficient grounds for banning such a practice.

Interestingly at least the German penal law has a very similar clause (§228), which prohibits acts that "offend good morals" (*wider die guten Sitten*) or, in the Latin phrase, are *contra bonos mores*. This clause explicitly excludes consentient acts as well, i.e. is directly opposing the *volenti* maxim. Obviously it needs to be specified what good morals might be and consequently the Bavarian Higher Regional Court decided in 1999 that "an offence against good morals is present, according to established court rulings, if an act is in opposition to the sense of decency of all equitably and justly minded people."[12]

A very similar version of this strategy to justify interventions, which is better known in bioethics, is Leon Kass's idea of the "wisdom of repugnance".[13] Instead of relying on an ideal, i.e. the right-minded person, Kass somewhat infelicitously refers to the infamous "man or woman in the street" (Kass 1997, p. 19). Nevertheless, Kass has a clear sense of where we stand in most modern societies: There are no reliable common normative boundaries to individual behaviour, save for core rights; there is a practice of normalisation in society – what used to be outrageous yesterday is common practice today; there is no shared sense of decency in our multicultural societies; there is widespread individualisation. We even regard ourselves, our lives and our bodies as projects to be shaped by our desires. Obviously, Kass raises all these issues with a sentiment of disquiet, but his description seems correct. This is simply the situation of modern pluralist, mostly Western, societies, if you like it or not.

What should we make of the mentioned paternalist strategy? It seems obvious that despite the many differences in normative viewpoints many people find surrogate motherhood or extreme body modifications repugnant. But is this sufficient reason for a ban? The main problem I see with this argument is not so much its

---

[10] Devlin's approach is not straightforwardly paternalistic, because he does not justify intervention by the resulting improvement (either morally or in terms of their prudential good) of the intervenee. Rather, he claims that all immoral behaviour might pose a threat to the integration of society. This reasoning, if successful, would be in line with the liberal harm principle.

[11] It is therefore unfair by Hart to replace the notion "public morality" by "popular morality" when describing Devlin's thesis (Hart 1963, p. 19).

[12] "Ein Verstoß gegen die guten Sitten liegt nach ständiger Rechtsprechung vor, wenn eine Handlung dem Anstandsgefühl aller billig und gerecht Denkenden zuwiderläuft."

[13] This approach and similar ideas are sometimes called "yuck factor"; see, for instance, Glover (1999).

philosophically disappointing basis – no reasons, but simply feelings or intuitions are put forward – but its reference to a common evaluative point of view in relation to either a particular society or mankind universally. The very fact that we live in pluralist societies, a fact that Kass explicitly recognises, is responsible for the lack of such a common basis of substantial values. What is more, even if we could identify examples of universal disgust, say in relation to human reproductive cloning, it would not be good enough for the examples we are discussing here, because surrogate motherhood, extreme body modifications, or voluntary active euthanasia are all assessed differently by different people and in different cultures. So I find the second strategy of the paternalist – or willing intervener, if we don't see him as a paternalist – unsuccessful, as there is no such thing as commonly shared values in modern societies.[14]

The third strategy of the paternalist against certain services is more closely related to a particular feature of indirect paternalism, namely that it involves more than one party, of which one is an assistant. As I said earlier, the services we are considering, such as surrogate motherhood or voluntary active euthanasia, require assistance by other people. Now, there seems to be an important difference between preventing a person directly from doing something and preventing another person to offer a requested service to that person, even where the very same actions are involved. In one word, indirect paternalism – which implies preventing assistance – might well be normatively different from direct paternalism.[15] This is mainly due to the fact that there does not seem to be an entitlement to be offered assistance, whereas a person usually is deemed to have the right of self-ownership. The latter allows persons to do lots of things to themselves.[16] So what we need to look at now, when considering a possible justification of contested services, is whether the assistant might have a moral claim, after all, to be allowed to offer the service, or if the person seeking assistance has a claim to such a service.

I believe there is indeed a normatively significant difference between direct and indirect paternalism, which would call for much more detailed reasoning than I can provide in this essay. There is very little that has been written so far on the topic of

---

[14] There is a variant of the idealised "right-minded person" approach, which I should mention, though I do not have the space to discuss it in this paper: If we do not interpret the approach as aiming at an empirical abstraction, referring to existing people, if not simply to the majority, we might still be able to make sense of the approach. We could interpret the notion of good morals – or the good for human beings more generally – as itself an ideal fuelled by philosophical argument, not merely by feelings of existing people. This is perfectionism. It has its own problems, but is still a viable option. Indeed, in another paper (Schramme 2009), where I discuss perfectionism in relation to paternalism, I endorse a form of perfectionism that I call negative perfectionism. It is negative, because it only addresses things that make a life bad.

[15] This is an important insight that is often ignored, for instance by Feinberg (see von Hirsch 2008; du Bois-Pedain 2010)

[16] Surely one may want to insist that the consent given by a person to the service of the assistant is normatively sufficient to justify providing assistance. Indeed, this seems to follow from the logic of the *volenti* principle. But one aspect of my paper is to show why assuming a normatively different status of indirect paternalism is plausible and that the *volenti* principle cannot be an absolute principle.

indirect paternalism (but see especially von Hirsch 2008; Simester and von Hirsch 2011, p. 166 ff.). It seems that one can be an adamant anti-paternalist yet allow for indirect paternalism. Although in the following I will reject indirect paternalism *in some cases*, I do not deem this to amount to a rejection of the rationale per se.

7.3   Objections to Indirect Paternalism

There seem to be at least two circumstances where it would be implausible to argue that the difference between intervention and preventing assistance has any normative significance: When a person has a justified claim, or entitlement, to a service,[17] or when the service itself is not dubious for moral or other reasons, for instance reasons that deem a service imprudent. One aspect regards the person seeking a particular service, the other aspect is concerned with the nature of the service provided.

The latter case applies to services such as selling sweets. Although we might have a paternalistic interest in banning it, because people tend to eat too many sweets, with well-known effects on their health, the service itself – offering a product in exchange for money – is neutral.[18] The possible negative consequences are due to the service users.

If the service is itself harmful, for instance because the offered good contains hazardous ingredients, a ban might amount to avoiding third party harm and hence not be an issue of paternalism. Examples of these cases might be well-known health and safety measures we find in many legal requirements regarding production, sales, trades and services. A service or offered good might also be itself harmful, yet something a customer wants anyway. This differs from the situation just mentioned, where we can assume that people do not agree with certain harms, especially where they are not known or cannot serve as means to other purposes. If we remember the difference between impersonal and personal harm, we could say that a service such as killing is an impersonally harmful service, but it might not always be deemed a personal harm. Very often it is of course not easy to say whether a service is as such harmful in the way that is of significance for its normative assessment, namely in terms of posing personal harm. Obviously this makes many cases, where paternalistic intervention is considered, so difficult to assess.

---

[17] I take 'claim' to be a moral notion here. It can be seen as a moral right, but I avoid the terminology to prevent confusion with legal rights. A person might have a legal entitlement to all kinds of morally dubious services, but these contracts are not my concern here, rather whether those contracts should be allowed. I also take 'claim' as to imply a duty of others to refrain from interference, so it is not just a 'liberty', in the Hohfeldian sense (Hohfeld 1923), where a person has permission to do something and hence is not doing something wrong. A justified claim, or entitlement, as such, does not include a duty of others to provide necessary means to pursue a goal, but I want to consider later how far such provision might indeed be morally required.

[18] Though we might want to introduce bans on, say, aggressive marketing of sweets.

Regarding one of the mentioned aspects that would undermine the justification of indirect paternalism we can therefore conclude that services that are in themselves morally and prudentially neutral may not be banned. This seems straightforward enough in theory, as there is no harm involved – so nothing we can protect a person against – but there are complications in practice, which are due to the potential accumulative harm of such services. These might be caused by misconsumption and overconsumption for instance. I have to leave this important issue for another occasion, but in general it seems that in order to target accumulative harms, which might well be personal harms after all, paternalism does not seem to be the right strategy. A more fitting intervention would be to warn people of the dangers of consuming certain services, but otherwise leave it to their assessment whether they want to use it or not. Regarding services that are in themselves harmful in a certain respect, such as explanting organs or killing another person, we need to ask whether they are of a type people would normally try to avoid, hence could be deemed general personal harms. Again, this obviously poses many more questions that cannot be discussed here, such as whether a general ban that prevents all potential service users from gaining access to the service can ever be justified. After all, there might always be at least one person for which this impersonally harmful service is not personally harmful. It seems that this is a problem of the normative assessment of general rules, such as legal bans, as opposed to individual, single case interventions, hence they point at a possible normative difference between interpersonal and legal paternalism.

The other condition undermining the rationale for indirect paternalism is fulfilled when a person has a moral claim or entitlement to a service, even where it could result in personal harm. Consider the case of parental education. Although we know that many parents raise their children in atrocious ways, we still respect children's right to be raised by their parents. Hence even if we find a particular service dubious for moral or other reasons, we might still be adamant that it should be allowed, even judged from a paternalistic point of view.

It would also be wrong to argue that any service failing on both criteria, i.e. that is deemed problematic for moral or other reasons and that does not involve entitlement to the service, should therefore be banned. After all, we need to balance the good of individual liberty against such a ban. I rather want to argue that there seems to be more scope for the paternalist in indirect paternalism than in direct paternalism. Concerning direct paternalism, there is a kind of presumption of entitlement to do many things to oneself in virtue of self-ownership, but this does not automatically apply to the same actions performed by another person on request. This is probably best seen by the example of suicide as opposed to assisted suicide. Although there might be a justification of the latter practice after all, the onus of justification is on the side of the defender of assisted suicide, whereas there seems to be presumption of the moral legitimacy of suicide (cf. Bergelson 2010).

I take it that many of the services considered for the purposes of this paper are indeed morally or prudentially problematic, hence fail on one (part) of the criteria: These services are not wholly neutral. They come along with at least impersonal harms. As I just said, this would not alone justify a ban, as on balance a legislature might want liberty to prevail. It seems unlikely, though, that balancing alone would tip the scales in favour of all services under consideration. People who would like to argue against a ban on, say, surrogate motherhood need a more principled argument. One way would be to consider more closely the second criteria just mentioned. We therefore need to scrutinise whether people might have a justified claim to the services under consideration.[19] If they do, the normative difference between indirect and direct paternalism breaks down and the third paternalist strategy fails.

7.4   Moral Claims to Contested Services

There are two special problems for a defence of entitlements to use contested services as applied to medicine particularly. The first one is the idea of traditional goals of medicine, which might stand against those claims. The second one is the possibility of conscientious objections by medical personnel. Both problems are of some importance, as it is certainly important to also normatively assess the situation of an assistant and not just the person requesting assistance. Nevertheless, I want to quickly establish why these points should not cause too much trouble for the antipaternalist, before moving on to the main question, whether controversial services can be justifiably claimed at all.

As far as the goals of medicine are concerned it is of course correct to state that health care's primary task is to cure disease and to alleviate suffering; it has not been introduced to improve people's lives over and above a negative threshold of impairment. Normally, health care treatments are offered only when they are indicated, i.e. when a health issue, usually a disease, is present. In addition, at least some of the services that are requested from medical personnel, such as cognitive enhancements, used to be tasks of other institutions, for instance education. These aims become more and more medicalised. Yet, although not completely beside the point, this argument cannot by itself establish why medicine should stick to its traditional goals. As long as services are being paid for by customers themselves, there does not seem to be a general reason against offering medical skills and knowledge for the desired use of healthy people. In fact, medicine has always and traditionally offered at least a few services that were not treatments of disease, for instance abortion. In fact, the whole profession for a long time was a paid service. So why should it not offer the whole range of its possible services, as long as clients choose them

---

[19] Some people might want to say that I am conceding far too much to the paternalist, as they would maintain that service users always have a claim right, if not to the provision of services, but to purchase services on a free market. In addition they might want to say that service providers have the right to sell their services, as long as these are neither immoral nor illegal. But this argument relies a) on the ideology of the free market, a topic I would like to avoid, and b) on a liberal reading of what might be regarded as immoral – namely only services that cause personal harm to others. My aim here is to scrutinize the paternalist strategy in relation to indirect paternalism without begging the question in favour of a strongly liberal, or even libertarian, point of view, although I have of course already hinted at certain aspects of a liberal viewpoint that seem to me unavoidable.

freely and other people do not suffer any harm because of these services?[20] It is notoriously difficult to establish goals internal to the tradition or actual practice of medicine.

Individual medical professionals must not be compelled to perform particular services, though. They might object due to conscientious reasons. Again, this is a valid point, and it seems all the more plausible regarding additional medical services than in the case of core treatments, where there is an extended debate about the justification of the conscientious objection. However, as long as there are some medical professionals who are willing and able to offer a service, such as killing a person on request, the possibility of conscientious objectors has no practical impact.

But why should clients have a claim to the services under scrutiny? Note first that to have a claim to have particular services offered is not the same as getting these services for free, or even as having a guarantee that they will be offered, for instance by introducing state services if nobody wants to offer them on a private market. I am only interested in the option of particular services, i.e. whether people have a legitimate claim to demand that a particular service might be performed by willing assistants, not whether everybody should be in a position to use the service.[21] My argument here proceeds in two steps: Firstly, in the remainder of this section I will argue that respect for autonomy, a core principle that even paternalists agree with,[22] needs to be seen as a demand to secure real or effective liberty,[23] which, again, means to offer enabling conditions for important individual life choices. Secondly, I will explain, in the following section, why the contested services are in congruence with established, widely accepted, services. This is a kind of normalisation argument, which proceeds by drawing analogies to uncontested practices. Its aim is to undermine the status of contestability of many services, such as voluntary active euthanasia. This is not a strong argument, but I nevertheless see it as an important element in undermining the indirect paternalist stance.

People regularly need assistance when leading their lives. This might be due to all sorts of reasons, for instance vulnerability, lack of capability, lack of time, or simply laziness. These reasons might differ in their normative significance of course. People also differ widely in what they regard as valuable activities and pursuits. Everybody has his or her own individual and sometimes idiosyncratic life plan or idea of the good life. I have argued at the beginning that people ought to have the freedom to do what they want as long as they do not cause harm to others. This is the traditional liberal stance. Obviously, paternalists would disagree and maintain that people ought to have the freedom only to pursue what is really worthwhile. This

---

[20] Some enhancements, for instance, might lead to disadvantages of other, unenhanced, people.

[21] Since some of those services will very likely only be affordable for rich people, this might lead to injustice. I disregard this issue in this paper.

[22] Many paternalists support intervention into choices only where it enhances autonomy. Some paternalists have a particular, more demanding, reading of the concept of autonomy, which does not comply with the interpretation I endorse (but see Cholbi 2013 for an important alternative).

[23] I prefer the term "real liberty", because it has been used in related discussions, especially in Philippe van Parijs's book *Real Freedom for All* (1995). Occasionally "positive liberty" is also used in the debate, but it might cause some confusion with another notion of positive liberty that was discussed in a famous essay by Isaiah Berlin (2002).

is a very basic quarrel between paternalists and anti-paternalists that I will need to ignore. But the point we have reached in this paper is a slightly different one: We want to consider whether indirect paternalism might be an option, even when direct paternalism regarding the very same outcome, such as desired death or biological reproduction, is not justified. So the paternalist would agree that direct paternalism would not be justified in the cases under consideration, hence the very basic point about worthwhile options does not apply. The issue then really is whether the introduction of services leading to the same result might legitimately be prohibited or otherwise prevented after all.

I stated earlier that the normative difference between indirect and direct paternalism hinges on the question whether service seekers have a claim to have a service introduced. This, again, is different from asking the question whether service providers have a claim to offer assistance. The latter question relates to the justification of a free market, the former is a question about the relevance of assistance for leading one's own life. It is a question what we mean when we say that we are free to do something, especially whether it requires the necessary means to be able to do it.

Liberty, self-determination, and autonomy are terms that are often used interchangeably, and indeed they are surely closely related. Liberty to do what one wants to do,[24] as long as one does not wrongfully harm other people, is a premise that is taken for granted at this stage of the argument. Where individual liberty has been granted, i.e. where we are allowed to do things ourselves, respect for autonomy implies that we are not hindered by others to pursue our aims. But liberty is not effective where we rely on the assistance of other people to pursue these aims and where they are hindered to offer their assistance. For example, to say a person is free to gain knowledge, where there are no teachers or books allowed, is making shambles of the notion of liberty to education. To be really free we constantly need the assistance of others. To respect autonomy therefore means to offer enabling conditions for services which support people in the pursuit of their individual lives (cf. Oshana 2003, p. 104; Möller 2009, p. 758).

To be sure, this way of understanding autonomy as real liberty raises some problems. For instance, there seems to be a potential confusion between liberty itself and having the means to make use of one's liberty. Indeed, a person who is not hindered to buy books might, in some important sense, be deemed free, even when she cannot afford these books. But note that here we are considering a different case, where in fact the assistance provided is banned from being provided. The analogy would not be to only call those people free to educate themselves who *have* the necessary means, but the analogy is drawn to those who *have access* to the necessary means. Conversely, we cannot call someone free to gain knowledge where books and teachers are banned, in the same way as we cannot deem an infertile couple free to procreate where IVF or other reproductive technologies are banned.

In addition, the point about necessary means seems to apply only in cases where at least some people have the means already, so we do not normally say that people

---

[24] This formulation is less complicated than it should be. We might want to add that people should only be at liberty to do what they want to do when sufficiently informed, when no coercing influences are present, and so on.

lack the liberty to jump over buildings because they do not have access to the necessary means to do so, which would in fact be artificial means of course. Again, this
has a kind of normalisation aspect to it: The more common a certain liberty becomes,
for instance because of technological development, the more we see liberty undermined where people do not have access to these means. Two centuries ago nobody
lacked the liberty to fly, because nobody had the means to do it, but when people
today are prevented from using aeroplanes, they lack the respective freedom.

So far, I have mainly used the notion of contestedness when referring to the
medical services under scrutiny. This has a certain empirical aspect: These services
are in reality contested, due to value judgements by real people. Yet, we might also
ask whether there are good reasons for these judgements. This is an issue of scruti-
nizing a feature of the practices, not an issue of finding out about the perception of
these practices in real people. That is why I now talk about the contestability of
these practices.

So the second step in this part of my argument consists in pointing out that the
ends that people pursue by using the services under consideration are decent and
understandable: People who request active euthanasia or physician-assisted suicide
want to end their suffering; surrogate motherhood often serves the purpose of having children at all[25]; body modification is done to pursue an individual aesthetic
ideal. Indeed, the sought services are merely extremes of widely accepted practices
and they are often the only means available: In almost all societies, we offer services
to alleviate suffering where we can and we usually allow people to die, even by their
own hands. Many liberal societies also accept the use of IVF and they allow many
services helping parents to have children. Again, many societies encourage people
to train and shape their bodies and do not usually have any problem with tattoos or
piercings. Hence, the services under consideration are in congruence with common
practice in many countries.

Similarly, the intentions and goals of the assisting parties are generally morally
valid. They want to help suffering people. Obviously this might not always be the
case, for instance if the provided assistance is merely performed on grounds of
financial reward. Here we might want to reconsider a ban on certain ways to provide
services, but this does not amount to a general case against contested medical services, nor does it normally apply to the examples that were the main examples in
this paper, especially voluntary active euthanasia.

7.5   Conclusion

Our discussion of the possible normative difference between indirect and
direct paternalism has brought us to the conclusion that, although it has normative
significance, it is inconclusive as regards the services under consideration. If an

---

[25] Obviously, it could be sought for other reasons, for instance because a potential mother finds
pregnancy too inconvenient. In these cases we might consider a ban, because it is not in congruence with accepted practices.

individual is allowed to do certain things that are deemed morally or prudentially problematic – though she might not actually be able to perform it herself – then services that offer the very same results should also be allowed. I therefore conclude that it does not matter, for these cases, whether an action of a person or a related service by someone else is hindered or banned for paternalistic reasons. If we oppose direct paternalism, we should also oppose indirect paternalism in parallel cases. This might still leave the paternalist with a strategy, but it is then a strategy that is not specific to indirect paternalism. It concerns whether the person who seeks assistance voluntarily agrees with it. This is similar to the question whether the self-harming person acts voluntarily. Yet, where there is a justified claim to assistance and no personal harm involved, voluntary consent is sufficient to justify the use of a service. Hence the *volenti* maxim is still in place, though in a slightly more complicated way, because not all assistance cases seem to be solved simply by applying this maxim.

References

Bergelson, Vera. 2010. Consent to harm. In *The ethics of consent: Theory and practice*, ed. Franklin G. Miller and Wertheimer Alan, 163–192. Oxford/New York: Oxford University Press.

Berlin, Isaiah. 2002. Two concepts of liberty. In *Liberty: Incorporating four essays on liberty*, ed. Henry Hardy, 166–217. Oxford: Oxford University Press.

Cholbi, Michael. 2013. Kantian paternalism and suicide intervention. In *Paternalism: Theory and practice*, ed. Christian Coon and Michael Weber, 115–133. Cambridge: Cambridge University Press.

Devlin, Patrick. 1965. *The enforcement of morals*. Oxford: Oxford University Press.

Du Bois-Pedain, Antje. 2010. Die Beteiligung an fremder Selbstschädigung als eigenständiger Typus moralisch relevanten Verhaltens – Ein Beitrag zur Strukturanalyse des indirekten Paternalismus. In *Paternalismus im Strafrecht: Die Kriminalisierung von selbstschädigendem Verhalten*, ed. Andreas von Hirsch, Neumann Ulfrid, and Seelmann Kurt, 33–56. Baden-Baden: Nomos.

Feinberg, Joel. 1986. *Harm to self*. Oxford: Oxford University Press.

Glover, Jonathan. 1999. Eugenics and human rights. In *The genetic revolution and human rights*, ed. Justine Burley, 101–124. Oxford: Oxford University Press.

Hart, Herbert Lionel Adolphus. 1963. *Law, liberty and morality*. Oxford: Oxford University Press.

Hohfeld, Wesley N. 1923. *Fundamental legal conceptions as applied in judicial reasoning*. New Haven: Yale University Press.

Kass, Leon. 1997. The wisdom of repugnance. *New Republic* 216(22): 17–26.

Kleinig, John. 2010. The nature of consent. In *The ethics of consent: Theory and practice*, ed. Franklin G. Miller and Alan Wertheimer, 3–24. Oxford/New York: Oxford University Press.

Möller, Kai. 2005. *Paternalismus und Persönlichkeitsrecht*. Berlin: Duncker & Humblot.

Möller, Kai. 2009. Two conceptions of positive liberty: Towards an autonomy-based theory of constitutional rights. *Oxford Journal of Legal Studies* 29(4): 757–786.

Oshana, Marina. 2003. How much should we value autonomy? *Social Philosophy and Policy* 20(2): 99–126.

Radcliffe Richards, Janet. 2010. Consent with inducement: The case of body parts and services. In *The ethics of consent: Theory and practice*, ed. Franklin G. Miller and Alan Wertheimer, 281–303. Oxford/New York: Oxford University Press.

Schramme, Thomas. 2008. Should we prevent non-therapeutic mutilation and extreme body modification? *Bioethics* 22(1): 8–15.

Schramme, Thomas. 2009. Political perfectionism and state paternalism. *Jahrbuch für Wissenschaft und Ethik* 14: 147–165.

Schramme, Thomas. 2013. Rational suicide, assisted suicide, and indirect legal paternalism. *International Journal of Law and Psychiatry* 36: 477–484.

Simester, A.P., and Andreas Von Hirsch. 2011. *Crimes, harms, and wrongs: On the principles of criminalisation.* Oxford/Portland: Hart Publishing.

Van Parijs, Philippe. 1995. *Real freedom for all: What (if anything) can justify capitalism.* Oxford: Clarendon.

Von Hirsch, Andrew. 2008. Direct paternalism: Criminalizing self-injurious conduct. *Criminal Justice Ethics* 27(1): 25–33.

Chapter 8
Paternalistic Care?

Roxanna Lynch

8.1 Introduction

This paper will examine the compatibility (or otherwise) of hard paternalism[1] and care. This examination of the possible compatibility of paternalism and care will ask whether acts of paternalism could simultaneously be acts of care and what effect (if any) paternalism might have on the quality of care.

The issue of the potential compatibility of paternalism and care is considered important to address because it is assumed that instances of givers and recipients of care having conflicting ideas about what would promote and constitute successful care for the recipient would not be infrequent. Because of the likelihood of this conflict occurring, it is plausible to suggest that givers and recipients of care would wish to understand whether and, if so, when it could be justifiable for a care giver to override the wishes of a recipient of their care, for that recipient's own sake, i.e. to act paternalistically towards them. This paper, then, seeks to examine whether care givers can act paternalistically towards the recipients of their care whilst continuing to provide care and, if so, whether that care would be at a level that could be defended as adequate.

In order to address the issues that form the focus of this paper, it must first be defined what is meant by paternalism and what is meant by care. Dworkin's (2010) definition of paternalism shall be used as a working definition of what is meant by the term. The definition and understanding of care utilised is the author's own (Lynch 2014). Once what is meant by paternalism and care has been outlined, the paper will then go on to see to what extent these two concepts could be compatible.

---

[1] Henceforth, 'hard paternalism' will be referred to only as 'paternalism'.

R. Lynch (✉)
College of Human and Health Sciences, Swansea University, Swansea, UK
e-mail: roxannajesselynch@gmail.com

© Springer International Publishing Switzerland 2015   115
T. Schramme (ed.), *New Perspectives on Paternalism and Health Care*, Library
of Ethics and Applied Philosophy 35, DOI 10.1007/978-3-319-17960-5_8

It is concluded that paternalism and care are not *necessarily* incompatible. However, paternalism is argued to pose a potential threat to care by both threatening the success of caring relationships and by directly opposing the aims of care. It is argued that though care givers may sometimes have to act paternalistically in order to give care, acts of care that are paternalistic nevertheless represent more 'risky' (in terms of their likelihood of success) acts of care.

8.2   Paternalism and Care Defined

Following Dworkin (2010), paternalism is defined as follows:

X *acts paternalistically towards* Y *by doing* (*omitting*) Z:

1. Z (or its omission) interferes with the liberty or autonomy of Y.
2. X does so without the consent of Y.
3. X does so just because Z will improve the welfare of Y (where this includes preventing his welfare from diminishing), or in some way promote the interests, values or good of Y.

From the above definition it can be seen that, according to Dworkin, paternalistic interventions are those in which one person interferes with another's autonomy against their will but for their own good.

Care is defined as follows:

Care is the successful promotion of some or all of the conditions necessary for the flourishing of the cared for, for the cared for's own sake (Lynch 2014).

The understanding of human flourishing that is utilised within this definition is argued to be necessarily constituted by the four generic goods of health, choice, pleasure and knowledge. This characterisation of flourishing understands the concept to be one that comprises both objective and agent relative aspects. Flourishing, and thus the generic goods that constitute it, is argued to be an objective good for human beings. This is because the constituents of flourishing are argued to represent aspects of the natural function of human beings and it is assumed that fulfilling that function is a good in itself, for all people. Flourishing is also understood to be agent relative, however, because ideally it is individuals themselves who determine the priority that they ascribe to the generic goods and thus the form that their flourishing will take.

As a minimum criterion for an act counting as an instance of care, it is argued (in line with the above definition of care) that an act must promote some or all of the conditions necessary for the flourishing of the cared for, for the cared for's own sake to *some extent*. That is, in order to count as an act of care, an act must promote the conditions necessary for at least one of the aspects of human flourishing to a recognisable extent.

Claiming that care involves promoting at least one of the conditions necessary for human flourishing to a 'recognisable' extent raises the question: 'recognisable to whom?' This issue is discussed at length elsewhere (see Lynch 2014). For the purposes of this paper the response to the question 'recognisable to whom?' shall be summarised in the following. In order to count as such, an act of care must promote at least one of the conditions necessary for human flourishing to an extent that is recognisable to the recipient of care or, should the recipient of care be deemed incapable of validating acts of care as such, e.g. because they lack mental capacity, a relevant third party. Relevant third parties could include a family member of the cared for or a suitably qualified professional.

If an act fails to fulfil any of the demands of the definition of care, or if it in some way impedes flourishing overall, then it cannot be said to count as an act of care. *Which* aspects of flourishing care givers should focus on promoting is argued to be dependent upon such things as their role responsibilities, the preferences of the cared for and the situation in which care is taking place.

Unlike other prominent understandings of care (see, for example, Noddings 2003, p. 35) that describe the *manner* in which care is given, e.g. in a responsive way, as being a necessary feature of care, the above stated definition of care is not similarly committed. Instead, this definition of care makes the more minimal claim that though care that is given in some ways, e.g. respectfully, is often more likely to achieve the aims of care, and, conversely, that care that is given in other ways, e.g. begrudgingly, is often less likely to achieve the aims of care, this is not necessarily the case. In order for an act to be caring on this definition of care it must only successfully promote some or all of the conditions necessary for the flourishing of the recipient of care, for that recipient's own sake.

What is meant by paternalism and what is meant by care has now been defined. In the following section it will be asked whether acts of care could simultaneously be acts of paternalism.

8.3 Can 'Care' Be Paternalistic?

In the previous section it was described both what is meant by care and what is meant by paternalism. When these definitions are taken together, it appears that *prima facie*, care and paternalism are not incompatible. This is because there is nothing in the definition of care used in this paper that is necessarily incompatible with Dworkin's understanding of paternalism. For example, in the above definition of care it was stated that recipients of care should *ideally* be free to determine (within the confines of their circumstances) the extent to which they access each of the generic goods that constitute their flourishing. However, stating that it should ideally be recipients of care who determine the method and manner of their flourishing does not prohibit other people, e.g. care givers, from sometimes contradicting or interfering with those choices if they feel that the choices that the people in their care have made will not be conducive to their flourishing overall. This is because ultimately the aim of care is to promote some or all of the conditions necessary for the flourishing of the cared for, for that cared for's own sake. It is argued that, typically, this aim will best be served if care is given in line with the preferences of the recipient of care. That said, it is accepted that on occasion this end may best be served by disregarding the preferences of the cared for, if the preferences of the cared for would clearly prevent their flourishing.

This paper argues, then, that care and paternalism are not *necessarily* incompatible and that sometimes, in order for care to take place, care givers *should* behave paternalistically towards the recipients of their care.

However, arguing that care and paternalism are not necessarily incompatible because there is nothing in the definition of either term that contradicts or is inconsistent with claims made in the other term does not fully answer the questions raised at the start of this paper. At the start of this paper it was stated that this paper seeks to address the issues of whether acts of paternalism could simultaneously be acts of care and what, if any, the effects of paternalism on the success of care would be. Though it has already been claimed that, in theory, acts of paternalism could be acts of care, and that sometimes care givers should act paternalistically if care is to take place at all, this claim does not fully address the second issue regarding what effects paternalistic acts may have on the *quality* of care. The anxiety is that even though care and paternalism may not be necessarily incompatible, i.e. paternalism does not prevent the possibility of care *per se*, it remains possible for paternalism to affect care in other ways. The remainder of this paper seeks to explore what these other ways might be.

The following section will examine the effect that paternalism (either in individual acts or at an institutional level) may have on the success of care in terms of its possible influence on relationships of care. As it has already been claimed that paternalistic behaviour may have positive effects on caring outcomes, e.g. by enabling care, this section shall focus on the possible negative effects that paternalism could have on care and caring relationships.

8.4   The Caring Relationship

In spite of the fact that the manner in which care is given is not argued to be a necessary feature of care, it was nevertheless conceded that the manner in which care is given is likely to have an effect on the success or otherwise of care. This is because care typically takes place in relationships of care. Caring relations are comprised of care givers and recipients of care. The definition of care articulated above implies that a good *caring relationship* will be one in which one party (the giver of care) successfully promotes some or all of the conditions necessary for the flourishing of the other party (the cared for) for that other party's own sake.

From the above description of the caring relationship it is inferred that a 'good caring relationship', will be one that is characterised by (amongst other things) trust between givers and recipients of care. Trust is considered to be important to caring relationships because it is argued to be important in enabling and maintaining good communication between givers and recipients of care. This claim is based on the assumption that people communicate best, e.g. most openly, with those whom they trust the most. It seems plausible to assert that, if a care giver is to decide correctly what conditions are necessary for the flourishing of the recipient of their care, then they will need to have a degree of understanding of what the recipient of care takes those conditions to be.[2] It is assumed that in order to achieve a level of understanding that will suffice to inform care givers of what it is that the recipients of their care need in order to flourish, there must be communication between givers and recipients of care.[3]

It is argued that in order to enable adequate understanding between givers and recipients of care, the communication that takes place between them must possess certain qualities whilst being free from others.

It is argued that two of the most important qualities that communication between givers and recipients of care should possess in order to be adequate are the qualities of honesty and transparency.

By 'honest', it is meant that communication should be truthful. Information exchanged between parties should be as accurate as possible and as detailed as is necessary to enable good decision making. By 'transparent', it is meant that such things as the motivations and biases of all parties to the interaction should, as far as possible, be made known to all parties to the relationship. These features of communication between givers and recipients of care are believed to be important. This is because, if they are absent, it is not clear how recipients of care would (1) reliably communicate to the givers of care what they understand the conditions most necessary for their flourishing to be or when those conditions have been met, or (2) how care givers and receivers could come to know whether they would be compatible partners to a relationship of care.[4]

Assuming that a reasonable level of communication is in principle possible between givers and recipients of care, e.g. there is a common language between the care giver and the recipient of care, the most significant threats to adequate communication between them are argued to be epistemic injustices and imbalances of power.

Epistemic injustice includes such things as 'testimonial injustice' and 'hermeneutical injustice' (as described by Fricker (2007)). Testimonial injustice, according to Fricker, "occurs when prejudice causes a hearer to give a deflated level of credibility to a speaker's word" (2007, p. 1) whilst hermeneutical injustice "occurs at a prior stage, when a gap in collective interpretive resources puts someone at an unfair disadvantage when it comes to making sense of their social experiences" (2007, p. 1).

---

[2] Particularly given that this author endorses a partially agent relative conception of human flourishing.

[3] Or, if this is not possible, with a 'next best alternative' individual – for example an advocate of some sort.

[4] For example, it is plausible to suggest that someone who is overtly racist may not be the best choice of care giver for a person of a different ethnicity to themselves. This is because people's interests and biases may in practice interfere with their ability to provide successful care. Leaving aside the issue of whether people should or should not have certain biases, the point remains that transparency in both givers and recipients of care is important to facilitate the optimal matching of partners in caring relationships.

It is claimed that epistemic injustices can pose a threat to the open communication that is necessary to the creation of understanding between givers and recipients of care. This is because epistemic injustices can prevent parties to caring relationships from really 'hearing' (because they have discredited the testimony of the speaker) and 'understanding' (because the necessary tools for interpreting experiences are lacking) what is being said to them. If one party to a relationship is unable or unwilling to hear and understand what the other party is saying, it will be unlikely that the relationship will move forward in the desired direction, i.e. by achieving care, because communication will have been compromised.

A further threat to the caring relationship exists as a result of the fact that most relationships potentially contain imbalances of power. Imbalances of power, e.g. in terms of strength or knowledge, pose a potential threat to caring relationships as they can serve to facilitate abuse by making it easier for one party to abuse the other. Abuse is taken to be contrary to the aims of care.

It is plausible to assert that imbalances of power can be exacerbated in situations in which certain parties to the dialogue are in some way rendered more vulnerable than others. In the context of care giver and recipient of care relationships, the issue of the vulnerability of the cared for is significant. This is because recipients of care are often such by virtue of a *known* weakness, e.g. impaired mental or physical health. It is not claimed that such impairments necessarily render the person so impaired vulnerable.[5] However, it is claimed that the likelihood of recipients of care being more vulnerable in some way than the average person, and of their being especially vulnerable to their care givers (because their care givers are likely to know their weaknesses, and because of their possible dependence upon their care givers) is higher and this suggests that relationships of care can easily become abusive and thus fail to count as caring.

Though the issue of the vulnerability of recipients of care is significant, the issue of the vulnerability of the care giver is also significant. For example, care givers can be actively and intentionally misled by the recipients of their care, e.g. regarding the severity of certain symptoms, and, as a result, the care they provide may fail through no real fault of their own.

Having discussed both what is important to and what can threaten successful caring relationships, attention will now turn to the possible threats posed by acts of paternalism to the development of such relationships.

8.5 Paternalism's Threat to Care

It follows from Dworkin's definition of paternalism that paternalistic interventions would contravene the expressed desires of recipients of care, but would claim to do so in the recipient of care's best interests. Given this understanding of

---

[5] Some people, for example, describe gaining great inner strength from illnesses in the form of greater perspective, patience and forbearance.

paternalism, it is argued that paternalistic acts could threaten the success of care in two distinct ways.

Firstly, paternalistic acts that conflict with the conditions necessary for the cared for's individualised conception of flourishing *could* be straightforwardly at odds with the aims of care. This is because it can plausibly be asserted that so called acts of care that are not compatible with the recipient of care's individualised notion of flourishing – even if they are claimed to be performed in accordance with the best interests of the cared for – will not be conducive to care because they will fail to promote the conditions necessary for the flourishing of the cared for.

For example, it can be seen that there could exist an *a priori* incompatibility between acts of paternalism and the aims of care because of the place ascribed to *choice* within the concept of flourishing. As an ability to choose has been claimed to be constitutive of human flourishing, it would seem that care and paternalism would be in tension. This is because paternalism, by definition, involves contravening someone's choices, whereas the flourishing that is the central aim of care involves people being allowed to make their own choices. Taken together, these two claims appear to point to a significant tension between care and paternalism. It will be argued, however, that the tension between choice and paternalism – though present – is not as significant as it first appears to be.

It has been argued that the generic goods that are constitutive of human flourishing are health, choice, knowledge and pleasure. Choice is understood as referring to the ability of people to control their lives, bodies and surroundings in a manner that is compatible with their accessing the other generic goods and other people being able to do the same. On such an understanding of choice, then, choice is both a good in itself, i.e. it is good *per se* to be able to choose, and is also instrumentally valuable to individuals being able to access other goods in a way and to an extent that is in accordance with their preferences. Given the place of choice within the account of flourishing endorsed by this paper, at this stage of the analysis it could still be plausibly inferred that care and paternalism could not be compatible.

However, though choice has been argued to occupy a central place within human flourishing, it must be remembered it is not claimed that in order to care, care givers must ensure the *actual* flourishing of the recipients of their care. Instead, it is only argued that care givers must successfully promote some or all of the *conditions* necessary for the flourishing of the recipient of their care, for that recipient's own sake. Additionally, it is claimed that in order to count as successful care, care givers must only promote *some* of the conditions necessary for flourishing, to some extent. It was argued that ideally it is individuals themselves who should determine the extent to which they access each of the generic goods, but that sometimes it will be more consistent with the broader, i.e. not just focussing on choice, aims of care for such decisions to be made by someone else.

Given the fact, then, that the account of care endorsed holds that carers must only promote the conditions necessary for some of the aspects of human flourishing to some extent, and that the extent to which they should be promoted can be determined by people other than the recipient of care, it is evident that in spite of the central role that choice plays within human flourishing, care and paternalism could nevertheless be compatible. This is because it could plausibly be argued that, *overall*, the conditions necessary for the flourishing of the cared for could sometimes be most reliably/successfully promoted by overriding some of the choices of the cared for.

It must be remembered, though, that choice has been argued to be constitutive of human flourishing. Because of this claim, if someone were to *always* be treated paternalistically and, subsequently, were to have no opportunity for choice in their lives, they would not be able to flourish. Constant paternalistic intervention by care givers, or paternalistic interventions that prevent a recipient of care's future ability to choose, would not, then, be compatible with care.

In conclusion, it is argued that paternalism and care could be compatible in spite of the threat posed to the good of choice by acts of paternalism as long as the act of paternalism only temporarily overrides an individual's ability to choose and does so in a manner that does not compromise the other generic goods. The claim that acts of paternalism could only be justified if they temporarily interfered with an individual's ability to choose, relates to the second type of threat that paternalism has been identified as posing to the success of care, namely threats to the caring relationship.

The second threat identified is that the very fact that a care giver could behave paternalistically may damage caring relationships by preventing good communication between givers and recipients of care. It is this second challenge posed by paternalism to the concept of care on which the discussion will now focus. The second challenge relates to a more subtle, and less researched, way in which acts of paternalism could pose a threat to caring relationships and thus to care.

It is argued that paternalism could interfere with good communication in a number of ways. It was argued above that good communication required honesty and transparency between all parties to the relationship as well as an absence (as far as is possible) of epistemic injustices and imbalances of power. It is argued that paternalistic acts have the potential to adversely affect all of these four aspects of good communication.

In the first instance, if a recipient of care believes that their will may be contravened by a care giver, they may be less likely – if given the choice – to enter into a caring relationship at all. This may be because prospective recipients of care have reasonable anxieties about having their beliefs and preferences disregarded. Anxieties about having their beliefs etc. disregarded may cause some people to choose not receiving care over receiving 'care' that they find upsetting in some way. On this level paternalism is not interfering with the good communication that is important to the achievement of care. Instead, it is preventing the possibility of care taking place outright.

In the second instance, if a recipient of care does decide to enter into a caring relationship with a care giver who they know to be permitted to behave paternalistically, the recipient of care may choose to hold back information from this care giver because, for example, they do not trust what the care giver may use the information for. As a result of this mistrust of the care giver, inadequate information may be given to them for them to care successfully. Subsequently, their 'care' may fail because it is misguided or inappropriate. This second scenario is an example of paternalism threatening care by decreasing the chances of the honest and transparent communication that is necessary to care givers coming to understand accurately what conditions are necessary to the recipients of their care's flourishing.

Turning now to the threats to good communication and care that were highlighted above, i.e. epistemic injustice and imbalances of power: it is argued that within the context of care, paternalism does not offer a challenge to care and caring relationships that is distinct from these threats. However, it is argued that the presence of paternalism is likely to serve to exacerbate the negative effects of these threats.

For example, care givers who seek to provide care in the knowledge that they may, on occasion and if they see fit, override the autonomy of the recipients of their care, may come to treat their patients in a manner that is not conducive to the development of a successful caring relationship. This is because they may come to either ignore or discredit wishes that ultimately they do not have to respect. In such situations, care givers may fail to make the effort to ascertain what the necessary conditions for the flourishing of the cared for are and instead choose to proceed as they think best. In such circumstances care is unlikely to be achieved. In this example the sanctioning of paternalism has exacerbated the imbalance of power that often exists between givers and recipients of care, possibly facilitating abuse. Additionally, it could be argued that the sanctioning of paternalism increases the chances of epistemic injustice because the very act of such sanctioning could be interpreted as signalling that the testimony and wishes of the recipients of care are often misplaced and/or of little worth.

So it can be seen that paternalism plausibly poses a threat to the success of care. This is because paternalism poses a challenge to the development of the good communication between givers and recipients of care that has been argued to be important[6] to the success of care. It should also be stated, however, that the nature of the threats posed by paternalism to care is not obviously particular to acts of paternalism or settings in which paternalism is sanctioned. For example, people can decide not to trust each other for various reasons, e.g. because of previous bad experiences or biases unrelated to the threat of paternalism. Furthermore, the fact that someone is permitted, e.g. by an institution such as a hospital, to act paternalistically does not mean that they will (and, conversely, the fact that someone is not permitted to act paternalistically does not mean that they won't). Though paternalism plausibly poses a significant threat to the success of care, then, it is not possible at this point to claim that it or its sanctioning poses a *unique* challenge to relationships of care. At this stage, all that can be plausibly asserted is that paternalistic acts signify a more 'risky'[7] approach to care giving.

---

[6] It has been argued that good communication is important to the success of care because it is likely to be part of what enables that success. It has not been argued that good communication is *necessary* to the success of care. This is because it is felt that – though unlikely – on some occasions care may take place even in the absence of good communication between givers and recipients of care.

[7] A 'risky' caring approach is understood to be one that is less likely to lead to successful care than an alternative approach.

However, a further possibility in terms of uncovering a unique threat posed by paternalism to relationships of care may lie in an exploration of what paternalistic behaviour could do to a specific group of people who for whatever reason, e.g. because of a damaged sense of self and self-worth, are more susceptible to the risks it poses. For example, McLeod and Sherwin argue:

> The exercise of paternalism is especially problematic when applied to patients whose autonomy is reduced by virtue of their history of oppression. Oppression involves unjust distributions of power, and healthcare settings are sites of very uneven power differentials. If HCP's,[8] especially physicians, further consolidate their already disproportionate power in relation to patients, especially those from oppressed groups, they exacerbate a problematic power differential and further reduce the already limited autonomy of their patients. Moreover, they are unlikely to be in a position to know what is ultimately in the best interests of patients whose life experiences are very different from their own; hence, they are unlikely to be in a position to exercise paternalism wisely (2000, p. 267).

This passage seems to highlight a point that, though obvious, is significant to a discussion of the potential threats to care posed by paternalism: that different recipients of care are likely to experience different types of act very differently. A paternalistic intervention that overrides the autonomy of a confident, capable individual may or may not count as an instance of care. But, importantly, it may also do no lasting damage to the cared for's ability to receive care. Confident individuals may well 'bounce back' from acts of paternalism. But a paternalistic intervention that overrides the autonomy of an individual who is oppressed and who already struggles to have meaningful dialogue with care givers may do lasting damage to that person's ability to receive care. This is because, if they are continually disregarded, such individuals may simply cease trying to speak up. In such circumstances, paternalism does appear to pose a uniquely insidious challenge (if not a logically unique one) to the success of care. The fact that paternalistic acts are performed for the recipients 'own good' may mean that recipients come to lose sight of what they themselves believe their 'own good' to be. In such instances paternalism would represent a highly risky act of care given the threat it would pose to both the cared for's present and future ability to receive care.

8.6   Conclusion

The main ways in which paternalism could compromise caring relationships have now been discussed. It has been argued that though acts of paternalism could clearly represent significant threats to the success of care and caring relations, it is neither the case that acts of paternalism would necessarily be harmful to the success of care nor that the threats posed to care by acts of paternalism are unique to paternalism (and the sanctioning thereof). However, given the fact that paternalism does plausibly pose risks to the success of caring relationships, within the context of care it is claimed that paternalistic acts should be deemed as 'risky'.

---

[8] 'Health Care Professionals'.

Given the risk that paternalistic acts pose to the success of care – especially given the fact that the risks posed by paternalism may be quite subtle and insidious – it seems plausible to assert that, if paternalism could be avoided within care, then it should be. If it is felt that an act that is paternalistic must be performed, then such an act would merit more substantial justification than other acts that aim at care.

Bibliography

Dworkin, G. 2010. Paternalism. In *The Stanford encyclopaedia of philosophy* (Summer 2010 edition), ed. E.N. Zalta. URL http://plato.stanford.edu/archives/sum2010/entries/paternalism/.

Fricker, M. 2007. *Epistemic injustice: Power and the ethics of knowing*. Oxford: Oxford University Press.

Lynch, R.J. 2014. *Care: An analysis* (Unpublished doctoral dissertation). Swansea: Swansea University.

McLeod, C., and S. Sherwin. 2000. Relational autonomy, self-trust and healthcare for patients who are oppressed. In *Relational autonomy*, ed. C. Mackenzie and N. Stoljar, 259–279. Oxford: Oxford University Press.

Noddings, N. 2003. *Caring: A feminine approach to caring and moral education*. Berkeley: University of California Press.

Chapter 9
The Bite of Rights in Paternalism

Norbert Paulo

9.1 Paternalism and Prima Facie Norms

Paternalism is roughly understood as an interference with a person against their will because this interference will benefit the person. We talk about *hard (or strong) paternalism* where a fully or at least substantially autonomous[1] decision of the person is not regarded as being decisive. We call *soft (or weak) paternalism*, in contrast, interferences in cases where the person is less than substantially autonomous. Most liberal philosophers nowadays are anti-paternalists in the sense that they hold hard paternalism to be unjustifiable. But some still do hold hard paternalism to be justifiable under certain conditions. Let us call them paternalists. A well-known defender of paternalism in that sense is Tom Beauchamp. He believes that anti-paternalists are not sensitive enough to the particular situations in which physicians are confronted with the question whether or not to act against the patient's autonomous wishes. Anti-paternalists, he argues, put too much emphasis on the patient's autonomy, especially when the infringement of her autonomy is minimal relative to the risk of harm she is facing.

Consider the *side rails case*[2] where a 23-year-old athlete who is scheduled for hernia repair, after receiving preoperative medicine but still with clear mind and full understanding of the alternatives and dangers, explicitly says that he does not want the side rails at his bed up. The nurse, after discussing the issue with the patient and

---

I am indebted to many people for comments on earlier versions of this paper. To the participants of the spring school on paternalism at Hamburg University; to an audience at Erasmus University in Rotterdam; to Ulrich Gähde, Daniel Groll, Doug Husak, and, especially, Tom Beauchamp.

[1] For the notion of "substantially autonomous decisions" see, e.g., Beauchamp (2009: 83).

[2] For discussions of this case see Silva (1989), Beauchamp (2009), and Chap. 10.

N. Paulo (✉)
Department for Social Sciences and Economics, Legal
and Social Philosophy, University of Salzburg, Salzburg, Austria
e-mail: norbert.paulo@sbg.ac.at

© Springer International Publishing Switzerland 2015   127
T. Schramme (ed.), *New Perspectives on Paternalism and Health Care*, Library
of Ethics and Applied Philosophy 35, DOI 10.1007/978-3-319-17960-5_9

against his explicit wish, nevertheless decides to put the side rails up in order to prevent the patient from falling out of the bed once he becomes drowsy from the medication. From her experience she knows that drowsy patients are at a high risk of falling out of bed and thereby harming themselves. The unit in this hospital is very busy and does not have the capacity to monitor the patient the whole time. In this case of hard paternalism we have a minor interference with the patient's autonomous wish versus a potentially much higher benefit for him (protection from harm). Beauchamp believes that "the lowered risk makes such actions plausible candidates for status as justified in the circumstances" (2009: 88).

This form of paternalism is commonly criticized with slippery slope-like arguments. The danger of medical paternalism, once it is (again) institutionalized, is so big that all tendencies towards paternalism are viewed with great skepticism, especially in times of a coexistence of various ethical convictions in liberal societies. Further, such cases should be seen as calling for maximal effort to convince patients to agree to the proposed action, thus turning it non-paternalistic. The idea is that no reasonable patient would refuse the side rails.[3]

But what if the patient, as in the example, does refuse? Anti-paternalists would have to say that there is no justification for putting the side rails up. If the patient later falls out of the bed and harms himself, so be it, even if the harm is serious. As a paternalist, Beauchamp takes another stance because he does not view autonomous decisions as being decisive and ruling out other considerations. On his view, together with Jim Childress developed since 1979 in seven editions of the seminal book *Principles of Biomedical Ethics* and described as "the *locus classicus* of the regnant bioethical paradigm" (Arras 2001: 73), commonly called *principlism*, four clusters of principles—respect for autonomy, nonmaleficence, beneficence, and justice—cover the whole field of biomedical ethics. The four principles are not absolute, they only have prima-facie character, i.e. none of them is generally prior to any other, none of them generally rules out consideration of the other principles. In principlism, ethical problems are conceived as conflicts between principles. In the side rails case, principlists would detect a conflict between the respect for autonomy (the patient's explicit wish) and the principle of beneficence (saving the patient from potential harm).[4] The conflict is then solved by balancing considerations of autonomy and beneficence. The point is that in principlism—and other normative systems that only include prima facie norms—autonomous decisions are necessarily open for being overridden by other considerations.

---

[3] See, e.g., Chap. 10. Note that Schöne-Seifert's own position is somewhat ambiguous. She first finds the position convincing that hard paternalism is never justifiable (Chap. 10: 112). Later she highlights that hard paternalism should be banned from medicine as a practice in society, leaving open the possibility to justify hard paternalism in particular cases (Chap. 10: 113). Since she does not develop this idea and does not provide any criteria as to when it might be justified I cannot really see the difference between her view and Beauchamp's.

[4] In principlism nonmaleficence only includes negative prohibitions of action, which is why the side rails case calls for the application of the principle of beneficence.

9.2   Constraining Conditions

This does not, of course, mean that one does not take autonomy seriously. There might, in fact, be very few cases where balancing yields justifications of hard paternalism. But this, then, depends on how one adjusts the relative weights of autonomy and beneficence in particular cases. Beauchamp and Childress restrict paternalism by offering special conditions. On their account, hard paternalism is only justified when

> 1. A patient is at risk of a significant, preventable harm. 2. The paternalistic action will probably prevent the harm. 3. The projected benefits to the patient of the paternalistic action outweigh its risks to the patient. 4. There is no reasonable alternative to the limitation of autonomy. 5. The least autonomy-restricting alternative that will secure the benefits and reduce the risks is adopted (2013: 216).[5]

These conditions are meant to restrict hard paternalism and taking the worries of anti-paternalists seriously, and I suppose they are efficient in doing so.[6] But, before I turn to severe problems with hard paternalism in principlism, let me briefly hint at parallels between these constraining conditions and the proportionality test that is used in law to limit legal rights. These parallels might be helpful to refine the use of the constraining conditions. Julian Rivers offers a lucid and concise summary of the legal proportionality test:

> State actions which limit the enjoyment of a right must be capable of achieving the end desired (suitable), it must be the least restrictive means of doing so (necessary), and it must be justified given the 'cost' to the right in question (proportionate). ... It is important to see that necessity and proportionality (in the narrow sense) are different tests: A measure may be the least intrusive means to achieve a certain end, and yet even the least intrusion necessary may be too high a price to pay in terms of the interference with other legally recognized interests. The test of suitability can thus be subsumed under the test of necessity. Any state action which is necessary, in the sense of being the least intrusive means of achieving some end must, by definition, be capable of achieving the end in the first place. It has to be suitable. Nevertheless, the test of suitability serves a practical function as an initial filter. Any state action which is not even capable of achieving a given end is unlawful, regardless of the existence of other alternative means. The test of proportionality in the narrow sense also has its threshold counterpart. Proportionality presupposes that the state action in question is directed towards the pursuit of an end which is generally legitimate. If the end is illegitimate, then no limitation of any right is justifiable (2002: xxxi f.).

Proportionality, thus, consists of four requirements: (1) a legitimate end; (2) suitable means to achieve this end; (3) that these means are the least intrusive to achieve the end (necessity); and (4) that the means are proportionate in the narrow sense (balancing). The very same structure is to be found in the justification requirements for paternalism within principlism.[7] In ethics, such conditions are merely introduced

---

[5] Note that Beauchamp (2009: 85) uses slightly different conditions that do not include suitability (condition 2 above), but focus on necessity (4 and 5) and balancing (3).

[6] For very similar conditions see Chap. 5.

[7] Beauchamp and Childress's condition 1 resembles the legitimate end requirement in the proportionality test; condition 2 calls for suitability, 4 and 5 for necessity, and condition 3 for balancing.

in an ad hoc manner, and their use is not very well developed. This is very different in legal theory, where proportionality attracts much attention.[8] I do not have the space here to discuss all implications of the proportionality test for ethics. Let me therefore only note that there is no point in balancing risks and potential benefits before asking whether the measure is necessary in the sense of being the least intrusive means to achieve a certain end.[9] Towards the end of the paper I will return to the proportionality test.

9.3   Principlism as a Convergence Theory

No matter how useful or problematic the constraining conditions may be, there is a much deeper problem in principlism. The problem is that principlism in fact not includes the four principles, but four "clusters of principles", i.e. clusters of considerations under the heading of autonomy, nonmaleficence, beneficence, and justice. Within these clusters are more specific rules, various rights, virtues, and ideals, all of which are, as already mentioned, "not wooden standards that disallow compromise" (Beauchamp and Childress 2013: 14), but are to be understood, following W. D. Ross, as prima facie binding only.

So what is the function of these different kinds of norms within principlism? One should distinguish between ideals on the one hand and principles, rules, rights, and virtues on the other; ideals do not impose obligations, the other norms do.[10] Within the realm of obligation-imposing norms the differentiation between principles, rules, rights, and virtues does not do much work in principlism, for all are supposed to be more or less interchangeably.[11] Principles and rules are only different in their abstractness. Rights are understood as correlative with obligations (which follow from rules and principles) and can potentially be present on all levels of abstractness. Most obligations are also translatable into virtues.

I suggest seeing the inclusion of different types of norms into principlism as just another step in the overall project of converging different approaches to ethical theory in a fruitful way. Exactly what principlism does regarding the foundation(s) of morality—not following one of the traditional schools in classical ethics such as utilitarianism, Kantianism, or virtue ethics but endorsing a *common morality* as a set of shared moral beliefs independent of time and culture—happens with the types of

---

[8] For recent book-length accounts of proportionality see Barak (2012) and Klatt/Meister (2012).

[9] Another point worth noting is that Beauchamp and Childress also provide requirements for the balancing process (2013: 23), which are effectively the same as the criteria for the justification of paternalistic acts. This is confusing, especially because condition 3 refers to balancing and would thus not serve any function. A comparison with the legal proportionality test could help to get the concepts straight.

[10] Virtues have some kind of mixed character. Some impose obligations, others do not.

[11] This is, of course, not to say that they are synonymous. They clearly follow different semantic rules.

norms, too. It is the attempt to provide a terminological framework for an under-standing of ethical debates. Although Beauchamp and Childress primarily talk about principles and more specific rules, they leave the door open for other approaches. This way of reasoning has the advantage that it keeps principlism in discussion with alternative approaches. The basic assumption behind principlism is that most of the alternative approaches capture some important aspect of morality and that one should aim at converging these approaches instead of understanding them as being mutually exclusive. The ultimate advantage of principlism is, thus, that it is meaningfully linked to many major traditions in ethical theory.

9.4   The Puzzle with Rights

This convergence theory, of course, also has its flaws. It can be seen as mere cherry-picking. It is easy to criticize principlism by insisting that one cannot converge the valuable insights of different ethical traditions without also accumulating all of these theories' problems and the conflicts between them. I will not pursue this path of criticism here. Instead, I will focus on the notion of rights in principlism and show that it is hard to make sense of rights talk in principlism. This discussion leads to the explication of the structure of rights. Once this structure is understood, it becomes clear that rights of individuals necessarily exclude hard paternalism.

In principlism, a

> right gives his holder a justified claim *to* something (an entitlement) and a justified claim *against* another party. Claiming is a mode of action that appeals to moral norms that permit persons to demand, affirm, or insist upon what is due to them. 'Rights,' then, may be defined as justified claims to something that individuals or groups can legitimately assert against other individuals or groups. A right thereby positions one to determine by one's choices what others morally must or must not do (Beauchamp and Childress 2013: 368, their italics).

Although this sounds as if rights were absolute and as if the right holder could ultimately determine what is morally right to do—within the scope of the right—one should recall that principlism does not contain any absolute norms. Thus, it also does not contain any absolute rights. The resolution of conflicts between rights works just as every other conflict between norms in principlism:

> [A] *prima facie* right … must prevail unless it conflicts with an equal or stronger right (or conflicts with some other morally compelling alternative). Obligations and rights always constrain us unless a competing moral obligation or right can be shown to be overriding in a particular circumstance (Beauchamp and Childress 2013: 15).

For Beauchamp and Childress, rights and obligations are correlative in the sense that rights talk is always translatable to obligations talk. To say "X has a right to do or have Y" is translatable to "some party has an obligation either not to interfere if

X does Y or to provide X with Y" (Beauchamp and Childress 2013: 371).[12] They seem to leave open the question whether rights or obligations are primary, i.e. if one has a right because the respective obligation already exists or if the obligation only exists because someone has the right—although there is some tendency towards the primacy of obligations.[13]

The question then is, if rights are fully correlative with obligations, do they serve any function within principlism? Obviously, the use of rights talk does some work in linking principlism to the debate in rights-based theories in ethics as well as to the debates in law and in politics. But is there any genuine gain from the inclusion of rights?[14] Beauchamp and Childress highlight merely practical and political considerations for the use of rights talk:

> No part of our moral vocabulary has done more in recent years to protect the legitimate interests of citizens in political states than the language of rights. ... We value rights because, when enforced, they provide protections against unscrupulous behavior. ... A major reason for giving prominence to rights in moral and political theory is that in ... practice ... they have the highest respect and better shield individuals against unjust or unwarranted communal intrusion and control than any other kind of moral category. ... By contrast, to maintain that someone has an obligation to protect another's interest may leave the beneficiary in a passive position, dependent on the other's goodwill in fulfilling the obligation (2013: 375).

This suggests that there is something special to rights that does not really fit with the correlativity thesis. But what exactly is it that makes rights talk stronger and more efficient than obligations talk? What is the bite of rights? Although Jeremy Waldron is right in saying that "[n]ot only do philosophers differ about what rights we have, they differ also on what is being said when we are told that someone has a right to something" (1989: 503), the easiest way to make sense of the bite of rights talk in comparison to other normative languages is captured in the traditional *will (or choice) theory* of rights. According to this theory the right holder is, to borrow H.L.A. Hart's words, "a small scale sovereign" (1982: 183). The metaphor of sovereignty in the personal realm is, of course, well-known in the paternalism debate from the work of Joel Feinberg:

> On my land ... I and I alone am the one who decides what is to happen. ... [But i]f we take the model of national sovereignty seriously, we cannot make certain kinds of compromise with paternalism. We cannot say ... that interference with the relatively trivial self-regarding choices involves only 'minor forfeitures' of sovereignty whereas interference with the basic life-choices involves the virtual abandonment of sovereignty, for sovereignty is an all or nothing concept; one is entitled to absolute control of whatever is within one's domain however trivial it may be (1986: 54 f.).

---

[12] For criticism of the correlativity thesis see, e.g., Raz (1986: 170 f.). Against Raz, see Kramer (1998: 23 ff.).

[13] Cf. Beauchamp/Childress (2013: 368, 373). This is the question of "rights-based" theories, where rights are the ultimate source of value; see Dworkin (1977) and Mackie (1984).

[14] Since principlism generally draws on W.D. Ross's notion of prima facie norms, one might hope for clarification by turning to Ross. In his discussion of rights, Ross (1930: 48 ff.) examines different versions of the correlativity thesis; but, unfortunately, he does not address the issue raised here.

For will theorists,[15] rights have the function to protect and foster individual autonomy. An essential feature of having a right is to have the power to enforce and to waive enforcement of the right, i.e. to have control over another's duty. This is what Hart and Feinberg aim to capture with the sovereignty metaphor. Within the will theory there is no room for inalienable rights (i.e., rights that the right holder cannot waive). But there might be room for absolute rights, i.e. rights that can never be limited.[16] This way of conceiving rights makes perfect sense regarding their juxtaposition to consequentialist moral theories,[17] which notoriously have problems to save important individual concerns—or, indeed, individuals—from being weighed-off against other peoples' interests or preferences. This is also why Robert Nozick conceptualized his idea of rights as side-constraints as personalized claims that are untradable across persons, and why Ronald Dworkin invoked the metaphor of rights as trumps over other considerations—especially public interests—, which highlights not only the barrier between different persons but also between different goods (or other considerations).[18] The same idea stands behind Jürgen Habermas' metaphor of rights as a "firewall" (1996: 254), Frances Kamm's insistence on "inviolability" (1996: 272), and many of natural law-inspired arguments against all balancing approaches.[19]

The main rival to the will theory is the *interest (or benefit) theory*.[20] According to the interest theory, the function of having a right is to further the right holder's interests, to make the right holder better off. The interest does not need to be an actual interest of a particular right holder. Instead, "interest" is to be understood as an interest people generally have. This is the reason why interest theorists have no problem to make sense of inalienable rights. I should note here, though, that inalienable rights are usually thought of as being only very few and concerning very central aspects of human life—such as dignity, basic liberty, and life itself. The right to bodily integrity, as in the side rails case, would certainly not count as inalienable.

The puzzle is that the understanding of rights within principlism is, as it stands, neither compatible with the will nor with the interest theory of rights. The point of the will theory is that some choices of individuals are protected. This does not necessarily mean that they can never be overridden or weighed up against competing interests. But strong versions of the will theory hold for a very limited number of instances, such as not being sacrificed for a greater good, that a right can never be

---

[15] What I call *will theory* here is, of course, a family of theories, which only share some basic features. Will theorists are philosophers as diverse as Immanuel Kant, H.L.A. Hart, Hans Kelsen, and Hillel Steiner.

[16] Cf. Finnis (2011: 223 ff.).

[17] Some even tell the whole story of the endorsement and use of rights as attempts to refine or oppose consequentialism, see Edmundson (2012) and Jones (1994).

[18] See Nozick (1974), Dworkin (1977: 90 ff., 1984); and Pettit (1987) on the relation between these two conceptions of rights.

[19] The most recent was put forward by Urbina (2012).

[20] Proponents of the *interest theory*—which is, of course, also a family of theories—are, e.g., Jeremy Bentham, David Lyons, Neil MacCormick, Joseph Raz, and Matthew Kramer.

overridden. This is clearly incompatible with principlism's emphasis on the prima facie character of all norms. I shall note, though, that such an account of absolute rights can only apply to very narrowly constructed rights since otherwise different rights would conflict (without any chance of conflict resolution when all rights are absolute). Such strong versions, thus, do not really reflect how we usually conceive rights.[21] But principlism is also incompatible with weaker versions of the will theory.[22] These weaker versions do not include absolute rights but nevertheless highlight the need to protect the right holder's choices in two ways. First, by reserving very high abstract weights for rights in the balancing process—thereby making it very hard to overturn them.[23] Principlism has no room for such abstract attributions of weight that are the same in every particular case. The second way is by limiting the kind of interests that can be invoked to limit the right (on the first step of the proportionality test), requiring that not all interests may be taken into account and summed up against the right, but only such interests that are sufficiently important relative to the right. Principlism is also unable to allow for such a limitation. Summing up, principlism is not compatible with any version of the will theory.

Similarly, principlism cannot subscribe to the interest theory, because it is committed to balancing all relevant considerations *in particular cases*. There is no room for an abstract ordering of values or rights. Degrees of importance (the weight) of particular considerations can, according to principlism, only be determined in particular situations and for particular agents. There is thus no way to assign higher relevance for certain rights for all persons as interest theorists do, especially with their notion of inalienable rights. Furthermore, as will become clear shortly, not even the interest theory allows for the kinds of hard paternalism that principlism holds to be justifiable. Last but not least is the interest theory commonly—though not necessarily—bound to natural law with its strong emphasis on absolute norms, i.e. the opposite of principlism's prima facie norms.

9.5   The Bite of Rights

Besides the differences in emphasizing autonomy and interest, there are far reaching similarities between the will and the interest theory.[24] There is, for instance, considerable agreement[25] in the literature on both, the will and the interest theory,

---

[21] See the discussion in Waldron (1989).

[22] For a discussion of "strong", "medium", and "weak" models, see Klatt/Meister (2012: 15–44).

[23] I am referring here to accounts such as Robert Alexy's (2002), who developed an idealized "weight-formula" that includes, inter alia, the abstract weights of rights (which are the same in every case) and the intensity of interference with the right (in the particular case).

[24] For an exceptionally clear and deep discussion of these and other aspects of the rival theories see Kramer (1998), Simmonds (1998), and Steiner (1998).

[25] Notable exceptions are Raz (1986) and Nozick (1974). Raz does not follow Hohfeld at all, Nozick only allows for a very limited set of rights (as side constraints), which are agent-relative

that the form of legal and moral rights can best be analyzed in terms of the four Hohfeldian incidents (privilege, claim, power, and immunity).[26] Another similarity that is much more important for paternalism is how rights protect peoples' choices or interests *structurally*. Once we leave behind us the problem-ridden idea that rights are inalienable and absolute we see that rights can be limited. Within certain boundaries, they can be weighed against other rights or other interests. Weaker versions of the will theory and the interest theory only differ in how generous they are in allowing competing interests to limit rights. There are no differences between all these ways to conceive rights when it comes to the *structure of the limitation of rights*. This structure is the proportionality test—consisting of a legitimate end, suitable means to achieve this end, necessity, and balancing—that I already introduced above. The crucial point for paternalism is this: On the first step of the proportionality test, only the protection of *rights and interests of others* are legitimate aims to limit individuals' rights, or so I shall argue.

An example will help to illustrate the point being made. A man who had been in security detention for more than ten years was to be treated with neuroleptics. The treatment would aim at curing his delusions, making him less aggressive, and generally re-socializing him, which is the core requirement for being discharged. The carer from the detention unit believed that this would benefit him, but the man refused the treatment for several reasons. Primarily he feared that he might be harmed by others when he was treated this way because, under this medication, he would not be able to protect himself anymore. Thus he preferred staying in detention over being cured (and potentially released). He finally went to court.[27]

In deciding whether or not the man is to be treated with neuroleptics, courts have to take several steps. The start is, of course, to detect the rights affected. Here is the first, though not very surprising, problem: In cases of paternalism some right of the cared-for (here, the detainee) is affected.[28] In this case the treatment certainly interfered at least with the detainee's right to bodily integrity; the detention interfered with his right to personal freedom. These rights generally protect what falls within their scope. But they can be specified with regards to other rights and interests, i.e.

---

and only negative in character, i.e. they only require to refrain from something (to the result that they can never conflict); see Waldron (1989).

[26] These are named after the American jurist Wesley Hohfeld who developed them as an analysis of legal rights. For a very clear outline of the Hohfeldian incidents—much clearer, indeed, than Hohfeld's own writings—see Wenar (2011) or Jones (1994). Wenar (2005) and Kramer (1998) provide refined versions of the incidents. On how the Hohfeldian incidents can be used to analyze moral rights, see Jones (1994: 47 ff.).

[27] This is the simplified version of a case the German Constitutional Court decided in 2011 (BVerfG, 2 BvR 882/09 of March 23, 2011). The German text of the decision and an English press release are available on the court's website (www.bverfg.de). On that decision see Bublitz (2011).

[28] This holds obviously for law, but also for ethical theories that include rights of individuals, just take Gert, Culver, and Clouser's definition of paternalism (Gert et al. 2006: 238) which excludes every action that does not violate one of Gert's ten moral rules (which is—on their account—the same as a moral right) or principlism (where, as already pointed out, patients have rights, which are correlative to obligations originating from rules or principles).

they can be limited. The different theories of rights do not differ in this respect. Rather, they disagree on the relevance they ascribe to various competing interests. But, again, the crucial point is this: in every model, these competing interests are always *interests or rights of others*. Only interests and rights of others limit rights.[29] What courts normally do is try to balance rights of different persons. But our case of coerced psychiatric interventions in security detention uncovers the second problem with paternalism for every normative account that includes rights: Only the rights of one person are affected—*his right* to bodily integrity concerning the treatment and *his right* to personal freedom concerning the detention.[30] This is not surprising for it only mirrors the typical structure of paternalistic actions: They interfere with a right of the cared-for in her own interest, which is almost always represented by another right of the very same person.[31]

At this point it becomes evident that my argument is only about hard paternalism, where an (at least substantially) autonomous decision of the cared-for is not regarded as being decisive.[32] The need to protect people against themselves mostly occurs in cases of soft paternalism, where the cared for is not substantially autonomous. To put it into the rights-terminology: Soft paternalism does not have the same structural problem with rights as hard paternalism.[33] In the former, the right holder is not regarded as competent to exercise her right and therefore someone else, a proxy (e.g., parents for their young children) exercises the right. It should be noted that, once we distinguish further between surrogate decision making and soft paternalism, interesting cases of soft paternalism are very rare. In fact, most instances of parents making decisions in the interest of their children are surrogate decision-making.[34] A case is only a case of soft paternalism when the cared-for is opposed to that very decision.[35] This is not to say, of course, that the problems of surrogate

---

[29] For my point I do not need to distinguish between rights of others and interests of others; neither between individuals' interests and interests like "national security" or "public welfare" (I believe that "interests" in the latter sense are almost always an abbreviation for accumulated interests or rights of individuals).

[30] As in most cases of hard paternalism we need a strong empirical assumption here: that the man will, once treated and released, not be a danger for other people.

[31] It is quite common to first acknowledge this role of rights but then weaken it again to account for actions as being paternalistic although they do not share this rights-structure; see, e.g., Shiffrin (2000: 218 f.).

[32] Recently, Daniel Groll (2012) made a parallel point concerning the rights of the cared-for, which in his terminology are *structurally decisive* authoritative wills or demands of the cared-for, and surrogate decisions, where in his terminology the will of the cared-for might be *substantially decisive*. But for Groll this distinction of roles the will of the cared-for might play in decision making does not necessarily rule out hard paternalism.

[33] But note that something structurally very similar remains when we interfere with surrogate decisions, e.g. the father's refusal to give his son a blood transfusion, because he (the father) is a Jehova's witness. I will not go into that here.

[34] Cf. Buchanan/Brock (1989).

[35] To give an example: Imagine a father who takes his 4-year old son Billy to the dentist. When the doctor now treats Billy this is justified because the father exercises Billy's rights and consents to the treatment, thereby waiving Billy's right to bodily integrity. There is no need to take the rights

decision making and soft paternalism are not important. Indeed, those are pressing moral problems, but they do not seem to share the specific structure of paternalism.[36] Thus, I take the rights terminology to be a very natural way to conceptualize paternalism and to be exceptionally clear in highlighting what is distinctive about it: (1) An act is a case of hard paternalism when the carer infringes a right of the cared-for and tries to justify[37] this interference with an interest or right of the cared-for. (2) An act is a case of soft paternalism when the carer interferes with a right of the cared-for—who is not sufficiently autonomous to have power over her rights—and tries to justify this interference with the cared-for's proxy having waived the cared-for's right, although the cared-for opposes the interference; there is, thus, no infringement of the cared-for's right. (3) An act is surrogate decision making when the carer interferes with a right of the cared-for—who is not sufficiently autonomous to have power over her rights—and tries to justify this interference with the cared-for's proxy having waived the cared-for's right, where the cared-for does not oppose the interference.[38]

In the security detention case the lower courts—wrongly—did the same most paternalists in ethics do in justifying paternalistic actions: They somehow balanced the rights of the cared-for and considered whether it is better for him to stay detained or to be treated. It was argued that he cannot waive his right to personal freedom and thus has a duty to respect and exercise this right by allowing the treatment with neuroleptics. What happened is that a right that normally imposes a duty on other individuals or the state was turned into a duty of the right-holder against himself. It was, thus, without much reasoning, claimed that personal freedom is an inalienable right, i.e. that the cared-for cannot waive it by preferring to stay detained. The result was that the right to personal freedom was invoked to limit the right to bodily integrity. Going through the proportionality test, the problem already occurs on the first step, which says that the interference with a right can only be justified with a legitimate end. No matter which theory of rights one endorses, legitimate ends to limit a right are the protection or promotion of the rights or interests of others. In the security detention case, clearly only the detainee's rights are at stake. There is thus no legitimate end and the right cannot be limited.[39] There is no way to proceed to the fourth step (balancing). This is true of all hard paternalistic actions; one would never pass the first step, because there is no legitimate end—represented by another

---

of others into account. This is a case of surrogate decision making. It turns to soft paternalism when Billy explicitly refuses the dentist's treatment.

[36] Note that my approach here is in line with Beauchamp's claim that hard paternalism is the only interesting form of paternalism, since only this form has to do with the limitation of autonomy. He argues that "it is easy to justify the conclusions of [soft] paternalism independent of any mention of a principle of paternalism" (2009: 82).

[37] As I will argue in turn, the carer can only try because hard paternalism is unjustifiable.

[38] I am, of course, focusing here on paternalism-like cases. We also talk about surrogate decision making where a proxy decides for someone else without any carer (or someone else) interfering with the cared-for's rights.

[39] Some try to avoid this problem by talking about "future persons" (Gerald Dworkin) or "different selves" (H.L.A. Hart), thus constructing something like a "harm to others" situation.

person's right—to protect or promote. The bite that we ascribe to rights talk in comparison to other normative language is, inter alia, this structural demand. The point is, thus, that rights of individuals as part of any normative theory necessarily exclude hard paternalism.

In passing I noted that not only can rights never be limited with reference to the same person's rights, but that to do so one would also have to claim that the limiting right is inalienable. Otherwise, the right holder could just waive it. I suppose that the burden to explain or even justify something like inalienable rights is difficult to overcome. One would have to make sense of rights entailing duties against the right holder. Even if one accepts inalienable rights, on all accounts I know of, only very few rights are so important that they can plausibly be inalienable. Personal freedom might be one of these. But recall the side rails case, where the patient's right to bodily integrity is at stake. This right is certainly not inalienable; otherwise we could not consent to any medical treatment. Provided one accepts inalienable rights and the particular case is within their scope, still one runs into the proposed structural limitation.

9.6   The Limitation of Rights

So far I have left open the question why rights can only be limited with reference to the rights or interests of other persons. I readily admit that I do not have a conclusive argument for that claim. Rather, my reasons are indirect and manifold:

1. The structure for the limitation of rights that I have outlined above absolutely protects the right holder against hard paternalism, which reflects the liberal anti-paternalistic intuition.
2. This structure thereby makes sense of a core content of a right; a minimal content that is not up for grabs. This also resembles the will theories' deep intuition of personal sovereignty.
3. It resembles this intuition without referring to absolute norms, trumps, firewalls, or inalienable rights. It is thereby compatible with moral theories, such as principlism, that are committed to prima facie norms and makes sense of the power of rights talk even in these theories.
4. The structure rests on a very broad basis. It is compatible with all theories of rights. It is compatible with strong will theories for my claim is only conditional (*if you limit a right, this is how to do it…*); if all rights are conceived as absolute, this is in accord with my account. It is compatible with weak will theories for it says nothing about the kinds of competing interests one can take into account (as long as these are someone else's); it further does not say anything about how to balance or how to adjust weights. It is also compatible with the interest theory for, even if one endorses inalienable rights one runs into the same structural problem; outside the scope of inalienable rights or when the right is limited with

reference to another interest of the right holder (and not to her rights) the situation is exactly the same as in the will theory.

5. It is very simple and clear; despite its simplicity it does provide clear demarcation lines between hard paternalism, soft paternalism, and surrogate decision making.

Together, these five reasons make a strong case for my claim. Indeed, everyone who does not accept this claim will have to explain why the literature on the limitation of rights does not mention the possibility to limit rights with reference to rights or interests of the same person. Further, one will have to show that an alternative also fits within the current conceptions of rights, is compatible with different normative theories, and still has the same explanatory power (in making sense of the special power of rights talk).[40]

9.7 Some Implications

Throughout the paper I referred to Beauchamp as a famous hard paternalist. I explained why principlism must—in Beauchamp and Childress's own understanding—hold that hard paternalism is justifiable. Trying to make sense of their inclusion of rights as norms in principlism I invoked traditional theories of rights to the result that the principlist understanding of rights is incompatible with all these theories. This result holds even if one includes my structural claim that rights can only be limited with reference to the rights or interests of others. But, if my claim is correct, then at least the inclusion of rights would serve a distinct function in principlism, which is what Beauchamp and Childress clearly wanted them to do. The special power of rights talk could at least be attributed to this claim and the thereby guaranteed core content of a right that is not up for grabs. This is not much, but it is better than nothing.

The most important implication is the abolishment of hard paternalism for all normative theories that include rights of individuals in their normative set. If my claim is correct, every set of legal or ethical norms that includes rights necessarily guarantees the right holders a certain normative core, some domain that cannot arbitrarily be limited. Especially can it not be limited with reference to the right holders' rights or interests. And since precisely this infringement of a right of a person combined with the attempt to justify the interference with an interest or right of the very same person is constitutive of hard paternalism, the structural limitation rules out every instance of hard paternalism.

I already noted that conceptual analysis does not solve the problems in dealing with paternalism. Not only is my claim neutral to soft paternalism and to surrogate decision making. Even when it comes to hard paternalism my claim does not

---

[40] I assume that coherence with currently accepted positions is a plus; it is, of course, still possible to develop a much better alternative theory of rights.

exclude something like 'provisionary paternalism', as some find it in John Stuart Mill's *On Liberty*, i.e. the interference with a person who is about to do something that potentially results in serious harm to the very same person in order to figure out whether the person is (at least substantially) autonomous. But my claim certainly forbids the continuation of the interference once the person's autonomy is beyond doubt—no matter what the balance between potential harm and benefit.

References

Alexy, R. 2002. *A theory of constitutional rights*. Oxford: Oxford University Press.

Arras, J. 2001. Freestanding pragmatism in law and bioethics. *Theoretical Medicine and Bioethics* 22: 69–85.

Barak, A. 2012. *Proportionality. Constitutional rights and their limitations*. Cambridge: Cambridge University Press.

Beauchamp, T.L. 2009. The concept of paternalism in biomedical ethics. *Jahrbuch für Wissenschaft und Ethik* 14: 77–92.

Beauchamp, T.L., and J.F. Childress. 2013. *Principles of biomedical ethics*, 7th ed. Oxford: Oxford University Press.

Bublitz, J.C. 2011. Habeas Mentem? Psychiatrische Zwangseingriffe im Maßregelvollzug und die Freiheit gefährlicher Gedanken. *Zeitschrift für internationale Strafrechtsdogmatik* 8: 714–733.

Buchanan, A.E., and D.W. Brock. 1989. *Deciding for others*. Cambridge: Cambridge University Press.

Dworkin, R. 1977. *Taking rights seriously*. London: Duckworth.

Dworkin, R. 1984. Rights as trumps. In *Theories of rights*, ed. J. Waldron, 153–167. Oxford: Oxford University Press.

Edmundson, W.A. 2012. *An introduction to rights*, 2nd ed. Cambridge: Cambridge University Press.

Feinberg, J. 1986. *Harm to self. The moral limits of the criminal law*, vol. 3. Oxford: Oxford University Press.

Finnis, J. 2011. *Natural law and natural rights*, 2nd ed. Oxford: Oxford University Press.

Gert, B., C.M. Culver, and K.D. Clouser. 2006. *Bioethics. A systematic approach*, 2nd ed. Oxford: Oxford University Press.

Groll, D. 2012. Paternalism, respect, and the will. *Ethics* 122: 692–720.

Habermas, J. 1996. *Between facts and norms*. Cambridge: Polity.

Hart, H.L.A. 1982. *Essays on Bentham*. Oxford: Clarendon.

Jones, P. 1994. *Rights*. New York: St. Martin's Press.

Kamm, F.M. 1996. *Morality, mortality*, vol. 2: Rights, Duties, and Status. Oxford: Oxford University Press.

Klatt, M., and M. Meister. 2012. *The constitutional structure of proportionality*. Oxford: Oxford University Press.

Kramer, M.H. 1998. Rights without trimmings. In *A debate over rights*, ed. M.H. Kramer, N.E. Simmonds, and H. Steiner, 7–111. Oxford: Oxford University Press.

Mackie, J. 1984. Can there be a right-based moral theory. In *Theories of rights*, ed. J. Waldron, 181–186. Oxford: Oxford University Press.

Nozick, R. 1974. *Anarchy, state, and utopia*. Oxford: Blackwell.

Pettit, P. 1987. Rights, constraints and trumps. *Analysis* 47: 8–14.

Raz, J. 1986. *The morality of freedom*. Oxford: Clarendon.

Rivers, J. 2002. A theory of constitutional rights and the British constitution. Translator's introduction. In Alexy, xvii-li.

Ross, W.D. 1930. *The right and the good*. Oxford: Clarendon.

Shiffrin, S.V. 2000. Paternalism, unconscionability doctrine, and accommodation. *Philosophy and Public Affairs* 29: 205–250.

Silva, M.C. 1989. *Ethical decision-making in nursing administration*. Norwalk: Appleton and Lange.

Simmonds, N.E. 1998. Rights at the cutting edge. In *A debate over rights*, ed. M.H. Kramer, N.E. Simmonds, and H. Steiner, 113–232. Oxford: Oxford University Press.

Steiner, H. 1998. Working rights. In *A debate over rights*, ed. M.H. Kramer, N.E. Simmonds, and H. Steiner, 233–301. Oxford: Oxford University Press.

Urbina, F.J. 2012. A critique of proportionality. *American Journal of Jurisprudence* 57: 49–80.

Waldron, J. 1989. Rights in conflict. *Ethics* 99: 503–519.

Wenar, L. 2005. The nature of rights. *Philosophy and Public Affairs* 33: 223–252.

Wenar, L. 2011. Rights. In *The Stanford encyclopedia of philosophy* (Fall 2011 Edition), ed. E.N. Zalta. http://plato.stanford.edu/entries/rights/.

Part III
Paternalism in Psychiatry and Psychotherapy

Chapter 10
Paternalism: Its Ethical Justification in Medicine and Psychiatry

Bettina Schöne-Seifert

10.1 Introduction

Paternalistic treatment purposely hinders or undermines the desires, decisions or actions of another person for his or her (presumed) welfare. Under some circumstances such conduct appears appropriate, under others objectionable, and under still other circumstances we tend to be uncertain, or even of two minds, in our evaluation. To systematize and justify these intuitions, or else to question them by looking at them more closely, is the task of ethics. And this is not a marginal task: The problem of paternalism touches upon fundamental questions of our moral self-understanding and our ethical, legal and political standardization. Its thorough treatment requires, among other things, a conceptual and axiological clarification of the opposing notion of self-determination (autonomy). At the present occasion such an undertaking cannot be made. It should be made clear, however, that autonomy in this context means the power of decision-making in questions of life plans and ways of life. In the specific context of medicine it most often means the right of veto when it comes to treatment – something that the law and ethics of Western societies regard as the substance of fundamental rights, and specifically as the object of a legitimate moral claim.

Paternalism comes across to us, on the other hand, as a practical phenomenon, or as a problem in completely different contexts of different actions, and should be decided upon accordingly. The requirement to wear a safety belt, the enforcement of helmet laws, or the prohibition of drugs are some prominent examples from legal policy. In modern Western medicine since the early 1970s, taking into account varying local and cultural specific practices, medical paternalism has been seen and

B. Schöne-Seifert (✉)
Institute of Ethics, History and Theory of Medicine,
University of Münster, Münster, Germany
e-mail: schoeneb@ukmuenster.de

© Springer International Publishing Switzerland 2015   145
T. Schramme (ed.), *New Perspectives on Paternalism and Health Care*, Library
of Ethics and Applied Philosophy 35, DOI 10.1007/978-3-319-17960-5_10

discussed more and more as an ethical problem: to begin with, the subject, as handled in the US-American literature (Beauchamp 1995; Childress 1982), has in the meantime become part of medical ethics of almost all Western countries.

With its specific paternalistic questions, psychiatry has now taken up a more specialized area, which places itself in connection with compulsive hospitalization and coercive treatment. To put it succinctly, after hundreds of years of an often encroaching, violent and – in today's view – inhuman psychiatry, Western psychiatry since the 1960s has become more liberalized, more transparent and brought more in line with patient rights. It lies just as much in the nature of the handling of psychologically and mentally ill people that these patients often will not or cannot consent to what can be offered as diagnostically promising, appropriate and – from the perspective of a third party – rational treatment. The questions of whether, when, and for what reason cases may be treated paternalistically for this reason belongs to the ethical questions fundamental to the psychiatric profession. It is of considerable importance to answer these questions plausibly and consistently with ethical norms that apply outside of psychiatry – for the trust of the patients and their dependents, for the self-assurance of the psychiatrists and their teams, and for society's trust in psychiatry.

In comparison, questions of paternalism in psychiatry have until now been rarely discussed on the side of philosophical ethics. This is just as true in general as it is for the German-speaking region. On the other hand, psychiatric problems currently offer differentiated illustrative material, suggesting the idea of drawing together the general ethical debate on paternalism and the problems of psychiatry – with anticipated gains in both theory and practice. The following reflections are just a few steps along this path.[1]

10.2   Concepts of Paternalism

The concept of paternalism stems from the 19th century and, first of all, was used in the social and economic sciences. In ethics this issue had indeed already been explicitly discussed by John Stuart Mill, and yet the term itself only entered into ethical terminology in the 1970s, where up till now it has been used only inconsistently (Garren 2006). Specifically, there is a wide and narrow concept of paternalism as well as – and partly related to this – a value-neutral and a negatively connotative concept.

It is beyond dispute that paternalistic action aims at the welfare of a person for whom the power of decision will be denied. By virtue of this aim, it can either be targeted to benefit or safeguard the person's well-being, or it can be to stave off an existing harm or to prevent the threat of harm. In the following, all of these variations will be covered by the phrase "for the person's welfare." A caring motivation

---

[1] This article takes up a project that I had the opportunity to pursue years ago at the University of Zurich when working for the chair of philosophy with Anton Leist.

for action is therefore not yet a sufficient description of paternalism, since one can indeed treat a person beneficently without conflicting with the person's own will. Hence in addition there must be the conscious impeding or undermining of existing desires or, in special cases, the denial of the power to decide, before the respective desires have even been developed. Caring alone, as a motivation for paternalistic action, affords no justification for such conduct: It could aim at a false understanding of welfare, or else it could be that well-being does not morally counterbalance the reality of such an imposition.

In the next step, a decisive decision regarding the definition is made by a more specific characterization of those preferences ignored and actions impeded by paternalists: The *narrow* conception of paternalism, as championed by Beauchamp or Quante (Beauchamp 2009; Quante 2002, ch. 8), holds that these need to be autonomous preferences or actions, whereas the *wide* conception, which is supported by Childress, VanDeVeer or Hodson (Childress 1982; Hodson 1982; VanDeVeer 1986), and that I will also endorse in the following paragraphs, implies *no* autonomy of the "paternalized" or of his or her actions.

These alternative definitions are also expressed with the help of a conceptual difference, which Joel Feinberg brought to the debate, namely, the differentiation between *hard* (or *strong*) paternalism on one side and *soft* (or *weak*) paternalism on the other side (Feinberg 1971).[2] Both variations are differentiated in that the first one undermines autonomous wants or decisions, whereas the second does not. In this way the narrow conception of paternalism is congruent with the hard variant, whereas the broad conception also encompasses weak paternalism.

Everyday examples of *strong* paternalism in legal policy are compulsory insurances or the enforcement of helmet laws for motorcyclists, insofar that they are justified by reference to the welfare of the affected individual. An example of strong paternalism in medicine is something like lying out of compassion to a terminally ill patient about his or her prognosis, although the patient had made an explicit and well-considered request for truthful information.

Weak paternalism, which concerns a person's non-autonomous wants, decisions, or neglect or undermining of actions, is a ubiquitous phenomenon – if only because it applies to dealing with children: The mother who brings her child to be vaccinated despite his or her bawling protest, or the father (*pater* – the namesake of paternalism) who holds his child back from jumping into deep water because she cannot swim, count – unspectacularly from a philosophical point of view – as weak paternalists. Other examples deal with people, who, through ignorance or overexcitement or even through mental impairment or mental illness, cannot make autonomous decisions, and whose autonomy deficits are supposed to be *compensated* through caring intervention. Essential requirement for these variations of paternalism are presently expressed or cognizable wants or intentions

---

[2] Feinberg himself later changed the terminology without substantial changes: "Strong" paternalism became "hard" paternalism and "weak" became "soft" paternalism. But the standard nomenclature leans towards the original terms (Beauchamp 1995). Incidentally, Feinberg believes that the weak version should not even be called paternalism at all.

(following Hodson, I will call them "empirical"; Hodson 1977, 67),[3] which are not autonomous. Again, this requirement does not yet constitute an adequate justification. Nevertheless, there is most likely no one who holds weak paternalism to be categorically inadmissible.

Although the decision between the narrow and wide concept of paternalism is, in the end, purely terminological, and one can make it either way, I believe the wide conception is preferable for two pragmatic reasons. Firstly, I hold weak paternalism – in contrast to Beauchamp (2009) – as neither generally trivial nor uninteresting from an ethical point of view. Secondly, from a systematic point of view it would be unwise to place the caring undermining of the wants of non-autonomous persons in a completely different "drawer of discussion" than strong paternalism, for the very reason that the way we draw the boundaries between autonomy and non-autonomy should play a central role in the debate over paternalism.

Whoever chooses the wide conception of paternalism for these or other reasons has apparently – against common linguistic usage – at the same time settled for a value-neutral interpretation of the concept, which encompasses the entire spectrum from inadmissible to morally imperative paternalistic actions. But even supporters of the narrow definition of paternalism can agree to such value-neutrality, if they hold strong paternalism under certain circumstances as acceptable (Beauchamp 2009).

A special form of weak paternalism is the so-called Ulysses-paternalism, where one undermines the empirical wants of another, because the affected person has, by earlier directives, given the paternalist the power to do so.[4] In order to listen to the entrancing singing of the Sirens, which irresistibly leads sailors to wreck their ships at the coast, Homer's Ulysses has his crew tie him to the mast of his ship after stuffing their ears with wax. When he hears the singing he is spell bound and demands to be set free, but the crew – obeying his precautious command – only tie him down more tightly. After surviving this deadly danger he thanks them. Just as Ulysses had instructed to have his later *empirical* (non-autonomous) wants to be undermined, so do many people competent of decision-making legitimate and specify a paternalistic undermining of preferences in the case of a later deficiency of autonomy. Typical examples of this are provisions for cases of (recurring) manic episodes or the progression of an Alzheimer patient's dementia. The difference, as compared to the other forms of weak paternalism, lies in the individual authorization to a paternalistic intervention, which is also specified in its content.

---

[3] Yet here we would have to ask whether, for instance, a requirement to wear a helmet falls under the concept of paternalism, if such a rule consequently results in a situation wherein many people do not even develop a preference for driving a motorcycle without helmet. We have to broaden the definition so that bans to further the well-being of affected parties can also be counted as paternalistic if they prevent the development of respective preferences.

[4] Admittedly some authors deem Ulysses contracts to be strongly paternalistic and criticize them accordingly; see Spellecy (2003).

In the following I will briefly discuss strong paternalism and more thoroughly weak paternalism – the latter especially in the context of psychiatry. I cannot address Ulysses-paternalism, even when, specific to the context of provisions for dementia, it brings up interesting and topical philosophical questions (cf., for instance, Davis 2002; Quante 2002; Brudney 2009).

10.3 The Justification of Strong Paternalism in Medicine

Strong paternalism in medicine, understood as the ignoring or undermining of the autonomous wants of patients, has had a long tradition. The "compassionate lie," "doctor knows best," or talking about the hope that one should never take away from the patient, had, for more than half of the last century, never been cast in a negative light or considered an item of criticism (Beauchamp 1995; and, regarding medical informational paternalism, Wear 1993, ch. 2). In the end, however, the long march forward in the Western culture of self-determination has not halted in front of this bastion of socially, uncritically accepted heteronomy.

In the meantime in medical ethics strong paternalism has been seen by many as "politically incorrect," and is no longer held, at least in the context of consent, as compatible with law, which grants the authority of consent to all patients capable of autonomy. But except for the fact that opinions about whether an action is strongly or weakly paternalistic can be, in concrete cases, markedly divergent (for more detail see Schöne-Seifert 2007, 52–54), the question of paternalism in practical ethical discussions seems to be answered more often with reference to basic law, intuitions, or with extreme reluctance, than with principled ethical arguments.

Essentially the question here is one of the clarification and substantiated relative evaluation of autonomy and well-being – including, if nothing else, the subsequent fundamental question of what makes the self-determination of patients valuable (see Feinberg 1986, 57–62, for a differentiation of the options). Even though this question and the task to answer it are quite complex, there are at least two positions concerning strong medical paternalism that are both well grounded and *prima facie* plausible:

According to the first account, strong paternalistic actions in individual cases are justified if a patient's gain in welfare that can (only) be brought about by these actions is both large and undisputable, and if at the same time the transgression of autonomy is minimal. A common example is the short-term, postoperative use of a protective bed rail against the preoperative wish of the patient. The theoretical justification appeals ultimately to the principle of proportionality in the evaluation of values that are not lexically prior to one another; application of this principle is assumed to lie in the power of moral judgment (Beauchamp and Childress 2009, 216f; Beauchamp 2009).

A second position, which denies that strong paternalism in medicine is legitimate in any single case, emphasizes the aspect of general trust in consistently respected self-determination of patients within the health care system: As soon as one makes exceptions – even in harmless single cases – to the unconditional respect for the autonomy of a patient capable of making decisions, *every* patient has grounds to mistrust the system (Tannsjö 1999, ch. 1).

I myself find the second position to be more convincing than the first for three reasons. Firstly, it is – in our contemporary society with its pluralism of values regarding questions of the good life and its anonymously practiced and highly specialized medicine – hardly to be expected that patients would trust the differentiating moral judgment of their doctors in these sensitive questions. Secondly, the danger of a doctor's hidden paternalistic behavior towards a patient is, for structural as well as substantial grounds, so big that it seems better not to give any leeway. Thirdly, the usual examples of supposedly justified, moderately strong paternalism are altogether more eligible to attempt to bring patients to a modification of their potentially harmful preferences. Which competent patient (and such a patient is assumed by strong paternalism for conceptual reasons) would not, prior to surgery, allow a protective bedrail to be attached to the bed after the operation, and be it for the sake of argument that the patient could otherwise not be operated upon? Strong paternalism, I conclude, should be completely and visibly banned from the field of practical medicine, understood as a socially responsible and, in its normative rules, transparent practice. In this way it remains open (even here) whether strongly paternalistic actions might be justified in single, concrete cases outside of this practice (cf. Scoccia 2008 for this problem in the context of legal rules regarding (assisted) suicide).

10.4 Justification of Weak Paternalism in Medicine

Weak paternalism, as I have described above, serves the caring *compensation* of insufficient autonomy – here in the context of patients' preferences and decisions. For whoever exercises self-determination must be able to take responsibility for these actions. In order to do so persons need particular cognitive preconditions and adequate understanding of what is going on. If these prerequisites are not fulfilled, then it could potentially be harmful for patients, if they simply had their way. In the same way that a father ought to prevent his non-swimmer son from jumping into deep water, and in the same way that we ought to prevent a friend who is set to accidentally drink algae instead of water from doing so, a doctor should also treat a patient against her expressed wishes under certain circumstances. *Not to* treat weakly paternalistically can, in certain situations, infringe upon doctors' ethical obligations of caring, and would therefore in medicine be just as inadmissible as the prevention of self-determination through strong paternalism. So much is, basically, uncontested.

Yet it is generally underdetermined under which terms weak paternalism (in medicine) is admissible or even imperative. Thereby questions arise, on the one hand, concerning the conceptualization and assessment of a threshold of autonomy deficits, and on the other hand concerning the required additional conditions.

10.4.1   Exculpatory Autonomy Deficits

The most comprehensive bio-ethical analyses and foundations of the principle of autonomy in the tradition of liberalism originate from Beauchamp, Faden and Childress (Faden and Beauchamp 1986; more concise but substantially the same Beauchamp and Childress 2009, ch. 4). The following remarks owe a lot to these three authors, although I do not follow them all the way.

10.4.1.1   Autonomy of Action or of the Person?

Because the systematic place of self-determination in bioethics lies with the decisions of human test subjects and patients, Beauchamp, Faden and Childress have focused their attention explicitly only on the autonomy of such concrete decision-making situations. They do not concern themselves with the theory of the "autonomous person," especially since even autonomous people might not act autonomously in some cases – perhaps due to ignorance or coercion (Beauchamp and Childress 2009, 100).

This restricted view – as one might put it – concerning isolated instead of global autonomy has not remained unopposed. According to Michael Quante, for instance, one aspect of autonomy cannot be explicated without reference to the other (Quante 2002, ch. 5; cf. Schöne-Seifert 2007, 56f.). This much can be said about this controversy: If I see it correctly, Beauchamp et al. are not denying the factual and conceptual connection between isolated and global competences to decide and reason (for instance, if the latter are completely missing, so are the former). Instead they circumvent the notoriously difficult question of what, globally speaking, defines an autonomous person, and content themselves with the theoretically simpler and easier to operationalize requirement of isolated autonomy. It is obvious that this autonomy must anyway be presented specific to content and situation, and possibly tested. Whether the necessary competences are limited to the competence of reason and decision-making, as with Beauchamp et al., or whether they are more all-encompassing (see below), has certainly not yet been concluded.

10.4.1.2   Components of Autonomous Action

According to the influential account of Faden, Beauchamp and Childress,[5] a patient's decisions must (1) be made deliberately in order to count as genuine actions at all. Additional to this comparatively trivial requirement, for decision-making autonomy

---

[5]The following is a significant simplification of the detailed account in Faden and Beauchamp (1986), ch. 7–10, or Beauchamp and Childress (2009), ch. 4. I believe the simplification is still true to the core of the original thoughts. For instance the mentioned authors list seven elements for a valid consent of patients, yet some of them can be seen, I believe, as either preconditions of the three main requirements (e.g. "information and "advice" as preconditions of understanding) or as specific descriptions of actions ("deciding" and "authorizing").

to be realized, further necessary and together sufficient conditions need to be fulfilled regarding an acting person and their conduct: (2) a relevant decision-making competence, (3) an understanding of what is involved, and (4) freedom from external control. Thereby it is emphasized that components (2) through (4), and therefore also the resulting generic notion, are *gradual* terms. None of these requirements has to be completely fulfilled: In order to autonomously give consent and thereby legitimize treatment, patients must neither possess outstanding intellectual competences nor understand the suggested treatment, its ramifications, its risks and its alternatives in all details; nor do they need to make this decision without any interference from other people. It is in fact enough if they have these abilities in "adequate" amounts – and in this way decide with adequate autonomy. According to Faden et al., thresholds are necessary to mark this boundary, which need not be in agreement with a particular theory but ultimately with the everyday understanding of adequacy.

Surely no participant in the associated discussion wishes to require less than these minimal conditions for self-responsible action, – but doubtlessly some want to go beyond. The debates about this possible surplus go under the catchwords *rationality* or *authenticity requirement* of autonomy and essentially revolve around the question of whether actions/decisions can only be regarded as adequately autonomous when they are *well-considered* or substantially *coherent* with the wants, beliefs and value judgments of the affected person. Both requirements are at the very least closely related: Whoever decides after good consideration will do so in the light of her beliefs; though value coherence appears to be possible through intuitive judgment and decision making – and is therefore the more basic requirement. The subject of dispute between opponents (e.g. Feinberg 1986; VanDeVeer 1986) of such additional requirements and their advocates (e.g. Brock 1988; Scoccia 1990) is whether such requirements for value coherence or good consideration might express an intellectual, excessive demand far from everyday expectation, and whether they prescribe to the affected persons such value coherence in a paternalistic way without sufficient justification.

The first objection can be plausibly met, in that one admits a *prereflective* form of evaluative coherence, such as Michael Quante has developed (Quante 2002, ch. 5). The second objection, however, is hard to deny, at least from a liberal perspective: Admittedly, we know from our own and other's experience that value-coherent – and in this sense also authentic – action is subjectively important to many people. To live life in "one's own way" can be an important source for a love of life and a sense of self-worth; to accept this with others would be understood as an expression of respect for people. But that does not *have* to be the case: Many people would like to separate themselves from their "old" maxims and decisions – and no one would be morally justified in impeding them, as a consistent liberal must argue, by referring to their own interests. As a rule, but not in every individual case, respect for a person's autonomy will protect actions and decisions that they fittingly do against the backdrop of their own preferences and beliefs. This normative differentiation must also be expressed conceptually.

Yet, as even a liberal would admit, the dispositional *capacity* to decide in a value-coherent way belongs to our notion of autonomous action – such as David de Grazia and Michael Quante advocate, as opposed to Beauchamp et al. (Quante 2002; ch. 5;

DeGrazia 1994) This shows itself significantly in that decisions of cognitively "fit" patients, whose orientation on their own evolved values may have been misplaced or skewed by pathological emotional disorders, will very well be experienced as non-autonomous. Hence not the factual value-coherence of decisions, but indeed the capacity to bring about such coherence, seems to belong to the plausible preconditions even of the isolated autonomy of action. I cannot discuss, on this occasion, the advantages and difficulties of such a model, which is drawn from Dworkin and Frankfurt (Dworkin 1970; Frankfurt 1971).

After these sketchy remarks it should at least be clear that autonomy deficits – which are a conceptual precondition for weakly paternalistic intervention with another person – can appear on three levels: as absence or deficit of specific capacities or competencies in the person, as a relevant ignorance of the person, or as determination of the person by others. In medical contexts these deficits must be fixed as much as possible before one is allowed to consider the legitimation of weakly paternalistic interventions at all: If a patient lacks important information for an upcoming decision or if her family members put her under massive pressure, then these facts are generally no exculpation for a doctor to then make decisions on the patient's behalf – rather, they are reasons to change something about the deficits themselves. The doctor can only change very little about the capacities of the patient; hence it is no accident that Beauchamp et al. call them a gatekeeper requirement of autonomy of actions.

10.4.1.3 Capacity for Autonomous Action (Competence)

Self-determination in actions is accompanied by taking over responsibility for their consequences. In order to be able to carry this responsibility, the agent must possess specific cognitive and mental capabilities – in ethics this is just as uncontested as in law when it concerns a person's own legal capacity or their ability to consent or to make a will. Here, just as there, broadly the same preconditions will be required (Petermann 2008, ch. 4): The person concerned must be able to understand relevant information, be able to rationally process that information, and must be able to make a decision against this backdrop. Specific groups of patients – such as children, unconscious persons, or people with severe mental deficits – obviously lack these capacities; yet in normal cases these capacities are presumed. In the case of a supposed lack of adequate ability to reason, which is always to be assessed in concrete situations and with reference to concrete decisions, they are to be examined, potentially by professionals.[6]

Whereas Beauchamp et al. specifically analyze these cognitive elements, other authors have required and discussed further capacities. Without being able to introduce the details of these various suggestions and to compare them with one another, I would like to state the following: It appears rational to firmly turn the capacity of self-referential evaluation into a prerequisite of autonomous action. The position that this capacity is a purely cognitive faculty and that it is *eo ipso* covered by the

---

[6] Regarding the thresholds there are certainly various standards; cf. Helmchen (1996).

requirements sketched above is to be opposed on the grounds of the example of a depressed patient, who indeed understands everything respecting her treatment options, but who is, due to the illness, "completely indifferent" to everything (Charland 1998).

Another component of the capacity to act and decide is the ability to guide one's actions. Beauchamp et al. regard the problem of absent self-guidance only in the context of external control. Hence they are concerned with cases in which patients are steered or controlled to their decisions by third parties through coercion or illegitimate manipulation. At the same time, however, they ignore cases in which people *cannot* guide themselves in the first place. These can be patients (such as small children or people with severe mental disabilities) whose capacity to self-guidance is missing along with other cognitive or mental capabilities, so that the lack of guidance is not the special deciding factor when it comes to the judgment of the capacity of autonomy. On the other hand, it could also involve – especially in psychiatry – patients with *isolated* defects in their self-guiding capacity, for instance due to anxiety disorder or obsessive-compulsive disorder. It is therefore not by chance that the capacity to guide one's behavior or the "capacity to will" is discussed as a component in its own right both juridically and in the psychiatric debate regarding decision-making autonomy. In conclusion, it appears that Helmchen and Lauter's list of basic capacities for autonomy is reasonable. They identify four capacities a person must display in specific contexts in order to be able to act and decide autonomously:

1. Ability to understand
2. Ability to process
3. Ability to evaluate
4. Ability to guide oneself (Helmchen and Lauter 1995)

I have already mentioned a number of times, but not yet established, that the competence for decision-making – that is the batch of mentioned component parts – must be a "context sensitive" concept. The justification for this standard point of view (Buchanan and Brock 1989; Charland 2001; Helmchen and Lauter 1995; Petermann 2008; Wettstein 1995) is not merely grounded in the fact that competence deficits are gradually formed and can fluctuate. It is also due to their commonly isolated or sectorial existence (as can be seen, for instance, in case of the mentioned area of abnormal delusions). Therefore these deficits are only relevant if they stand in direct relationship to that over which it is to be decided.

10.4.1.4 Autonomy-Competence as a Gatekeeper Requirement

The foregoing considerations should have illustrated that competence – within the area under consideration – is a psychological term with a *normative* charge. Competency statements "set the stage" for how this or any patient *should* furthermore be handled (Beauchamp 1991; Charland 2001). Any determination of the necessary components of competence as well as, within these elements, every designation of thresholds between insufficient levels of characteristics and levels deemed adequate should serve as a double protection of the patient: On the one hand a protection from self-harm for which a patient cannot actually be accountable, on the other hand a protection from inadmissible caring compulsion by doctors or caregivers. This balance between too much and too little paternalism is being achieved mainly through the definition and the determination of competency.[7] This is due to the fact that both formal and informal judgments of the competence of a patient have a gatekeeper function. The other conditions of autonomy, which in a sense become relevant only after[8] this "sorting," serve as parameters for therapists, care-givers or relatives, and must first *be realized* mostly through their behavior – through comprehensive and conversationally competent information, through abstaining from manipulative behavior, and through sensitivity for psychological coercion by third parties. Even the setting of informational requirements and the interpretation of freedom of choice are normative balancing acts, but not ones between too much and too little paternalism.[9]

In this sense it is hardly surprising that competence is such an extremely contested concept in interdisciplinary literature. This is not due to any complex descriptive content, but to its serving as a decisive junction in the debate on paternalism. Thus, while the establishment of a *concept* of competence and its threshold always has normative consequences, the elaboration of suitable criteria is predominantly a psychopathological task, and the assessment of a particular patient in the light of these criteria is a purely clinical and empirical matter.

The previous result is theoretically modest: Weakly paternalistic action *may* be justified if patients have deficits relevant to autonomy. Besides children, patients with impairments of consciousness, or those with sufficiently severe mental disabilities, also patients with illness-induced, isolated impairments in competence can be considered as recipients of weakly paternalistic intervention. Who counts as such will ultimately be ascertained by the determination of competence thresholds.

Still, the respective paternalistic action must then be legitimized through yet another condition.

10.4.2 Acting for a Patient's Well-being – Subjectively Constructed

Deficits of autonomy in a patient P are not a carte blanche for the well-meaning compulsion of P at one's own discretion – so much is undisputed. Rather, P's treatment should be adjusted to P's assumed will, when there is adequate evidence, and

---

[7] An important question in this context, which cannot be treated here, concerns the sliding-scale conception of competency: According to this conception the requirements for decisional autonomy have to be more demanding in congruence with increased risk of harm. This position has many supporters, e.g. (Drane 1985; Wilks 1997). For a critical assessment see Demarco (2002).

[8] This must not be understood in a strictly chronological way: Often judgments regarding competency are made *during* patient briefing.

[9] They are rather balancing acts between mistakes due to ignorance vs. excessive demands on patients and doctors, or else between a decision by third parties vs. leaving patients on their own.

only secondary and as a substitute to the intersubjectively understood well-being of the patient. That is, at least, the liberal credo of medical ethics, which I want to follow[10] and which *generally* applies to the justification of surrogate decisions. Yet for paternalistic action there is a special feature: The fact that it is performed *against* the wishes of a patient and not merely *without* the consent – which might not be possible to obtain – makes its acceptance more disturbing both when anticipating and watching it – be it because of an unjustified thinking in stereotypes or because of a justifiable fear of extension of such practices. Not only strong, but also weak paternalism will easily be felt as encroaching, violent or disrespectful – more than the treatment of a critically ill patient who is unable to express herself. This must be taken into account in the regulation and management of such cases.

10.4.2.1 Hypothetical Consent of the Patient with Counterfactual Assumption of Capacity for Autonomy

A suggestion at first glance would be to permit weakly paternalistic action only if it prevents serious harm (Murphy 1974; Wolf 2000). But this condition would rule out that one might spare someone mere inconveniences through paternalistic intervention, which seems intuitively implausible. For instance, what fundamentally opposes preventing a confused patient from eating a moldy piece of bread, from which she could "only" get a stomach ache?

More convincing is therefore the claim that the patient P would hypothetically agree to this specific intervention, supposing, counterfactually, that she were competent (Hodson 1977; Scoccia 2008; against this point of view: Wolf 2000). If, however, one considers how unlikely it can be to have *positive* evidence for this, because patients have simply not thought or spoken about it, then one could see good reason to reverse the burden of proof: It could be enough to have no evidence of opposition – an option that cannot be discussed here.

It is only subordinate to these conjectures, whether positive or negative, that is, when they have to remain inconclusive for principled or contingent reasons, that an inter-subjective standard of interest may be used. This requirement corresponds – somewhat unsurprisingly – to what in bioethics is general consensus for legitimate surrogate decision-making: These decisions should, if at all possible, correspond to the presumed will of the patient and hence fit the patient's own beliefs and values she had when she was still autonomous. It is therefore aimed at a substituted judgment.

Well-grounded speculation about what a patient, who is no longer sufficiently capable of deciding autonomously, *might* have wished necessarily require reference points from times during which the patient could make autonomous decisions and articulate or indirectly express relevant beliefs or value judgments. Relatives, friends or trusted doctors who employ these conjectures are trying, in the view of such

---

[10] The subjectivism about values, on which this position bases, cannot here be justified in its own right.

evidence in a thought experiment, to determine how the patient would have decided hypothetically (Buchanan and Brock 1989; Brudney 2009). The goal to reach a value-coherent decision *for the patient* is, however, by all means not always so easy to realize in practice.

For reasons of the already mentioned special worry concerning the undermining of the opposing wishes of the patient, it is also important to make the aspect of apparently encroaching determination by others *itself* the subject of hypothetical approval – that is, to ask ourselves whether P (in a competent state) would have approved in principle of caring paternalism *as such*, or rather rejected it.

In some cases questions about P's hypothetical acceptance or rejection of a treatment are futile. That is, because the corresponding thought experiment assumes that P has shown, in a sense, a mere gap in her competence, which could be bridged in the manner discussed above. Thereby, for all patients whose "hypothetical competence" remains an empirically unsupported fiction – whether because they were never competent before (small children, severely mentally retarded patients), or whether because they will never be competent again (severely demented patients; Davis 2002; Brudney 2009) – this strategy for justification is a non-starter. The general consensus for surrogate decision-making demands that in these cases we revert to the standard of the intersubjectively determined best interests (Beauchamp and Childress 2009, 135ff; Buchanan and Brock 1989).

10.5   Paternalism in Psychiatry

10.5.1   Contextual Characteristics

Psychiatric paternalism is a very real phenomenon: Even among the contemporary conditions of a medically advanced, legally secure and human-oriented modern psychiatry, as has increasingly become the standard in Western societies, involuntary hospitalization and coercive treatment for "the well-being" of the patient are practiced. Even now patients are admitted to psychiatric clinics through the use of bodily force by police, tied to bed rails, and treated with psychotropic drugs against their will. Although these practices are legally permissible in Germany and other Western countries only under very strict conditions,[11] it is important to test their *ethical* appropriateness, to justify them expressly if necessary, and to make them

---

[11] By way of example, in Germany psychiatric sections and treatment are, according to the "law regarding support and security measures in case of mental illness" (*Gesetz über Hilfe und Schutzmaßnahmen bei psychischen Krankheiten*), only legal if due to mental illness or disability a patient poses a "present and significant threat for themselves or others" or if there is a danger of suffering serious health-related harm. Involuntary hospitalization is only allowed if limited in time and under court supervision. Coercive treatment may only be performed if the patient lacks capacity to consent or on the basis of consent of a person authorized by the patient or of a legal guardian (Dressing 2004; Koch et al. 1996).

transparent to all involved parties. In so doing there are some specific features of this medical subject to consider.

Firstly, psychiatry plays not only historically, but also at present, a sinister role in a number of totalitarian states in the forced treatment of political or otherwise "unwanted" people. These experiences understandably form the background of many perceptions of psychiatric coercion – even if such psychiatric abuse is (was) not at all, but is (was) at most cynically claimed to be, "for the well-being" of the people treated in this way – and thus cannot be categorized at all as paternalism and most certainly not as weak paternalism.

Secondly, there are always still patients who, whether correctly or incorrectly, see themselves as victims of disrespectful or even downright harmful coercive treatment, and enduringly suffer from this perception.

Thirdly, the rapidly growing possibilities of psychiatry to dramatically affect the feelings, moods and behaviors of people through medication or other measures understandably elicit particular fears of abuse.

Fourthly, many people unfortunately still consider treatment in psychiatry as such as defamatory, embarrassing, and even degrading. And ultimately the regular professional dealing with patients, who, induced by illness, may act "irrationally," can invite the creation of a generalized "paternalistic" attitude and its structural consolidation.

For these reasons, the importance of a proactive ethical analysis and justification for psychiatric coercive measures on societal, institutional and individual levels cannot be overstated.

10.5.2   Justifications of Weak Psychiatric Paternalism: Programmatic Desiderata

First of all, the psychiatric treatment of patients, as it is presently done in so many places, has to be repeatedly checked so that coercive measures are practiced as rarely as possible. Some "encroaching" methods of treatment can be achieved by replacing them with less aggravating ones, which require perhaps only a higher degree of attention and imagination.[12]

A second important point is a clear differentiation between caring and otherwise motivated coercive measures (for the protection of third parties, etc.), as it is generally laid down in law. Only the former fall under the concept of paternalism and it can only be admissible if the affected person is incompetent of decision-making in relevant ways and if the coercive treatment is considered likely to serve her personal well-being and to correspond to her hypothetical will. Coercive measures in the interest of third parties must be justified or criticized by other means.

---

[12] For an account of the profession to record and regulate psychiatric compulsion (see Kallert et al. 2005; Müller 2005).

A third desideratum is to always avoid a *generalized* assumption of psychiatric patients lacking competence in decision-making: Neither a mild mental retardation nor a circumscribed delusion, which does not relate to aspects of the proposed therapy, should stand in the way of an everyday, isolated decision of treatment.[13] Indeed, contrary to the perceptions of lay people, psychiatric patients as a collective whole are, as measured by clinically established competency tests, apparently no less competent to decide about their treatment than the comparative group of somatic patients (Grisso and Appelbaum 1998).[14]

Fourth, even legitimate judgments of incompetence have an appreciable dark side: They can aggrieve, degrade or stigmatize patients, or even act as self-fulfilling prophecies (Beauchamp 1991; VanDeVeer 1986, 415). Especially in times of the finally liberalized psychiatry, caveats against judgments of incompetence are only too understandable. Here the preferred strategy appears to be treating patients as far as possible as if they were – counterfactually – competent: Whenever possible and reasonable they should be informed, advised and consulted ahead of decisions about therapy. Doctors and their medical teams should knowingly accept to some extent the role of "reserve paternalists," who only have to jump in if the "as-if-competent" patient decides against the treatment that is recommended by medical opinion.

One additional point, which requires further analysis, is finally the question of how to validate in the psychiatric context the condition of hypothetical consent of the patient to potential coercive measures. An uncontroversial option for certain patient groups (such as addictions or periodic affective disorders) consists in encouraging Ulysses contracts (see above), in which paternalistic treatment measures are justified via anticipation. For many other patients there will be substantial biographical evidence for the alleged agreement to promising paternalistic treatments. Very difficult cases are the ones in which patients, in the absence of treatment, will face the threat of serious harm, and in which one has only very few or no indication of their attitudes towards psychiatric coercive treatment. Here it appears, as introduced above, probably to be correct to reverse the burden of proof. This means to assume in the face of missing indications of rejection that people as a rule would, if necessary against their actual will, agree to the treatment of severe psychiatric illnesses, if the treatment lives up to medical standards, is promising and is not disproportionally onerous.

An exemplary case would be a young patient with a high suspicion of acute psychosis who refuses any treatment. Suppose she feels controlled by aliens, has within just a few months given up her university studies and abandoned her social contacts, and only comes out of her parent's attic apartment in order to eat alone in front of the refrigerator. She confronts hostilely all pressure from her parents to see a doctor

---

[13] This is accommodated in Germany, Switzerland, and the USA (but not in all European countries; see Lauter 1996) by clearly distinguishing between legal capacity and (the less demanding concept of) capacity to consent to treatment.

[14] Yet it is generally admitted that there is a need for further psychopathological research in this area (see also Helmchen and Lauter 1995).

and is apparently without any awareness of her illness. According to accepted psychiatric law, i.e., its established interpretation, this behavior does not constitute a "clear and present danger" to the patient herself (interpreted mostly narrowly as suicidal tendency) or to third parties, and therefore would not lawfully be sufficient grounds for compulsive hospitalization for the purpose of psychiatric diagnosis and subsequent treatment. If it is however actually the case that for successful treatment of acute psychosis a quick onset of treatment is important, this would from an *ethical* perspective likely be a case of admissible, even required, paternalism – and probably also sufficient reason to systematically rethink the reasonableness of the existing law and practice.

A considerably different and typical case would be that of an older patient with chronic and therapeutically no longer well-amendable psychosis, who has lived his entire life as a partial beggar, and only occasionally allows himself to be taken up and admitted to a clinic in order to be pampered. After a few days he wants to "head out" and refuses any psycho-pharmacological treatment. Here hypothetical consent to appropriate coercive treatment appears doubtful.[15]

There remains much to be done in order to develop a plausible, coherent and, for medical practice, a helpful ethical structure of justification for paternalistic psychiatric treatment. It would be worthwhile to illuminate the argumentation sketched above in more detail and to test, refine and specify it in light of paradigmatic illustrations. Even in the best-case scenario no algorithm could emerge here that would replace the power of moral judgment in psychiatric teams. But this power of judgment, which is indispensible for making the right decisions in individual cases, could be trained so that transparent argumentation can be achieved.[16]

References

Beauchamp, T.L. 1991. Competence. In *Competency. A study of informal competency determinations in primary care*, ed. M.A.G. Cutter and E.E. Shelp, 49–77. Dordrecht/Boston/London: Kluwer.

Beauchamp, T.L. 1995. Paternalism. In *The encyclopedia of bioethics*, 2nd ed, ed. W.T. Reich, 1914–1920. New York: Macmillan.

Beauchamp, T.L. 2009. The concept of paternalism in biomedical ethics. *Jahrbuch für Wissenschaft und Ethik* 14: 77–92.

Beauchamp, T.L., and J.F. Childress. 2009. *Principles of biomedical ethics*, 6th ed. New York/Oxford: Oxford University Press.

Brock, D.W. 1988. Paternalism and autonomy. *Ethics* 98: 550–565.

Brudney, D. 2009. Choosing for another: Beyond autonomy and best interests. *Hastings Center Report* 39(2): 31–37.

---

[15] Müller (2005) contains evidence that at least in singular cases legal guardians and courts end up with highly problematic decisions.

[16] This paper was originally published in Jahrbuch für Wissenschaft und Ethik 14, 2009, 107–127. Translated from German by Andrew Fassett.

Buchanan, A.E., and D.W. Brock. 1989. *Deciding for others*. Cambridge/New York: Cambridge University Press.

Charland, L.C. 1998. Appreciation and emotion: Theoretical reflections on the Mac-Arthur treatment competence study. *Kennedy Institute of Ethics Journal* 4: 359–376.

Charland, L.C. 2001. Mental competence and value: The problem of normativity in the assessment of decision-making capacity. *Psychiatry, Psychology and Law* 8(2): 135–145.

Childress, J.F. 1982. *Who should decide? Paternalism in health care*. Oxford/New York: Oxford University Press.

Davis, J.K. 2002. The concept of precedent autonomy. *Bioethics* 16(2): 114–133.

Degrazia, D. 1994. Autonomous action and autonomy-subverting psychiatric conditions. *The Journal of Medicine and Philosophy* 19(3): 279–297.

Demarco, J.P. 2002. Competence and paternalism. *Bioethics* 16(3): 231–245.

Drane, J.F. 1985. The many faces of competency. *The Hastings Center Report* 15(2): 17–21.

Dressing, H. 2004. Compulsory admission and compulsory treatment in psychiatry. In *Philosophy and psychiatry*, ed. T. Schramme and J. Thome, 351–356. Berlin/New York: de Gruyter.

Dworkin, G. 1970. Acting freely. *Noûs* 4: 367–383.

Faden, R.R., and T.L. Beauchamp. 1986. *A history and theory of informed consent*. New York: Oxford University Press.

Feinberg, J. 1971. Legal paternalism. *Canadian Journal of Philosophy* 1: 105–124.

Feinberg, J. 1986. *The moral limits of the criminal law*, Harm to self, vol. 3. Oxford: Oxford University Press.

Frankfurt, H.G. 1971. Freedom of the will and the concept of a person. *The Journal of Philosophy* 68: 5–20.

Garren, D.J. 2006. Paternalism, part I. *Philosophical Books* 47(4): 334–341.

Grisso, T., and P.S. Appelbaum. 1998. *Assessing competence to consent to treatment*. Oxford/New York: Oxford University Press.

Helmchen, H. 1996. Common European standards and differences, problems and recommendations. In *Informed consent in psychiatry. European perspectives of ethics, law and clinical practice*, ed. H.G. Koch, S. Reiter-Theil, and H. Helmchen, 381–411. Baden-Baden: Nomos.

Helmchen, H., and H. Lauter. 1995. *Dürfen Ärzte mit Demenzkranken forschen? Analyse des Problemfeldes. Forschungsbedarf und Einwilligungsproblematik*. Stuttgart/New York: Thieme.

Hodson, J.D. 1977. The principle of paternalism. *American Philosophical Quarterly* 14(1): 61–69.

Kallert, T.W., M. Glöckner, G. Onchev, J. Raboch, A. Karastergiou, Z. Solomon, L. Magliano, A. Dembinskas, A. Kiejna, P. Nawka, F. Torres-González, S. Priebe, and L. Kjellin. 2005. The EUNOMIA project on coercion in psychiatry: Study design and preliminary data. *World Psychiatry* 4(3): 168–172.

Koch, H.-G., S. Reiter-Theil, and H. Helmchen (eds.). 1996. *Informed consent in psychiatry. European perspectives of ethics, law and clinical practice*. Baden-Baden: Nomos.

Lauter, H. 1996. Assessment and evaluation of competence to consent. In *Informed consent in psychiatry. European perspectives of ethics, law and clinical practice*, ed. H.G. Koch, S. Reiter-Theil, and H. Helmchen, 307–321. Baden-Baden: Nomos.

Müller, P. 2005. Zwangseinweisungen nehmen zu. *Deutsches Ärzteblatt* 101(42): 2262.

Murphy, J.G. 1974. Incompetence and paternalism. *Archiv für Rechts- und Sozialphilosophie LX* 4: 465–486.

Petermann, F.T. 2008. *Urteilsfähigkeit*. Zürich/St. Gallen: Dike.

Quante, M. 2002. *Personales Leben und menschlicher Tod*. Frankfurt a.M: Suhrkamp.

Schöne-Seifert, B. 2007. *Grundlagen der Medizinethik*. Stuttgart: Kröner.

Scoccia, D. 1990. Paternalism and respect for autonomy. *Ethics* 100(2): 318–334.

Scoccia, D. 2008. In defense of hard paternalism. *Law and Philosophy* 27: 351–381.

Spellecy, R. 2003. Reviving Ulysses contracts. *Kennedy Institute of Ethics Journal* 13(4): 373–392.

Tannsjö, T. 1999. *Coercive care. The ethics of choice in health and medicine*. New York: Routledge.

Vandeveer, D. 1986. *Paternalistic intervention: The moral bounds on benevolence*. Princeton: Princeton University Press.

Wear, S. 1993. *Informed consent. Patient autonomy and physician beneficence within clinical medicine*. Dordrecht: Kluwer.

Wettstein, R.M. 1995. Competence. In *Encyclopedia of bioethics*, vol. 1, ed. W.T. Reich, 445–451. New York: Macmillan.

Wilks, I. 1997. The debate over risk-related standards of competence. *Bioethics* 11(5): 413–426.

Wolf, J.C. 2000. Paternalismus und andere ethische Konflikte im Alltag der Amtsvormunde und Amtsvormundinnen. *Zeitschrift für Vormundschaftswesen* 55: 1–15.

Chapter 11
Informed Consent, the Placebo Effect and Psychodynamic Psychotherapy

C.R. Blease

11.1 Introduction

Psychobabble is not mere babble (David Jopling 2008: 157)

The study of paternalism arguably becomes most penetrating when the discussion moves to authentic case studies. In the medical setting, paternalism was – until recently – the *sine qua non* of professional excellence. Physicians were guided by the principle of therapeutic privilege: a physician's knowledge and training, it was gauged, trumped the right to patient choice. Nowadays in (Western countries) the medical profession eschews these ethical norms: patient autonomy and choice are now (in codified form, at least) principles to which physicians must legally adhere.

This paper examines the issue of paternalism and informed consent through the lens of psychotherapy: in particular the varieties of psychotherapy that go under the collective label 'psychodynamic psychotherapy' (such as Freudian approaches). I consider the extensive (and long-standing) evidence for the charge that psychodynamic psychotherapy does not work as a result of its theoretical underpinnings: rather, it is only effective because it elicits the 'placebo effect'. The term 'placebo effect' has a long history of ill-definition (and I will hold back from addressing this until later). The fundamental assumption in the medical ethics literature is that the use of the placebo effect infringes on patient autonomy since it necessitates deception by clinicians. The ensuing debate on the ethics of placebos has pivoted on whether it is ever justifiable to deceive patients about a treatment in order that patients might thereby therapeutically benefit. Given that psychodynamic psychotherapy is frequently recommended by physicians for many conditions (including childhood abuse, trauma, depression, and anxiety), I argue that physicians and

C.R. Blease (✉)
School of Philosophy, University College Dublin, Dublin, Ireland
e-mail: charlotte.blease@ucd.ie

© Springer International Publishing Switzerland 2015
T. Schramme (ed.), *New Perspectives on Paternalism and Health Care*, Library of Ethics and Applied Philosophy 35, DOI 10.1007/978-3-319-17960-5_11

psychotherapists are not currently providing adequate disclosure about how this form of therapy[1] works (namely, that it may work by triggering the placebo effect).

The paper begins with some background discussion on informed consent in clinical practice (Sect. 11.2.1), including a survey of 'standard' (and, as I later argue, empirically impoverished) ethical perspectives on the ethics of placebo use (Sect. 11.2.2). Next, I turn to psychodynamic psychotherapy: I describe the received view that psychodynamic psychotherapy is effective because it affords patients therapeutic, truthful insights into aspects of their psychological history (Sect. 11.3.1). I briefly explain the serious (and well known) scientific objections to the received view (Sect. 11.3.2), before considering the charge that psychotherapy only works (to the extent that it does) because it is placebogenic (Sect. 11.3.3). In the next section of the paper, I pause to observe how the term 'placebo' refers to a range of triggers in the healthcare encounter, and offer a definition of the term 'placebo effect' (Sect. 11.3.4). Finally, in Sect. 11.4, I diagnose the current failings among health professionals, with respect to adequately informing patients about how psychodynamic psychotherapy is thought to work. I also analyse Jopling's solution for 'open placebos' disclosure in psychotherapy and contend that his formulation depends on too narrow an understanding of the placebo effect, and what triggers it. I conclude that providing adequate disclosure for psychotherapy leads to some counterintuitive but perhaps unavoidable consequences.

11.2 Informed Consent

11.2.1 Background

Dworkin formulates paternalism as follows (2014): "X *acts paternalistically towards* Y *by doing* (*omitting*) Z:

1. Z (or its omission) interferes with the liberty or autonomy of Y.
2. X does so without the consent of Y.
3. X does so just because Z will improve the welfare of Y (where this includes preventing his welfare from diminishing), or in some way promote the interests, values, or good of Y."

Physicians are duty-bound to respect patient autonomy: medical ethics guidelines oblige physicians to tell the truth and to ensure that informed consent is obtained before undertaking any medical intervention. The code of ethics of the American Medical Association (AMA) states that, "withholding medical information from patients without their knowledge or consent is ethically unacceptable" (2006a: Opinion 8.082); and "The physician has an ethical obligation to help the patient make choices from among the therapeutic alternatives consistent with good medical

---

[1] I variously use the term 'psychotherapy' in this paper to refer to 'psychodynamic psychotherapy'.

practice" (2006b: Opinion 8.08). The UK's General Medical Council (GMC) asserts that physicians must obtain "consent or other valid authority" (2010: paragraph 36); physicians must "discuss with patients what their diagnosis, prognosis, treatment, and care involve", and "share with patients the information they want or need in order to make decisions" (2008). The American Psychological Association now obligates therapists to obtain consent for psychotherapy (1992); and the American Psychiatric Association also urges that "[a] psychiatrist shall not withhold information that the patient needs or reasonably could use to make informed treatment decisions" (1998: 24).

But what does 'informed consent' mean in clinical practice? It cannot entail providing exhaustive amounts of information on different medical options. Beauchamp and Childress understand 'informed consent' to comprise the following components: "(1) competence; (2) disclosure; (3) understanding; (4) voluntariness; and (5) consent" (2009: 120). On this analysis, patients must have the mental capacity to understand the information disclosed to them and to make a decision about their treatment options; there should be no coercion involved if the patient has received adequate information and understood her choices. In regard to disclosure, Beauchamp and Childress contend that information relevant to decisions includes, "those facts or descriptions that patients or subjects usually consider material in deciding whether to refuse or consent to the proposed intervention" and "information that the profession considers to be material" (2009: 121). These stipulations are sufficiently 'gappy' to provide some problems for physicians; however, we might summarise them as the need to provide patients with relevant information regarding current knowledge about success rates of interventions, side-effects, other benefits and risks. Physicians should also be prepared to provide information (on occasion) about how interventions are thought to work. How this information is to be determined also poses problems for physicians. If we can simplify by talking of heuristics the approach that might best forge a 'patient-centred' approach is one that combines the "reasonable person standard" and the "subjective standard" in information disclosure (ibid, pp. 122–124). The "subjective standard" involves physicians tailoring the information needs to each patient as best as possible, according to their belief set, their prior medical history, anxieties about a procedure, and so on. There is also a need to disclose a background benchmark of relevant information and this might be determined by estimating the kinds of information that a "reasonable person" would require in treatment decisions. Whilst there are certainly outstanding problems with conceiving an idealised rational patient for our purposes we can minimally defend the notion that, for many treatments, there will be prominent and perhaps significant facts about which patients ought to be informed.

This minimal assumption leads to the next consideration: the issue of patient understanding. Physicians are confronted by patients who vary in their aptitude, ability, and their prior beliefs – all of which can influence or impede the processing of information. In addition, a patient's condition may directly or indirectly impair his or her ability to grasp information. One problem that deserves special emphasis (for later discussion) involves what Beauchamp and Childress call "the problem of nonacceptance and false belief" (2009: 130). They argue that "A single false belief can invalidate a patient's or subject's consent, even when there has been a suitable disclosure and comprehension"; and assert that, "If ignorance prevents an informed choice, it may be permissible or possibly even obligatory to promote autonomy by attempting to impose unwelcome information" (2009: 130–1). Later in the paper, I argue that this is a significant consideration in psychotherapy: patients are inadequately informed about how psychotherapy works.

11.2.2 Autonomy and the Placebo Effect

The use of placebos in clinical practice has been understood to present a particular problem for informed consent: it has been argued that placebos necessitate intentional deception on the part of the physician. At the outset it should be emphasised that how we understand the term 'placebo' has pivotal bearing on the ethical consequences of placebo use [in Sect. 11.3.3 I will reflect on recent research on placebos including the possibility of 'open placebos']. For now, we can note that the overwhelming conception of placebos in the medical community (including among medical ethicisits) is that placebos are sham treatments that only work because of some kind of deception: for example, a physician might prescribe a medication that is known to have no pharmacological effect on a particular ailment but (it is assumed) if the patient *believes* that the medication may be palliative there is likely to be some symptomatic relief as a result (Raz and Guindi 2008). This definition appears to underpin professional medical ethics codes. For example, the AMA declares that, "the use of a placebo without the patient's knowledge may undermine trust" and "compromise the patient-physician relationship" (2006: Opinion 8.083). This reveals the common, "*a priori* empirical assumption"[2] that the AMA understands placebos as necessarily invoking deception if they are to elicit therapeutic effect. The AMA guidelines continue: "A placebo may still be effective if the patient knows it will be used but cannot identify it and does not know the precise timing of its use. The physician need neither identify the placebo nor seek specific consent before its administration" (2006: Opinion 8.083). This prescriptive claim appears to display some conceptual (and, more to the point, empirical) confusion. On a charitable reading there appears to be a consistent underlying commitment to the view that placebos necessitate deception; yet, the guidelines also seem to suggest that the only ethical way of harnessing the placebo effect (and avoiding the charge of non-disclosure) is by demanding that physicians reveal, in a sort of semi-covert manner, that placebos are being deployed. This stipulation appears to draw on the idea of authorized concealment (something that has been advocated in nocebo use (Colloca and Miller 2011)). The thinking behind this perspective is that the person is not deceived about placebos being use but merely about the particular tokens or timing of their usage. As we will see later this is a view that is peculiarly philosophical and impractical – it ignores the possibility of placebos which cause side effects, for

---

[2] This *mot juste* is owed to David A. Jopling (2008).

example. Indeed, it might be contended then that AMA clause calls to mind the proverbial case of having one's cake and eating it. On ethical grounds we might query whether the patient is able to make an autonomous choice about treatment if: (i) the timing the treatment is out of kilter with consent; and (ii) the placebo in question is hidden. In any case, we can avoid messy oversights by confronting a neglected issue: it is an empirical issue whether giving patients a placebo and disclosing this information to them undermines the placebo's therapeutic effects.

Ethical debate over the justification of the use of placebos in clinical practice has (overwhelmingly) turned on the conceptual assumption that placebos involve either partial disclosure or outright deception. On a deontological approach to this understanding the use of placebos constitutes an infringement of patient autonomy: regardless of therapeutic gain for the patient, the physician must always disclose to the patient information regarding treatment intervention (Brody 1980; Kleinman, Brown and Librach 1994). For example, taking the case of 'sugar pills' prescribed for pain relief it is argued that a reasonable person would desire to know this fact: namely, that the pills do not work because of any pharmacological properties *per se*. The argument continues that failure to inform the patient (to intentionally withhold this information) invokes flagrant deception and an infringement on patients' autonomous treatment choices.[3]

The debate from a utilitarian perspective is less clear-cut. On the one hand, it has been argued that the consequences of not informing patients about placebos will lead to a harmful "domino effect" on patient trust in the medical profession and that this will negatively outweigh any immediate therapeutic effects to the patient as a result of non-disclosure (Bok 1974; Beauchamp and Childress 2009: 124; Kanaan 2009; Schwab 2009). In this vein, Bok argues that, "to permit a widespread practice of deception…is to set the stage for abuses and growing mistrust" (1974: 23): patients may delay or avoid seeking orthodox medical treatment, or come to understand some medical interventions to be "inert" and therefore a "sham".

A different utilitarian defence is the view that, on balance, the deceptive use of placebos can be justified in certain circumstances (Rawlinson 1985; Lichtenberg et al. 2004; Foddy 2009). Foddy, for example, claims that disclosed placebos would diminish their effectiveness by lowering the expectations of patients. Adopting this line of reasoning, Rawlinson identifies the following conditions for placebo use (Rawlinson 1985: 415):

1. Placebos are only employed for the patient's benefit and not for some expedient reason, on the part of health professionals;
2. Placebos can only be used when there is weighty evidence that they are necessary;

---

[3] Note that most patients assume that painkillers work due to their specific chemical properties – so in this case, the doctor does not have an obligation to disclose anything *further* about how those chemicals work – rather, this information in itself is adequate because a reasonable patient already has adequate knowledge. Where the presumption of adequate information comes undone is if most reasonable patients have the wrong assumptions about how a treatment works and this false information has the potential to infringe on the patient's treatment choice (see Blease 2014).

3. The physician can make the case for the necessity of the deception to the satis-
faction of a reasonable observer;
4. The physician determines that the long-term dependence on the placebo effect
would not conceal the disorder to the patient;
5. The physician takes into consideration the values of the patient and whether
deception would undermine the patient's future relationship with the physician.

It appears that the dispute over the utilitarian justification for "deceptive placebo
use" pivots on differing commonsensical estimations of the impact of deception on
patient trust, in the long-term. This is something that can only be settled by empirical
evidence; moreover, the suggestion that a physician is in a position to judge (from
the armchair, as it were): (i) the reasonableness of placebo use for an "ordinary
bystander"; and (ii) individual patient's views on medical deception (Rawlinson's
clauses 3 and 4), shows an overestimation of our folk psychological capabilities.
Underlying these views is a naive psychological view of cognition: we can object
that it is surely an empirical matter whether informing patients explicitly about
placebos does, in fact, have a causal bearing on distrust in the medical profession in
the long-term. These are facts that can neither be discerned by the "reasonable"
person in the street; nor by the "reasonable" doctor in the surgery. It is a matter for
scientific psychological study to reveal if patient distrust is affected by either the
deceptive or the non-deceptive use of "placebos" (see Kaptchuk et al. 2010; Kelley
et al. 2012).

Other notable attempts to reconcile the problem of informed consent and placebo
use include O'Neill's argument that deceptive placebo use does not infringe on
patient autonomy (1984). O'Neill contends,

> In human contexts, whether medical or political, the most that we can ask for is consent to
> the more fundamental proposed policies, practices and actions... Respect for autonomy
> requires that consent be possible to fundamental aspects of actions and proposals, but
> allows that consent to trivial and ancillary aspects of action and proposals may be absent or
> impossible. (1984: 176)

O'Neill claims that autonomy needs to be reconceived: it cannot be defined as
the exhaustive opportunity of decision-making in any given domain; O'Neill
describes this as "idealistically autonomous". She argues that autonomous decision
making is only relevant when it comes to "fundamental" choices. And for O'Neill
disclosure of placebos by physicians does not constitute a fundamental aspect of
medical intervention – therefore deceptive placebo use does not jeopardise patient
autonomy.

O'Neill's argument moves too fast. First, contrary to O'Neill's speculation,
placebos may (in fact) be perceived by patients to be a fundamental aspect of their
treatment: we need empirical evidence whether patients *perceive* placebo use to be
a significant treatment in itself rather than some "trivial" therapeutic supplement.
Second, O'Neill's account (like so many others) depends on un-argued assumptions
about the nature of "placebos": it relies on a conceptually and empirically impoverished view of "placebos" (as something like *sugar-pills* that *necessitate deception*
in order to elicit therapeutic effect) but as I argue in Sect. 11.3.3, placebos refer to

more than just "sugar pills" and placebo treatments can have significant (and serious) side-effects (Blease 2013a, b); I also examine the evidence for 'open placebos' and the claim that deception may not be necessary in eliciting placebo effects.

First we need to understand the problem of informed consent in psychotherapy. In the next section, I describe the standard explanation of how psychodynamic psychotherapy works – what I dub "the received view" (Sect. 11.3.1); before summarizing the criticisms of this view (Sect. 11.3.2), and evaluating the hypothesis that psychotherapy is only effective because it *is* a placebo (Sect. 11.3.3) (this will necessitate a much more detailed account of what we mean by the term 'placebo').

11.3 Psychodynamic Psychotherapy

11.3.1 The Received View

In the late 1990s Shapiro and Shapiro estimated that there were around 500 formalized versions of psychotherapy currently used in clinical settings (Shapiro and Shapiro 1997); today the current estimate may be closer to 700.[4] These can be further classified into some major sub-groups which chiefly include psychodynamic, cognitive-behavioural, and person-centred versions of psychotherapy. The chief differences are as follows: psychodynamic psychotherapy is characterised by the goal of uncovering why the patient is feeling and behaving as she does; in cognitive behavioural therapy ('CBT') the objective is simply to change behavioural patterns by reflecting on undesirable behaviours, emotions and thought-patterns through a process of re-training; and in person-centred therapies ('PCT') the principal therapeutic goal of the therapist is to establish a non-judgmental 'therapeutic alliance', or empathetic relationship with the patient. Whilst the aim of psychodynamic psychotherapy is to analyse the patient's current problems and psychological history, CBT is only concerned with solving the patient's present-day problems, and person-centred therapies aim to create an environment that facilitates the patient's own reflection and problem-solving. Finally, psychodynamic therapy usually takes much longer than CBT and PCT (usually not less than 6 months of weekly hour-long sessions), compared to 3–4 months in the case of CBT, and variably but usually shorter time frames than PCT.

In this paper I will be only be concerned with psychodynamic psychotherapy for two main reasons. First, the theoretical principles of psychodynamic psychotherapy have been challenged even more extensively in the scientific and philosophical literature than CBT or PCT. This makes it a particularly salient form of psychotherapy to investigate from an ethical standpoint: if its basic theoretical principles are highly questionable, we need to consider the range of ethical problems pertaining to its use as a treatment. Second, psychodynamic psychotherapy involves the

---

[4] Estimate owed to Bruce Wampold (in conversation).

commitment of more time on the part of the patient and therefore a greater financial obligation by the patient or healthcare authority: for personal investment in cost and time it is a form of therapy that deserves serious ethical analysis.

The list of therapies that come under the label 'psychodynamic' [which I will hereafter refer to as 'psychotherapy' for brevity] include versions of therapy derived from the theorists Freud, Jung, Adler and Klein. These therapies are unified in their claim that patients undergo a process of exploration that leads to the uncovering of (*bona fide*) psychological insights about themselves. According to the received view, this process involves interpretation and self-reflection on the part of the patient as she is guided through what is often termed an "excavation" of her emotions, behaviour, and thoughts: the process is considered to be excavational because only through interpretive analysis can the patient discover hidden insights about her troubled psychology. This process of self-exploration, however, is dependent on the particular theoretical framework of each version of psychotherapy: that is to say, *each* particular theory of psychotherapy posits very *different* unconscious psychological processes that will are revealed during the therapy. So, depending on the therapist's particular theoretical purview, resistances, repressed memories, dreams, unconscious drives, displacement activities, repressed denials, neuroses, or inferiority complexes, may be revealed to the patient during guided dialogue with the therapist. Jopling carefully formulates the received view as comprising two epistemic features: first, the therapeutic exploration is "authentic and truth-tracking": the putative 'insights' are not considered to be mere fictions or artefacts of therapy; second, it is the process of putative self-discovery that produces therapeutic benefit to the patient (2008: 71ff).[5]

11.3.2   Criticisms of the Received View

The most serious objections to the received view come from cognitive and social psychology. The first is the charge that the ontologies and processes that comprise the theories of psychotherapy have simply not been assimilated, or vindicated by scientific psychology: for example, references to "oral", "anal" or "phallic" stages of development (as Freud's theory of analysis claims) have not found any analogues in scientific theories of infant development. The list could be developed to include Jung's notion of a "collective unconscious", Alder's "inferiority complexes" and so on; these terms receive no theoretical preservation within prevailing scientific psychological theories. This means that psychotherapists are not referring to entities or processes that are psychological "real" in their dialogue with patients.

---

[5] Interestingly there is some evidence that Freud equivocated about whether analysis was curative. "I often console myself with the idea, that even though we achieve so little therapeutically, at least we understand why more cannot be achieved. In this sense our therapy seems to me to be the only rational one", Gerhard Fichtner, ed, *Sigmund Freud/Ludwig Binswanger: Briefwechsel*, 1908–1938, cited in E. Shorter (1997: 152)

The second objection relates to how the patient purportedly obtains information about such entities and processes: this is the criticism from introspection. From the perspective of current science, the methods of psychotherapy are based on false claims about the epistemic access afforded by introspection (cf. Wilson 2002; Kurzban 2010). Psychotherapy depends on the assertion that patients can track and interpret the reasons and causes for their inner-thoughts and feelings – that we can have privileged access to the mechanisms that give rise to psychological states. But as experimental work in social and cognitive psychology shows, "People tell more than they can know" (Nisbett and Wilson 1977): we have no direct access through conscious reflection (or through the kind of conversational exchange that occurs in therapy) to the cognitive mechanisms that give rise to our psychological states. Evolutionary psychology sheds light on why this is the case and it is worth pausing to consider the significance of this field of research (which is still underappreciated by humanities and social science scholars) (Barkow 2006). The mind-brain, just like the rest of our physiology, has been shaped by natural selection – it has been adapted to solve a vast range of recurrent problems in our ancestral environment (ranging from navigation, finding mates, foraging, negotiation, detecting free-riders, and so on). Cognitive and evolutionary psychology understands the mind-brain to be an information-processor: it detects information in the environment and processes it in a way that elicits behavioural responses. Since natural selection works as a biological filter on genetic variation, it is a directionless, satisficing process which (as has often been stated) is only concerned with the four fs: feeding, fleeing, fighting and fucking. Thus, from evolutionary perspective there is no adaptive, functional reason for us to be privy to the non-conscious cognitive mechanisms underlying this information processing: in short, as Wilson contends, "The modern view of the adaptive unconscious is that a lot of the interesting stuff about the human mind – judgements, feelings, motives – occur outside of awareness for reasons of efficiency" (Wilson 2002: 8). The epistemic claim that we can access causal mechanisms giving rise to thoughts, feelings and behaviour via introspection or psychoanalytic excavation is deeply flawed: these mechanisms can only be revealed by scientific psychology and not first-person analysis.

There are other important consequences of these evolutionary considerations for psychotherapy and mental health. Natural selection is not a goal-directed, 'truth-tropic' process: therefore, it may be advantageous (because conducive to survival and reproductive ability) to select false beliefs and false belief-forming strategies, over true beliefs and truth-tropic cognitive processes (Churchland 1987; Barkow 1989; Kurzban 2010). In fact, research from social psychology shows that marginally over-estimating one's abilities, popularity, level of attractiveness, and even the future (the so-called 'Pollyanna principle') is linked to mental well-being (Taylor and Brown 1988; Wilson 2002; Kurzban 2010; Trivers 2011; Blease 2011). Indeed, individuals who are suffering from mild-depression, furthermore, have more realistic evaluations of themselves (Alloy and Abramson 1979; Kapci and Cramer 1988; Blease 2012b, 2015a).

11.3.3 The 'Psychotherapy as Placebo' Hypothesis

At the outset we can note that the evidence shows that psychotherapy works very
well for many patients who suffer from depression, anxiety, or trauma (Shapiro and
Shapiro 1997: 102). However, studies also show that the version of psychotherapy
(including CBT) appears to be irrelevant to outcome – there is no significant difference in effectiveness between varieties of psychotherapy (Sloan et al. 1975; Luborksy
et al. 1975; Sloan and Staples 1984; Wampold and Imel 2015). If we assume: (i) that
psychotherapy can be instrumental in treating patients; and (ii) defer to the scientific
majority that psychotherapy is explanatorily bankrupt; we need to enquire: What are
the components in common to different psychotherapies that appear to lead to beneficial effects? In response to this question, Frank and Frank have forwarded the "common factors" hypothesis. This is the view that it is the common features shared by all
versions of psychotherapy that are causally relevant in treatment (1991). One important shared feature of psychotherapy, it has been proposed, is that that all versions of
psychotherapy endow the patient with a narrative framework that provides a rationale
or sense of coherence with regard to his or her feelings, thoughts and problems
(Frank and Frank 1991; Jopling 2008). It may be that story-making affords patients
a means of organising and explaining problems in a way that produces some therapeutic benefits; studies of victims of trauma show that individuals who manage to
forge some sort of explanatory understanding about why the traumatic event happened and who believe that the experience has enhanced their lives as a result ("it
happened for a reason"), tend to recover best (Janoff-Bulman 1992; Pennebaker
1997). So, psychotherapists and their patients may forge "explanatory fictions"
(Jopling 2008) that may carry some beneficial import. We might therefore speculate
that in order for the patient to derive benefit from the interpretation of her life that is
being forged, she should find the version of psychotherapy to be plausible.

Other factors common to different versions of psychotherapy include the caring
context, the status of the psychotherapist (for example, authoritativeness), the social
prestige associated with a form of therapy, the psychotherapist's qualities (including
confidence, empathy, ability to listen), the psychotherapist's optimism, the patient's
expectations about the treatment, being given a diagnostic label for one's problems, and
being given a set of rituals or techniques to practice or work on between therapy sessions (Parloff 1986; Frank and Frank 1991; Kaptchuk 2002; Jopling 2008; Wampold
and Imel 2015). All of these factors have variously been grouped together under the
label 'placebos'. But how should we understand the terms 'placebo' and 'placebo
effect'? In the next section I will answer by elaborating on recent scientific findings that
pull away from conventional (yet established scientific) wisdom on the subject.

11.3.4 What Science Tells Us About Placebos

Focusing on scientific findings is especially important given the widespread
misconceptions about placebos including the claims that: placebos are "inert" and have
no "real" physiological or psychological impact on the patient's symptoms; that

placebos only have "non-specific" effects; that placebos involve only subjective or transitory relief from symptoms; and, (as noted) that if placebos are to produce any effect this must involve deception (see Raz and Guindi 2008).

If we are to embark on an ethical analysis on the use of placebos we need to have a clear definition of what we mean by 'placebo' and 'placebo effect'. At present the term 'placebo' is employed by empirical researchers in a variety of ways; it might appear that the term placebo is a placeholder for a plethora of very different therapeutic interventions that trigger 'the placebo effect'. Consider the following: today, empirical researchers contend that the placebo effect constitutes a signifi-cantly beneficial effect for specific disorders (including angina, asthma, anxiety, depression, pain, Parkinson's disease, irritable bowel syndrome). It is also claimed that placebo effects are highly specific effects, and inter-subjectively measurable including via blood sugar, cholesterol, and cortisol levels, and blood pressure (see Jopling, forthcoming). But there is still ambiguity in what is being referred to as a placebo effect and care is still required in deployment of these terms. To illustrate: 'placebos' (for example dummy pills) may not trigger 'placebo effects' but only provide a 'placebo response' – that is to say, a response to what has been dubbed 'a placebo'. Yet, specifying *'the* placebo effect' therefore necessitates the harder, scientific task of delineating the specific mechanistic pathways that induce particu-lar therapeutic effect(s) in patients.[6] How do we decide which therapeutic benefits arise from the placebo effect, as opposed to other unknown therapeutic aspects of an intervention? How do we decide what is a *bona fide* placebo from that which is not? In the long-term, these questions require greater focus and attention among placebo researchers. For now, we can note that there may be more than one mechanism of action for what researchers currently dub *'the* placebo effect'. And for the purposes of this paper, I will avoid the thorny, theoretical-cum-empirical issues about how best to define 'placebo effect'. I only note that this is (of necessity) an important work in progress at the philosophical-theoretical end of empirical research. In this paper I will tender a pragmatic (but undoubtedly short-lived) working definition of 'placebo effect' to encompass "positive care effects" where these include *beneficial effects to the patient which* are *incidental to the principle mechanism of action of the target biomedical or bio-psycho-social treatment* (Blease 2012a). Placebos, on this definition, are reliable triggers for such beneficial effects. It should also be noted that this definition of 'placebo effect' provides a 'moving classification': therefore, it is likely that some of the processes dubbed 'placebo effect' under this definition may later be expunged and re-defined as other therapeutic phenomena.

Examples will help to illuminate this definition. Medication for pain-relief has specific pharmacological properties which target pain-receptors in the nervous system: this is the principal mechanism of action for painkillers such as paracetamol.

---

[6] Explanations for the placebo effect include the claim that placebos are "meaning" responses since the responses vary according to different cultures (Moerman 2002); the trigger for conditioned responses with medical phenomena (e.g. pills); psychological "expectancy responses" (Kirsch et al. 2004; Benedetti 2005); and the claim that the context of care and communication style of healers can be placebogenic (Di Blasi et al. 2001; Kaptchuk 2002; Blease 2012a).

But pain relief can be augmented by the following factors: the pills are red in colour; the patient is told to take the pill four times per day (compared to twice daily, where the dosage is equivalent); the pills have a brand name; the pills are pricier than other varieties (Huskisson 1974; Branthwaite and Cooper 1981). Other factors can also induce or augment these analgesic effects including the mode of administration of the medication (for example, injections for pain relief are more effective than pills), and telling a patient that he is receiving medication compared to giving pain-relief medication surreptitiously (De Craen et al. 2000). These factors are incidental to the principal mechanism of action: the analgesic properties of the drug paracetamol.

A growing body of research also shows that individuals suffering from depression may be responsive to placebo effects as defined; indeed – and contrary to O'Neill's assumption that placebos are trivial aspects of care (1984) – it has been claimed that antidepressants, and even electroconvulsive therapy may wholly depend for their effectiveness on the placebo effect (Kirsch 2009; Blease 2013b). It has also been hypothesized that the numerous common side effects of antidepressant medication (e.g. dry mouth, drowsiness, low sex drive) make it a particularly potent placebo: individuals expect that they are receiving 'strong' medication, which somehow triggers palliative expectancy effects (Kirsch 2009); in the case of electroconvulsive therapy it has been hypothesised that the "theatre" of the intervention, the attention given to the patient, the side effects (including headaches and memory loss), the patient's belief in the effectiveness of the treatment (Blease 2013a, b) may trigger beneficial effects.

In addition, new research also appears to challenge the assumption that placebos necessitate deception: recent 'open placebo' research purports to show that disclosing to patients that they are receiving a placebo does not diminish the placebo effect (Park and Covi 1965; Sandler et al. 2008, 2010; Kaptchuk et al. 2010; Kelley et al. 2012). These studies used placebo 'sugar pills' and, (for example, in the Kapthcuk study, 2010) patients were informed that they were being given "placebo pills made of an inert substance, like sugar pills, that have been shown in clinical studies to produce significant improvement in IBS symptoms through mind-body self-healing processes." The conclusions of these researchers is that there is some (albeit limited) vindication for the compatibility of open disclosure and placebo use in clinical practice.

What does any of this mean for psychotherapy? According to my proposed definition of placebos – 'therapeutic effects which are incidental to the principle mechanism of action of the target treatment' – the following components of psychotherapy may be placebogenic: patient expectations, the cultural prestige of the therapy, the therapist's empathy, the lowering of inhibition, social contact with someone considered to be authoritative and trustworthy, the ritual of healing, the healing environment and its trappings, and even the expense of the treatment; we can add to this the construction of an explanatory narrative for the patient's problems.

Testing forms of psychotherapy for their effectiveness against a placebo intervention presents methodological problems. In order to test any treatment the control must satisfy the following conditions:

1. The placebo control contains all the relevant non-characteristic features of the test treatment t, to the same degree that they are present in the experimental treatment process;
2. The placebo control has no additional relevant features over and above the non-characteristic features of the experimental treatment. (Howick 2011: 82)

No study in psychotherapy has successfully fulfilled these conditions: in order to conduct comparative research of psychotherapy with a placebo, one would need to formulate a version of sham psychotherapy replete with psychotherapists who "believed in it" (since double-blinding is also a standard requirement in placebo studies) (Shapiro and Shapiro 1997: 108; Kirsch 2005).[7] One way to circumvent the problem of constructing a sham psychotherapy is to draw on comparison studies of different versions of psychotherapy whereupon the epistemological claims about the curative component of each theory cancel each other out; and, as noted, the evidence so far shows that no one form of psychotherapy is superior to another. Luborksy et al. use the words of the Dodo in Alice in Wonderland to sum up these findings, "everybody has won and all must have prizes" (common factors hypothesis in psychotherapy has since been dubbed the 'Dodo Bird Conjecture') (1975; see also: Rosenthal and Frank 1956; Smith et al. 1980; Sloane et al. 1975; Wampold and Imel 2015).[8] Some studies declare that the "therapeutic alliance" or "collaborative bond" between patients is the strongest predictive measure of success in psychotherapy (Brown 2013). These studies assess therapeutic alliance via measurements of patients' and therapists' contributions to dialogue in therapy; how freely patients feel able to talk; and patients' efforts to carry out tasks. However, this measure does not tease apart what causes this therapeutic alliance: it may be that the narrative aspect, for example, brings about this bond, or if the patient considers the therapist to be particularly authoritative or prestigious, that this fosters the alliance. In short, while these studies vindicate the 'common factors' hypothesis – that it is the *shared components* of all version of psychotherapy that are therapeutic – the term "therapeutic alliance" is still sufficiently vague as to underdetermine which of the placebo features (if any) is most significant.

---

[7] Arguably the closest that any study has come to providing a suitable sham comparison was Strupp and Hadley's study (1979) using a control group of empathetic college professors (with no training in any form of psychotherapy): the study found that there was no difference in patient improvement between "sham psychotherapy" and psychotherapy.

[8] Some critics of the explanation that psychotherapy works as a placebo argue that different versions of psychotherapy propose that different so-called placebogenic components are also necessary for successful recovery (Parloff 1986; Kirsch 2005): for example, that creating a secure environment, and the patient's conviction that the form of psychotherapy works, are important factors in successful treatment. However, it might be countered that the central therapeutic claim of different versions of psychodynamic psychotherapy is that the patient is afforded insights into her life: for this reason, the purported "insight-tracking" of psychotherapy (ideally) needs to be tested.

11.4 Psychotherapy and Disclosure: Current Failings

Given the claim that that psychotherapy involves fictional story-construction – what Wilson describes as "literary criticism in which we are the text to be understood" (2002: 163) – we need to consider the circumstances and preconditions in which psychotherapy might be ethically employed. This is no trivial matter. If psychotherapy is as effective as regularly talking through one's problems with a trusted friend, for example, then the mainstream provision of psychotherapy in health services needs to be assessed. On the other hand, it may be that psychotherapy also has negative psychological side-effects (including the forging of false memories).[9]

At this juncture it is also important to note that the hypothesis that psychotherapy works as a placebo stands in opposition to O'Neill's assertion that placebos are "ancillary or trivial" aspects of medical care (1984): the placebo explanation of psychotherapy renders the placebo effect as *the fundamental engine* of treatment. It is also important to reiterate that it is not yet known which placebogenic features of psychotherapy are most significant (and perhaps, they are all additively important). This means that the question of patient autonomy and adequate disclosure with regard to psychotherapy cannot easily be sidestepped. In order to tackle the question of informed consent, and to render the problem more manageable, I assume: (a) psychotherapy *only* works by harnessing placebo effects; and (b) physicians and psychotherapists are currently failing to disclose this information to their patients. With this in mind, we can proceed to evaluate the current failure to disclose relevant information. There are likely to be different reasons and motivations for why physicians and psychotherapists fail to provide adequate treatment disclosure in the case of psychotherapy – each reason warrants separate ethical evaluation.

The first consideration is that it is likely that the majority of physicians and psychotherapists are uninformed of the (growing) literature on psychotherapy and placebos (Raz and Guindi 2008). It might be argued that by ignoring this important empirical literature, health professionals are failing in their duty to keeping themselves up-to-date about research developments in the field of psychoanalysis. This is a significant failing in itself if we expect health professionals to keep medically informed about what is still a frequently used, medically 'orthodox' line of treatment for many mental health conditions, especially those believed to be rooted in childhood trauma (including depression, post-traumatic stress disorder, eating disorders, anxiety disorders). Grünbaum argues, most psychoanalysts consider what they do to have a valid scientific basis (1984). Physicians (certainly in the UK and USA) still recommend psychotherapy to patients but it is likely that they are as ill-informed as psychotherapists with regard to explanations for its effectiveness. Given research into psychotherapy and placebos has been ongoing for some 50 years, the blame for these failings rests both at an institutional and a health agency

---

[9] Whilst the problem of the psychological side-effects of psychotherapy is certainly relevant to the ethical use of this treatment, it is an issue that takes us too far from the concerns of this paper (the issue of placebos and informed consent) (see Jopling 2008)

level. Medical schools and psychotherapy training courses ought to inform students of this research; moreover, the current lack of regulation (and proliferation) of psychotherapists in the UK and USA provides further reason for urgent reform in the professionalization of psychotherapy – as Shapiro and Shapiro conclude, "With more than 250 different types of psychotherapy and hundreds of DSM... diagnoses, it may be necessary to establish a new governmental agency, a 'Food, Drug, and Psychotherapy Administration" (1997: 121).

It may even be the case that some physicians and psychotherapists are aware of the psychotherapy-as-placebo explanation for its effectiveness (or at least regard the received view with some cynicism) but singularly fail to provide adequate disclosure to patients or fail to ensure that patients understand this information because they do not consider this to be an important feature of their professional role. This is an issue that may be more prevalent in some healthcare systems than others (and it is also a problem that may be difficult to gauge); it is therefore essential that health professionals perceive the importance of respecting patient autonomy and ensuring that patients have adequate knowledge about how psychotherapy works. Indeed, with regard to the obligation to inform patients about psychotherapy, it has been argued that "practitioners still retain considerable latitude in defining what constitutes informed consent" (Beahrs and Gutheil 2001: 5). Furthermore, some physicians or psychotherapists may be fully cognisant of the importance of informed consent but fail to ensure adequate disclosure wholly out of expediency: this may be a more serious moral failing since apparently these professionals are fully aware that this decision violates patient autonomy (Blease 2014).

If physicians and psychotherapists are aware of the explanation that psychotherapy works as a placebo, should this information be disclosed to patients? And *what* should be disclosed? Perhaps it is possible to disclose to patients that psychotherapy works as a 'placebo'. This is a move that has been pioneered by David Jopling (2008). Following the recent evidence for the possibility of successful open placebos (Kaptchuk et al. 2010) Jopling urges that psychotherapy can be ethically employed if patients are adequately informed about how it works; he advises that patients could be informed that therapy involves the creation of "therapeutic fictions". In light of this, Jopling proposes that psychotherapists invoke the following stipulation:

> [This treatment] involves working with psychodynamic explanations, interpretations, and insights concerning your psychology, history, behaviours, feelings and personality that are not literally true, but more like explanatory fictions. It involves making no claims to the psychological and historical truth when exploring your problems and your past. When we work with these interpretations, we are working with the psychological equivalent of a sugar pill. They are not however fanciful, arbitrary or silly; but nor can we say that they are true... (2008: 263).

Jopling's proposal has much in common with O'Neill's purview with respect to informed consent and the placebo effect. On O'Neill's narrow (and problematic) definition of placebos we never need to disclose placebos because they are peripheral to the main therapeutic method of treatment. Similarly, Jopling advocates the disclosure of what he considers to be the fundamental beneficial component of psychotherapy: the therapeutic narrative component (which he also considers to be a placebo and which is in line with my broader definition of placebos). But whether the narrative feature of psychotherapy is its main therapeutic engine has not yet been determined. It will be recalled that it has been hypothesized that the placebogenic factors in psychotherapy include – not just the construction of therapeutic fictions – but the social status of psychotherapy, the prestige of the therapist, the 'ritual' aspects of therapy, the socio-emotional communication style of the therapist, the expense of therapy, the patient's expectations about the effectiveness of the therapy, and the commensurability of the patient and the therapist's beliefs about the version of psychotherapy. As noted, research into these component factors is ongoing (Wampold and Imel 2015). If we wish to calibrate our disclosure according to accurate information (and I think that is a given) then Jopling's proposal for informed consent falls short of adequate disclosure. Therefore, even if we were to concede with O'Neill that only the main method of treatment needs to be disclosed we have not thereby circumvented the problem *when the main course of treatment is placebogenic*.

In the case of psychotherapy, the question then becomes: What should we disclose to patients? Should we not inform the patient about all of the placebogenic features of psychotherapy? Perhaps Jopling's open placebo statement should also include the following: "In addition to the construction of therapeutic fictions, evidence shows that if I speak to you in a positive, empathetic and encouraging tone of voice, if you have a high opinion of me as a health professional, and if I charge you a reassuringly expensive hourly rate, this will lead to therapeutic mind-body effects. Do you consent to these aspects of care?" (Blease 2012a). This may seem somewhat counterintuitive but is that a good enough reason *not* to disclose this information? One rejoinder is to argue that some of these aspects of care are expected – for example, patients would not embark on psychotherapy if they did not believe it would be effective, and it is only common sense to expect a therapist to adopt a particular demeanour, and for the patient to have a high opinion of the therapist. But we might respond: Do patients routinely expect these features of care to be *the* engine of therapy *per se* (Blease 2015b)? We might argue that patients ought to be informed that if they do not have confidence in the therapy, or the therapist, that it is likely that the therapy will be less effective. In the same way, shouldn't patients have the right to know that their therapy will be more successful if the therapist consistently adopts a particular communication style, or if they don't feel that they have attained a 'therapeutic bond' with the therapist? Shouldn't patients be informed of the putative therapeutic consequences of the pricing scale for their hourly psychotherapy sessions? What would it mean, for example, to inform patients that the rate per hour is expensively pitched but that this is wholly (or partly?) for therapeutic reasons? As Beauchamp and Childress' stipulate, "a single false belief can invalidate a patient's consent" (2009: 130): patients have a right to know how therapy works if they are to make informed choices.

11.5   Conclusion

Ethical discourse on the question of informed consent and questions of patient autonomy need to pay much closer attention to ongoing empirical research. In the case of the placebo effect, when discussion is divorced from current scientific input the discussion floats free of applicable insight. Jopling's 'open placebo' proposal for psychotherapy provides a significant first step in the right direction. However, more detailed attention needs to be given to understanding how psychotherapy works – including whether psychotherapy *just is* a placebo. It could be that patient understanding of placebos diminishes its therapeutic returns, and we do not yet know whether disclosure in itself threatens (or enhances) patient trust in health professionals. When we have (even preliminary) evidence-based answers to these questions, the debate over the usage of placebos may only then be open to utilitarian challenges. In the meantime, psychotherapy, like any other treatment intervention, should be subject to adequate disclosure – even if this discomfits practitioners of the long tradition of psychodynamic 'talking cures'.

Acknowledgements   I would like to thank Thomas Schramme and the delegates at the Spring School on Paternalism, University of Hamburg for very helpful suggestions on a draft of this paper in March, 2012. I would also like to thank Jerome Barkow and Tom Walker for comments on a previous draft of this manuscript.

References

Alloy, L.B., and L.Y. Abramson. 1979. Judgment of contingency in depressed and nondepressed students: sadder but wiser? *Journal of Experimental Psychology: General* 108: 441e85.

American Medical Association. 2006a. Opinion 8.082: withholding information from Patients. http://www.ama-assn.org/ama/pub/physician-resources/medicalethics/code-medical-ethics/opinion8082.shtml. Accessed 25 Jul 2010.

American Medical Association. 2006b. Opinion 8.08—Informed Consent http://www.ama-assn.org/ama/pub/physician-resources/medical-ethics/code-medical-ethics/opinion808.page. Accessed 6 Jul 2012.

American Psychiatric Association. 1998. *The principles of medical ethics especially applicable to psychiatry.* Washington, DC: American Psychiatric Association.

Barkow, J. 1989. *Darwin, sex, and status: Biological approaches to mind and culture.* Toronto: Toronto University Press.

Barkow, J. 2006. Sometimes the bus does wait. In *Missing the revolution: Darwinism for social scientists,* ed. J. Barkow. Oxford: Oxford University Press.

Beahrs, J., and T. Gutheil. 2001. Informed consent in psychotherapy. *American Journal of Psychiatry* 158(1): 4–10.

Beauchamp, T., and J. Childress. 2009. *Principles of biomedical ethics.* Oxford: Oxford University Press.

Blease, C. 2011. Deception as treatment: The case of depression. *Journal of Medical Ethics* 37(1): 13–16.

Blease, C. 2012a. The principle of parity: The placebo and physician communication. *Journal of Medical Ethics* 38(4): 199–203.

Blease, C. 2012b. Mental health illiteracy? The unnaturalness of perceiving depression as a disorder. *Review of General Psychology* 16(1): 59–69.

Blease, C. 2013a. Electroconvulsive therapy: The importance of informed consent and placebo literacy. *Journal of Medical Ethics* 39(3): 175–176.

Blease, C. 2013b. Electroconvulsive therapy, the placebo effect and informed consent. *Journal of Medical Ethics* 39(3): 166–170.

Blease, C. 2014. The duty to be well-informed. *Journal of Medical Ethics* 40(4): 225–229.

Blease, C. 2015a. Too few 'likes', too many 'friends'? What evolutionary psychology tells us about Facebook depression. *Review of General Psychology* 19(1): 1–13.

Blease, C. 2015b. Talking more about talking cures: Cognitive behavioural therapy and informed consent. *Journal of Medical Ethics* (forthcoming)

Bok, S. 1974. The ethics of giving placebos. *Scientific American* 231: 17–23.

Branthwaite, A., and P. Cooper. 1981. Analgesic effects of branding in treatment of headaches. *British Medical Journal* 282: 1576–1578.

Brody, H. 1980. *Placebos and the philosophy of medicine: Clinical, conceptual and ethical issues.* Chicago: University of Chicago Press.

Brown, W. 2013. *The placebo effect in clinical practice.* Oxford: Oxford University Press.

Churchland, P.S. 1987. Replies to comments. *Inquiry* 29: 241–272.

Colloca, L., and F. Miller. 2011. The nocebo effect and its relevance for clinical practice. *Psychosomatic Medicine* 73(7): 598–603.

De Craen, A.J., J.G. Tijssen, and J. de Gans. 2000. Placebo effect and the acute treatment of migraine: Subcutaneous placebos are better than oral placebos. *Journal of Neurology* 247: 183–188.

Di Blasi, Z., E. Harkness, E. Ernst, A. Georgiou, and J. Kleijnen. 2001. Influence of context effects on health outcomes: A systematic review. *Lancet* 357(9258): 757–762.

Dworkin, G. 2014. stanford.library.usyd.edu.au, ed. E. N. Zalta.

Foddy, B. 2009. A duty to deceive: Placebos in clinical practice. *The American Journal of Bioethics* 9: 4e12.

Frank, J., and J. Frank. 1991. *Persuasion and healing.* Baltimore: John Hopkins University Press.

General Medical Council. 2008. Guidance on good practice. http://www.gmc-uk.org/guidance/ethical_guidance/consent_guidance_part1_principles.asp. Accessed 6 Jul 2012.

General Medical Council. 2010. Good medical practice, paragraph 36. http://www.gmcuk.org/guidance/good_medical_practice_relationships_with_patients_consent.asp. Accessed 25 Jul 2010.

Grünbaum, A. 1984. *The foundations of psychoanalysis: A philosophical critique.* Berkley: University of California Press.

Howick, J. 2011. *The philosophy of evidence-based medicine.* Oxford: Wiley-Blackwell.

Huskisson, E. 1974. Simple analgesics for arthritis. *British Medical Journal* 4(5938): 196–200.

Janoff-Bulman, R. 1992. *Shattered assumptions: Towards a new psychology of trauma.* New York: The Free Press.

Jopling, D. 2008. *Talking cures and placebo effects.* Oxford: Oxford University Press.

Kanaan, R. 2009. When doctors deceive. *The American Journal of Bioethics* 9: 29e30.

Kapci, E.G., and D. Cramer. 1988. The accuracy of dysphoric and nondepressed groups' predictions on life events. *Journal of Psychology: Interdisciplinary and Applied* 132: 659e70.

Kaptchuk, T. 2002. The placebo effect in alternative medicine: Can the performance of a healing ritual have clinical significance? *Annals of Internal Medicine* 136: 817–825.

Kaptchuk, T., E. Friedlander, J. Kelley, et al. 2010. Placebos without deception: A randomized controlled trial in irritable bowel syndrome. *PLoS One* 5(12): e15591.

Kelley, J., T. Kaptchuk, C. Cusin, S. Lipin, and M. Fava. 2012. Open-label placebo for major depressive disorder: A pilot randomized controlled trial. *Psychotherapy and Psychosomatics* 81(5). doi:10.1159/000337053

Kirsch, I. 2005. Placebo psychotherapy: Synonym or oxymoron? *Journal of Clinical Psychology* 61(7): 791–803.

Kirsch, I. 2009. *The emperor's new drugs: Exploding the antidepressant myth.* London: The Bodley Head.

Kirsch, I., S. Lynne, and S. Vigorito. 2004. The role of cognition in classical and operant conditioning. *Journal of Clinical Psychology* 60(4): 369–392.

Kleinman, I., P. Brown, and L. Librach. 1994. Placebo pain medication: Ethical and practical considerations. *Archives of Family Medicine* 3: 453–457.

Kurzban, D. 2010. *Why everyone else is a hypocrite*. Princeton: Princeton University Press.

Lichtenberg, P., U. Heresco-Levy, and U. Nitzen. 2004. The ethics of the placebo in clinical practice. *Journal of Medical Ethics* 30: 551e4.

Luborksy, L., B. Singer, and L. Luborksy. 1975. Comparative studies of psychotherapies: Is it true that "everybody has won and all must have prizes"? *Archives of General Psychiatry* 32: 995–1007.

Moerman, D.E. 2002. *Meaning, medicine, and the 'placebo effect'*. Cambridge: Cambridge University Press.

Nisbett, R., and T. Wilson. 1977. Telling more than we can know: Verbal reports on mental processes. *Psychological Review* 84: 231–259.

O'Neill, O. 1984. Paternalism and partial autonomy. *Journal of Medical Ethics* 10(4): 173–178.

Park, L.C., and L. Covi. 1965. Non-blind placebo trial. An exploration of neurotic patients' responses to placebo when its inert content is disclosed. *Archives of General Psychiatry* 12: 336–345.

Parloff, M. 1986. Placebo controls in psychotherapy research: A sine qua non or a placebo for research problems? *Journal of Consulting and Clinical Psychology* 54: 79–87.

Pennebaker, J. 1997. *Opening up: The healing power of expressing emotions*. New York: Guilford.

Rawlinson, M. 1985. Truth-telling and paternalism in the clinic: Philosophical reflections on the use of placebos in medical practice. In *Placebo: Theory, research and mechanisms*, ed. L. White, B. Tursky, and G. Schwartz. New York: Guildford Press.

Raz, A., and D. Guindi. 2008. Placebos and medical education. *McGill Journal of Medicine* 11(2): 223–226.

Rosenthal, D., and J. Frank. 1956. Psychotherapy and the placebo effect. *Psychological Bulletin* 55: 294.

Sandler, A., C. Glesne, and G. Geller. 2008. Children's and parent's perspectives on open-label use of placebos in the treatment of ADHD. *Child: Care, Health and Development* 34(1): 111–120.

Sandler, A., C. Corrine, and J. Bodish. 2010. Conditioned placebo dose reduction: A new treatment for attention-deficit hyperactivity disorder. *Journal of Developmental and Behavioral Pediatrics* 31: 369–375.

Schwab, A. 2009. When subtle deception turns into an outright lie. *The American Journal of Bioethics* 9: 30e2.

Shapiro, A.K., and E. Shapiro. 1997. *The powerful placebo: From ancient priest to modern physician*. Baltimore: The Johns Hopkins University Press.

Shorter, E. 1997. *A history of psychiatry: From the era of the asylum to the age of Prozac*. New York: Wiley.

Sloane, R., F. Staples, A. Cristol, N. Yorkston, and K. Whipple. 1975. *Psychotherapy versus behaviour therapy*. Cambridge, MA: Harvard University Press.

Smith, M., G. Glass, and T. Miller. 1980. *The benefits of psychotherapy*. Baltimore: John Hopkins Press.

Taylor, S., and J. Brown. 1988. Illusion and well-being: A social psychological perspective on mental health. *Psychological Bulletin* 103: 192e210.

Trivers, R. 2011. *The folly of fools: The logic of deceit and self-deception in human life*. New York: Basic Books.

Wampold, B., and Imel, Z. 2015. The Great Psychotherapy Debate: The Evidence for What Makes Psychotherapy Work. London: Routledge.

Wilson, T. 2002. *Strangers to ourselves*. Cambridge, MA: Harvard University Press.

Chapter 12
Paternalism in Psychiatry: Anorexia Nervosa, Decision-Making Capacity, and Compulsory Treatment

André Martens

12.1 Introduction

Decision-making capacity or mental competence is one of the most intensively discussed concepts in contemporary bioethics and medical ethics.[1] In this paper I argue that anorexia nervosa, an eating disorder primarily afflicting adolescent girls and young women, seriously challenges what I label the traditional account of decision-making capacity. In light of these results, it may in addition be necessary to rethink a certain popular type of paternalistic argumentation that grounds the justification of compulsory treatment, for example of anorexic persons who refuse treatment, on a lack of decision-making capacity.

In my conclusion I make the case for supplementing the list of abilities necessary for decision-making capacity with an explicitly evaluative-emotional though content-neutral element as suggested by several authors in recent years. Furthermore, I attempt to demonstrate that the justification of soft paternalism based on a lack of decision-making capacity is actually a more complicated task to do than often assumed. This is because, in my view, decision-making capacity should be regarded as a value-laden but, in itself, normatively impotent concept for justifying paternalism. Ascribing a lack of decision-making capacity (and therefore a lack of autonomy) is not already sufficient for the justification of compulsory treatment measures that override the treatment decisions of anorexic persons.

---

[1] There are deep terminological quarrels about how to distinguish 'capacity' from 'competence' (e.g. 'capacity' as a legal, 'competence' as a clinical term). An all-agreed definition seems out of sight. Although I principally acknowledge these concerns, for the sake of simplicity and following Charland (2011) I will use both terms interchangeably. Sometimes I use the term 'decisional capacity' instead of 'decision-making capacity' for stylistic reasons.

A. Martens (✉)
Department of Philosophy, University of Hamburg,
Von-Melle-Park 6, Hamburg 20146, Germany
e-mail: andre.martens@yahoo.de

© Springer International Publishing Switzerland 2015   183
T. Schramme (ed.), *New Perspectives on Paternalism and Health Care*, Library of Ethics and Applied Philosophy 35, DOI 10.1007/978-3-319-17960-5_12

This paper is structured as follows: The first section is dedicated to the traditional account of decision-making capacity. The second discusses an influential recent critique and possible supplements of the traditional account. In the third section, conceptual, psychological and normative issues surrounding decision-making capacity are brought together. Here it is that soft paternalism and its justification become the focus of attention.

12.2 Decision-Making Capacity

Contemporary medical ethics is dominated by a liberal stance. Respecting patients' autonomy can be seen as the central principle underlying much discussed concepts such as informed consent. Generally speaking, autonomous patients have the right to be self-determining in decisions about their treatment.[2] Not all patients, however, are autonomous at the time of important treatment decisions, as can be exemplified by comatose or brain-dead patients. But it is not necessary to come up with such extreme examples. At first appearance also some psychiatric patients seem to lack autonomy to a relevant degree.[3] Although some conditions for autonomy may be met by these persons, for instance being sufficiently informed about the consequences of treatment decisions and not being coerced by staff, family members and others, they nevertheless seem to lack decision-making capacity, which I regard in the following as a necessary but not sufficient condition for ascribing autonomy to a person. In my understanding, autonomy is a broader concept than decision-making capacity.

Traditionally, decision-making capacity is formulated in terms of certain abilities. Only if a person has all of those abilities can she be regarded as having decisional capacity. The majority of accounts of decision-making capacity accept the following set of four elements as necessary conditions:

1. **Understanding**: a patient has to be able to understand 'the factual information relevant to the decision she is being asked to make'.[4] Understanding includes the comprehension of the nature of one's mental disorder and of the treatment being recommended as well as its benefits and risks.[5] It is important to emphasize that only local understanding is needed, that is, the understanding of pieces of information related to and relevant for the focal treatment decision.

2. **Appreciation**: the second traditional element of decision-making capacity describes the ability to appreciate the consequences and significance of the focal

---

[2] Cf. Craigie (2009).

[3] This type of autonomy is often called personal autonomy. In the course of this paper, the relationship between autonomy as a normative term and autonomy as a set of abilities and preconditions for self-governance (a moral psychological conception that is the focal point of this section) will be further scrutinized.

[4] Culver and Gert (2004, 260).

[5] Grisso et al. (1995, 128).

treatment decision for one's own life. The patients must 'be able to apply the information abstractly understood to their own situation'.[6] Thomas Grisso, Paul S. Appelbaum and colleagues refer to appreciation as the 'patients' recognition that information given to them about their disorder and potential treatment is significant for and applicable to their own circumstances.'[7]

3. **Reasoning**: Some authors have argued that the former two elements need to be complemented by a rationality criterion which demands the patient to have the mental ability to engage in reasoning processes (such as weighing and comparing alternatives) and correct information manipulation.[8]

    I propose not to overstrain rationality constraints on decision-making capacity ascription, since overly demanding constraints will lead to counterintuitive consequences, namely, the need to regard many mentally healthy subjects in a variety of situations as mentally incompetent. This may be the case, for example, due to biases, superstition, time pressure, unconventional preferences and many more factors leading to putatively irrational decisions. There is a lot of literature on the topic of rationality constraints which I cannot discuss here in more detail.[9]

4. **Communication of Choice**: Finally, a person has to be able to express and communicate a choice in some way in order to be competent.[10]

Further basic abilities or mental functions presumably also important for decisional capacity, such as intentionality, long-term memory, the power of imagination and anticipation, the ability to execute mental time travel, belief ownership, introspection among others, cannot always be easily connected with the above categories. My proposal is to regard the four elements discussed above as the core of the traditional account, acknowledging that there are further abilities necessary for ascribing (personal) autonomy.

Noteworthy about the traditional account is its focus on cognitive or intellectual abilities as well as its fundamental neutrality regarding specific values. The term 'appreciation' already seems to have an evaluative dimension insofar as the 'application of relevant information to self' presupposes an axiology of the person doing so (in order to be able to appreciate the significance of a certain decision for oneself). Additionally, it is no news that reasoning and rationality are terms associated with certain types of norms (for example of coherence and consistency). Nevertheless, the first three elements of decision-making capacity sketched above are *basically* cognitive abilities concerned with information manipulation, its application and recognition. Furthermore, it is striking that the

---

[6] Appelbaum and Grisso (1995, 110).

[7] Grisso et al. (1995, 128).

[8] Charland (2011) gives an overview.

[9] See Bortolotti (2010) for an elaborated account of rationality constraints in the context of delusional disorders.

[10] Grisso et al. (1995, 129).

four elements described also exclude emotions in a strict sense from conceptions of decision-making capacity.[11]

For some time past there has been the call for supplementary evaluative or emotional elements in conceptions of decision-making capacity. In the next section I argue that recent research into anorexia nervosa has forcefully demonstrated the plausibility of this call.

12.3   In Pursuit of Thinness: The Challenge of Anorexia Nervosa

The traditional account of decision-making capacity sketched above enjoys great popularity in psychiatry nowadays which is mirrored in several attempts to make it productive for psychiatric practice. The *MacArthur Competence Assessment Tool for Treatment* (MacCAT-T) developed by Thomas Grisso and Paul S. Appelbaum counts as 'the most fully developed standardized method of assessing competence'[12] in a psychiatric setting, as it were the 'gold standard'[13] of decision-making capacity assessment, and is explicitly oriented by the traditional account.[14]

Jacinta Tan and her colleagues have recently challenged the traditional account of decision-making capacity by questioning its completeness and adequacy, grounding their reservations in empirical quantitative and qualitative research into anorexia nervosa.[15] They report that anorexics at different stages of their illness achieve good results in the MacCAT-T test, which, as previously mentioned, is usually regarded as implementing the traditional criteria for decisional capacity.[16] At the same time, anorexics often arrive at prima facie unreasonable and sometimes extremely harmful treatment decisions. They often refuse treatment even when they are severely ill. In practice it is usually a hard task for mental health professionals to obtain consent

---

[11] In my reading, Appelbaum and Grisso introduce 'appreciation' as some kind of cognitive function that links (connects) the axiology of a person with a decision-making process, without having an evaluative or emotional aspect *in itself*. But also see the controversy between Charland (1998b) and Appelbaum (1998). I think, however, that a full account of decision-making capacity needs not to focus exclusively on the linkage between axiology and decision-making process but also on the characteristics of specific values involved in the focal decision making, namely their form. I spell out this thesis at the end of this section.

[12] Tan et al. (2003b, 698). See also Breden and Vollmann (2004) and Vollmann (2006) for another critical perspective on the MacCAT-T.

[13] Vollmann (2006, 289).

[14] Cf. Grisso et al. (1995), Appelbaum and Grisso (1995), Grisso and Appelbaum (1995, 1998).

[15] Tan et al. (2003a, b, c). See also Tan et al. (2006a, b, 2010) and Hope et al. (2011).

[16] Tan et al. (2003b). Tan et al. interviewed ten female patients aged 13–21 years who met the DSM-IV criteria for anorexia nervosa or atypical anorexia nervosa. The median body mass index (BMI) was 17.10 kg/m^2 (roughly corresponding with a girl standing 1.65 m high weighing 46.5 kg), with a range from life-endangering 12.57 to near-to-average 19.62 kg/m^2.

to treatment measures.[17] In some cases no consent is to be obtained at all and civil commitment is considered.

Tan and her colleagues state: 'In terms of intellectual measures such as understanding and reasoning, even severe anorexia nervosa patients may be judged to be competent to make treatment decisions'.[18] They maintain: 'The current legal criteria of capacity, applied by the MacCAT-T test, failed to capture difficulties that were relevant to competence to refuse treatment in anorexia nervosa.'[19] According to Tan and her colleagues, it follows that criteria not captured by the traditional account are also relevant to decision-making capacity in anorexia nervosa and should be further scrutinized. In the following I support this view and defend the thesis that the traditional account needs to be supplemented by an evaluative or emotional element, rightly understood. Prior to the further development of this thesis, Tan et al.'s research and some problems raised by it should be examined carefully.

Despite the maintenance of certain cognitive abilities even in severe anorexia nervosa, Tan and her colleagues show that anorexia nervosa may still have a radical impact on persons suffering from it, not least on thinking processes (such as difficulties with concentration), attitudes (for example, towards death and disability), values (e.g. the importance of friendship, family relationships, academic success and, most importantly, thinness), and identity (anorexia nervosa is often perceived as becoming a pivotal part of one's personality).[20]

Tan and her colleagues suggest that these results may be highly relevant to conceptions of decision-making capacity. This claim appears to be based on the intuition that although anorexics may be regarded as competent according to the traditional account, still something seems to be 'wrong' with at least some instances of anorexic decision making, such as treatment refusal in terminal anorexia nervosa. I suppose that first and foremost conceptual intuitions lead Tan and her colleagues to arrive at this conclusion, which I share.

To summarize, Tan and her colleagues observed that often anorexic treatment refusal cannot be fully explained by means of cognitive or intellectual deficits. Instead, in many cases certain predominant values seem to be responsible for treatment decisions. They introduce the notion of pathological values and state:

> One possible way in which we can respect values of individuals while still considering values affecting competence is to trace the origins of values to determine whether they arise from an individual or a disorder. If a value or value system can be clearly determined to arise from a mental disorder rather than the person, then this value cannot be seen to be authentic to the person himself or herself, and, if it affects treatment decision making, should be considered suspect in terms of compromising competence. This determination is not easy in many cases, but may be feasible in others.[21]

---

[17] Tan et al. (2003b).

[18] Tan et al. (2006a, 279). Cf. Tan et al. (2003b, 701).

[19] Tan et al. (2003b, 706).

[20] Tan et al. (2003b).

[21] Tan et al. (2006a, 278).

Their thesis is that what compromises treatment refusals expressed by anorexic persons is sometimes the decisions' rootedness in values shaped or caused by mental disorder. That is, values influence decision making, and the origin (source) of these values is crucial to the issue of whether to respect treatment decisions based on them. In the following I try to systematize this view and, in doing so, make sense of the notoriously problematic notion of pathological values.

The research of Tan and her colleagues seems to support what I call the 'inclusion thesis'.

Inclusion thesis: *Any full account of decision-making capacity must include at least one evaluative or emotional element.*[22]

In other words: values and emotions are highly relevant to conceptions of decision-making capacity. They play a bigger role than captured by the term 'appreciation', as will become apparent in a moment. To begin with, I introduce three arguments supporting the inclusion thesis. In my view they differ in terms of argumentative strength.

1. **The Empirical Argument**: Values and emotions, as a matter of fact, can and do influence decision making. Therefore it is advisable to consider values and emotions in conceptions of decisional capacity.
2. **The Semantic Argument**: To have the capacity to make a decision *means* that a certain outcome (one's decision) is brought about by a process making active use of and presupposing some axiological background (value system), which, in addition, is deeply interconnected with certain emotions and similar mental states. Therefore it is advisable to consider values and emotions in conceptions of decisional capacity.
3. **The Mismatch Argument**[23]: One's 'self' is, partly, constituted by values and related emotions. Sometimes a decision being made and expressed seems to be at odds with these self-constituting values: there is some kind of mismatch. This is for example because it is somehow grounded in mental disorder and may therefore be regarded as inauthentic to the self. The decision-making process had no access to one's own deeply entrenched value system, perhaps due to some general psychic dysfunction. Or it simply did not make proper use of it (performance error). If authenticity is relevant to decision-making capacity as some authors suggest, values and related emotions are indirectly relevant as well. We have to compare the values and emotions involved in an actual treatment decision with the authentic values and emotions of a person we would normally

---

[22] Strictly speaking, Tan et al. do not comment on the importance of including emotions and other related mental states in conceptions of decisional capacity in what I regard as their main publications. This idea is rather prominently defended by Louis Charland (1998a, b). In a reply to Charland and other commentators, however, they seem to consider this view (Tan et al. 2006b). My intention is not to interpret these authors but to defend a position which is loosely inspired by them and interconnects values and emotions.

[23] This argument can possibly be seen as one way among others of fleshing out the Semantic Argument.

expect to influence her decision, in order to decide whether or not she has local decisional capacity. If a decision lacks this connection, values might be expressed that are inauthentic to the person who utters it.

I use the term 'value' in a very broad sense (not restricted to moral values), including a wide range of evaluative judgments. Liking strawberry ice cream and giving paramount importance to thinness as in anorexia nervosa are both values in this sense. Some emotions and related mental states seem to be deeply connected with certain values (leaving aside the question whether they function as a source or an expression of them).[24] The anorexic pursuit of thinness, for example, is deeply interconnected with an intense fear of gaining weight or becoming fat, as captured by the DSM-IV definition.[25] I suggest looking at the relationship between values and emotions as a form of nested co-occurrence relationship. This is why I propose to supplement the traditional account not only with an evaluative element but also to include emotions.

Tan and colleagues refer to both the Empirical Argument and the Mismatch Argument.[26] It occurs to me that the first two arguments, the Empirical and the Semantic, are more powerful than the third one, as I will explain shortly. I think they already suffice to make the inclusion thesis plausible. Note that so far I have said nothing about the exact role values (and emotions) and their features play in an extended account of decision-making capacity. But I hope to have shown that they necessarily play *any* role.

Why are both the Empirical Argument and the Semantic Argument stronger than the Mismatch Argument? The traditional account of decision-making capacity presumes that certain intellectual abilities such as understanding and reasoning play an important role in decision-making processes, that is, their dysfunction will affect and influence decision making. This is why they are included in the traditional account. But also values (for example the paramount importance of being thin in anorexia nervosa) and emotions (for example the anorexic's fear of gaining weight), as Tan and others' research has shown, influence decision making. I regard this mainly as an empirical question. If so (as the Empirical Argument suggests), conceptions of decision-making capacity should include some evaluative or emotional element.

---

[24] I personally support cognitive emotion theories that acknowledge an evaluative component in the formation process of emotions. The most important point for my argumentation, however, is that I regard values and emotions as necessarily intertwined, no matter what kind of emotion theory one favors.

[25] The *Diagnostic and Statistical Manual of Mental Disorders* of the *American Psychiatric Association* (APA 1996, DSM-IV, 307.1) associates anorexia nervosa with an '[i]ntense fear of gaining weight or becoming fat, even though underweight.' Note that DSM-5 has been recently published including some changes in the diagnostic criteria of anorexia nervosa. For example, Criterion A no longer contains the term 'refusal' in the context of weight maintenance and Criterion D (amenorrhea) was deleted.

[26] See esp. Tan et al. (2003b, 2006a, 2010) and Hope et al. (2011).

As for the Semantic Argument, to assume that there is something 'wrong' with the anorexic's decision to refuse treatment in terminal anorexia nervosa presupposes some kind of axiology (value system). Decisions are always made on the basis of such axiological backgrounds that provide us with reasons for one or another decision and, in the form of related emotional states, with motivations to act and decide – all terms broadly construed. This is simply what 'making a decision' *means*. By definition there are no entirely accidental decisions. Decisions do not just happen. Even if notions such as value system and reasons may be rejected as crude, unscientific folk psychology, they are necessary to make sense of notions such as decision and choice. Again, if so, why not include an evaluative or emotional element in conceptions of decision-making capacity?

I am somewhat reluctant to draw upon the Mismatch Argument, although it may have some intuitive appeal. This is primarily because it operates with the concept of authenticity which is, to say the least, problematic.

Above I quoted Tan and others as follows:

> If a value or value system can be clearly determined to arise from a mental disorder rather than the person, then this value cannot be seen to be authentic to the person himself or herself.[27]

What does 'authenticity' mean here? According to Demian Whiting, an inauthentic value could be a value that is either 'not really possessed by the patient' or a value that is 'possessed by the patient but for which something other than the patient is responsible'.[28] He criticizes both interpretations and finally rejects an authenticity condition on decisional capacity ascription. His main argument against the second (in my view stronger) interpretation reads as follows:

> I am not sure of the extent to which *anyone* is responsible for the values he or she has (thus, for instance, can it really be sensibly held that valuing pleasure over pain is a matter of personal *choice*?). Consequently, if an authentic valuation is one for which the patient is responsible then I am unclear about the extent to which any of a person's values can be judged to be authentic.[29]

This thesis may be too strong, but it is certainly true that at most only a few values are the result of conscious deliberative activity for which we can be held responsible due to some form of self-control. Evaluative judgments – such as liking sunsets, being mad for strawberry ice cream, among others – do not belong to this class.[30] Having and acting upon them (e.g. making the decision to go to an ice cream parlor) does not compromise one's decision-making capacity. Responsibility for one's values, then, is not what matters in terms of authenticity. But what does matter? This question cannot easily be answered.

---

[27] Tan et al. (2006a, 278).

[28] Whiting (2010, 343).

[29] Whiting (2010, 343).

[30] This, however, does not mean that we cannot influence them. We might read a well-researched article asserting that strawberry ice cream usually contains cancer-causing ingredients and consequently stop liking it most (possibly we even express emotions such as disgust when seeing strawberry ice cream).

Furthermore, it is not easy to make sense of the expression 'something other than the patient is responsible for certain (pathological) values', at least if we want to avoid some suspect ontology presupposing psychopathology entities that are both in some sense external to a person and nevertheless affect a person's decisions internally.

If, on the other hand, authentic value simply means something like 'representing a stable disposition that is integral part of our identity and self-concept', many values held by persons with long-term anorexia nervosa probably have to be considered as authentic and do not differ in this respect from stable values held by healthy persons. Many anorexic values are indeed extremely stable and regarded by concerned persons as 'self'-corresponding (ego-syntonic). Theories operating with first-order and second-order desires or values do not seem to work well in anorexia nervosa, and phenomena such as ambivalence and ambiguity in some anorexic decision making as observed by Tan et al. complicate the issue even more. These are some reasons why I think that the first two arguments are stronger and more elegant than the Mismatch Argument, although the (philosophical) jury may be still out.

The idea behind the authenticity discourse is not only to highlight the importance of values and emotions for conceptions of decision-making capacity, but also to explain what is intuitively wrong with some instances of anorexic decision making. That is, to give reasons supporting our intuition that anorexia nervosa can compromise decisional capacity.[31] I think there is a stronger way of doing so beside accounts drawing upon the notions of authenticity and pathological values.

I propose the following fifth condition for decisional capacity:

5) Compulsive values: *Decision-making capacity requires the absence of decision-affecting compulsive values as well as related emotions.*

My suggestion is to look at the form (or nature) of values rather than at their origins in psychopathology. This also enables us to give a solid though still content-neutral account of decisional capacity.[32] Not only the formation process or source of values, but also their evolution over time and functional impact on other mental states is important,[33] because it indicates those properties or features of values that constitute their form. Values evolve in a specific manner in the individual and have certain effects that are also contingent upon what and how they are. Stability, rigidity, and volatility are examples of such properties related to values' evolution. We should also think of certain properties (such as compulsiveness) that determine a value's interaction with other mental states and cognitive mechanisms.

I do not claim that my five elements account is complete, nor do I claim that the origin of values in psychopathology (its causal history and conditions of formation) might not play any role in conceptions of decision-making capacity. Moreover,

---

[31] Authenticity theorists' fifth condition for decision-making capacity could then read as follows: 'Decision-making capacity requires the absence of pathological values, that is, values that are inauthentic to the person.'

[32] Sometimes this position is called 'proceduralism', as opposed to 'substantivism' which looks at the content of values. Since my focus is on the form of values and not on their formation process, I do not use the term 'proceduralism' in this section.

[33] Values in this sense can be both activating and inhibiting mental states.

presumably not all pathological values are compulsive values. My account cannot capture *all* cases of compromised decision-making capacity. That would be presumptuous. I think, however, that at least compulsive values compromise decision-making capacity.

Compulsiveness renders nonsensical the idea of decision making. Not merely the notorious extreme stability of compulsive values (their rigidity, fixity) is a problem, or the fact that they often go along with distorted cognitive mechanisms that are responsible for value conservation, or that they are not open to revision even in the light of massive and continual counter-evidence.[34] In my view they compromise decisional capacity first and foremost because of their *impact* and *effect* on deliberation and related processes necessary for decision making. This idea is closely related to the concept of voluntary decision-making.[35] If a certain value renders impossible a space of deliberation, weighing alternatives etc., in other words, if it renders impossible potential revision, overturns alternative decisions that are incompatible with it from the outset and if it always leads us to the same 'decision', then – for conceptual reasons – we have not made a genuine decision and are actually unable doing so. While values may be formed without us even noticing it, decision making requires some space of self-control, volition or will power, which I regard as constituents of voluntariness. Decision making is a balance act between drawing upon our existing axiology and the ability to deliberate. Decision-making *capacity* is the ability to master this balance act. Compulsive values such as the paramount importance of thinness in anorexia nervosa clearly compromise deliberation and therefore decision making in general. This is basically a conceptual argument. It may as well be applicable to other mental disorders.

In the first two sections of my paper I was mainly concerned with conceptions of decision-making capacity and mental abilities necessary for its ascription. Including a fifth condition for decisional capacity referring to the absence of compulsive values in decision-making processes may help to justify why some anorexic persons indeed lack decision-making capacity while acknowledging Tan et al.'s finding that decision-making capacity is not just about cognitive abilities, narrowly defined.

I now turn to the relationship between decision-making capacity and soft paternalism and discuss a certain type of normative argumentation in support of the latter, which I regard as highly flawed.

12.4   Soft Paternalism and Decision-Making Capacity

In this section I defend two related theses. First, the conception of decision-making capacity is value-laden, but it does not prescribe any particular values (proceduralism); nevertheless, in itself it is normatively impotent. Second, some anorexics and other persons suffering from mental disorders may lack decision-making capacity;

---

[34] These are static features of compulsive values.

[35] See Nelson et al. (2011) for an elaborate account of voluntariness that rejects authenticity as a necessary condition of voluntary action, focusing instead on intentionality and freedom from controlling influences.

but this fact alone does not justify soft paternalism. The justification of soft paternalism is actually a far more complicated task to do than simply demonstrating that a person is lacking decision-making capacity and is therefore incapacitated to make autonomous decisions.

In the following I criticize a certain type of argumentation in support of soft paternalism that is grounded on a lack of decision-making capacity and enjoys some popularity. Treatment refusal in anorexia nervosa will again be the prime example of my analysis. Before these normative issues can be addressed, the term 'soft paternalism' needs to be introduced.[36]

Paternalism, according to a very basic and certainly contestable formula, indicates any interference with the will expressions of a person P, but for the good (benefit, welfare) of person P.[37] A beneficent motive, in some sense, must underlie the paternalistic interference.

In 1972 Gerald Dworkin gave a by now classic definition of paternalism:

> By paternalism I shall understand roughly the interference with a person's liberty of action justified by reasons referring exclusively to the welfare, good, happiness, needs, interests or values of the person being coerced.[38]

Dworkin focuses on a person's liberty of action, whereas my rather 'internalistic' focus is on a person's will and its expressions. Furthermore, Dworkin's definition already highlights the normative dimension of the concept of paternalism. Reference to the welfare (good, happiness etc.) of a person can normatively outweigh these person's actual choices. I think, in order not to get lost in conceptual analysis, these two definitions are a sufficient starting point for the following argumentation.

In recent decades, many conceptual differentiations and clarifications were made. Gerald Dworkin and Joel Feinberg played a prominent role.[39] Very important for the issue of decision-making capacity and its role in the justification of compulsory treatment is the differentiation between hard and soft paternalism. Soft paternalism, modifying my above formula, refers to any interference with the *non-autonomous* will expressions of a person P, but for the good (benefit, welfare) of person P. The basic idea is this: If a person lacks autonomy, say due to severe anorexia nervosa or dementia, and her will expressions are overridden for instance by the medical staff in order to promote her good (benefit etc.), this will be a case of soft paternalism.[40] Compulsory treatment of the severely mentally ill can therefore

---

[36] See several other papers of this volume for more detailed analyses of the many different types of paternalism.

[37] 'Will expressions' is intended to be a general term embracing more specific terms such as 'decision' or 'choice'.

[38] Dworkin (1972, 65). Note that Dworkin later changed his wording and does not refer to interference with action anymore. See also his contribution in this volume.

[39] See e.g. Dworkin (2005) and Feinberg (1986).

[40] Imagine the case where a person suffering from severe dementia utters the wish to go sailing although an awful hundred-year storm is gathering. Keeping this person away from accomplishing her endeavor can be regarded as a case of soft paternalism because overriding her will is to the good of the person (if saving her life is part of this person's good).

be regarded as a case of soft paternalism. Crucial here is the ascription of a lack of autonomy, for instance based on the negative outcome of a decisional capacity assessment test.

Some authors downplay or underestimate the difficulty of justifying soft paternalism. Tom L. Beauchamp is a prominent example. In a recent paper he maintains:

> Everyone supports altruistic beneficence directed at confused cardiac patients, ignorant consumers, frightened clients, and young persons who know little about the dangers of alcohol, smoking, drugs, and motorcycles. No caring person would leave these individuals unprotected, and no reasonable philosopher would defend a normative thesis that permits such outcomes. The knotty problems about the justification of paternalism lie not in these behaviors. They lie exclusively in strong paternalism, which takes over and overrides autonomy.[41]

Some leading psychiatrists, for instance J. L. T. Birley, former president of the *Royal College of Psychiatrists*, go even further and seem to treat soft paternalism as some kind of 'natural' institutional reaction to intricate life situations of the individual. Birley states: 'Every citizen should have the right to be admitted against his or her will, to be treated without loss of dignity, in a first class psychiatric service.'[42]

Soft paternalism and its justification are, however, by no means a trivial matter. Admittedly, if soft paternalistic measures such as compulsory treatment could be justified, it would facilitate the daily routine of psychiatric staff. Soft paternalistic measures can save lives and prevent severe harm. I do not deny this. We should, though, distinguish soft paternalism's utility and – in many cases – very positive outcomes from the challenge and necessity of its normative justification. And, not least important, we should take the latter challenge very seriously. Soft paternalism is anything but self-evident and unproblematic.

In the following I focus on a popular justification strategy of soft paternalism that eventually demonstrates how difficult and non-trivial its justification is.

The argument reads as follows: Given a lack of autonomy of the paternalized person, say due to a lack of decision-making capacity, and given a beneficent motive of the paternalist, soft paternalism is justified simply because the 'healthy' paternalist's will expression (resp. treatment decision) *outweighs* the non-autonomous will expression of the paternalized person. That is, medical staff, family members and other surrogates need not to respect the paternalized person's treatment decisions in situations where a lack of autonomy can be ascribed.[43]

---

[41] Beauchamp (2009, 82–83).

[42] Birley (1991, 1).

[43] Some might even argue that under certain circumstances we are obliged to override and disrespect the paternalized person's will, for example because we have a duty to help that obliges us to actively take over responsibility for other persons. I acknowledge that this view is not identical with the one I discuss in the main text. It should, however, be noted that such a 'duty to help' would apply to decisions of both autonomous and non-autonomous persons that we deem unwise or harmful. If promoting the good of a person is our unrestricted duty, then we no longer need to distinguish soft from hard paternalism.

In the case of anorexia nervosa we can often replace 'will expressions' not only by 'decisions' but also by 'values'. Treatment decisions, for instance the refusal to participate in weight-gain programs, express genuine values such as giving paramount importance to being thin, as discussed in Sect. 12.3. Also will-overriding surrogate decisions (by medical staff or family members) express certain values such as giving paramount importance to being alive and having a healthy weight. In anorexia nervosa, and possibly also in other mental disorders, the conflict between the paternalized person and the paternalist is best understood as a value conflict.[44] The paternalist's argumentation is based on the assumption that a lack of autonomy (decision-making capacity) entails a normatively decisive imbalance between both values. Respecting the will of another person is only obligatory if the 'scales' are about the same level. The *fact* that some person lacks autonomy changes the whole situation.

In effect, the soft paternalist relies on some kind of *superiority claim*: healthy, beneficent will expressions (resp. values) are superior to will expressions (resp. values) of non-autonomous persons.[45]

Let us have a fresh look at what exactly happens here. I think this type of argumentation in support of soft paternalism consists of four steps:

1. Initially, the decision-making capacity to consent to or refuse treatment, say of an anorexic person, is assessed (based on some conception of decision-making capacity).
2. This assessment then leads to the ascription of a lack of decision-making capacity.
3. This lack of decision-making capacity justifies overriding the anorexic person's will expressions (values).
4. The paternalist finally overrides the anorexic patient's will expressions.

Let us further assume that the decision-making capacity assessment has been based on my five conditions account that includes some content-neutral evaluative-emotional element. This assessment, then, seems to be a purely empirical task, if we take the presupposed account for granted. Up to this point, there is nothing problematic about the argumentation. Some anorexics indeed lack decision-making capacity. In the first two sections I tried to show why in many cases of even severe anorexia nervosa the traditional account with its cognitive-intellectual focus will not be sufficient for reaching this conclusion. But an extended account, or so I argued, will most likely do.

---

[44] I support the view that anorexic persons may express admittedly compulsive values (and, if Tan et al. are right, pathological ones) that are nevertheless still *genuine* values (not mere 'utterances' that cannot be truly regarded as values). Recall also my remarks on Whiting (2010).

[45] This 'superiority claim' is the content of the paternalist's normative conclusion. Note that hardly any soft paternalist will ascribe global incompetence to a person suffering from a mental disorder. Each individual case needs to be looked at separately. Anorexics may lack decision-making capacity with regard to important treatment decisions without lacking it with regard to other decisions.

What is problematic, however, is the 'jump' to step 3. This becomes clearer if we enrich the above example and assume that an anorexic person lacks decision-making capacity due to some compulsive values. In this case, the assessment of the *form* of the patient's values is crucial, I argued. But how can we ever reach a normative conclusion – the superiority of the paternalist's own values over the paternalized person's values and being therefore justified to actively override her values – that is purely based on decision-making capacity assessment?

At least two problems are coming up at this point.

(a) Ascribing (a lack of) decision-making capacity to a person is usually regarded as an empirical task based on some conception of decision-making capacity. In this argumentation, however, it seems to change the moral status of the assessed person.

(b) Inferring from a lack of decision-making capacity that a person's values can legitimately be overridden, is equal to jumping from a form assessment to a normative content conclusion ('the paternalist's values are superior to the paternalized person's values').

The first problem (inferring normative conclusions from empirical assessments) is already tricky and raises important issues of the relationship between empirical aspects and normativity. It may be solvable. Philosophers at least from David Hume onwards have tried. I will not comment on this matter.

The second problem (inferring a content conclusion from a form assessment) is already serious enough, and in my opinion hints at a fallacy. It resembles some kind of qualitative jump requiring additional justification that is actually not given. How come that the form of a value changes the normative status of a person or, in other words, that it has an influence on normative argumentations related to the content of values?[46] Form is not content. Any possible relationship between these two aspects of values needs additional arguments that are, in my view, insufficiently addressed up to date, if at all soft paternalism is regarded as something non-trivial.[47] This is why I regard decision-making capacity *in itself* as a normatively impotent concept.

It is a dilemma: On the one hand, modern liberal medical and bioethics usually want to defend content-neutrality regarding specific values. If doing so, we must accept that a value such as 'I want to be thin, regardless of whether I die' is not in itself (in terms of content) problematic. Problematic is rather its form. But we cannot simply jump from form to content when justifying paternalism. So we are thrown back to content: Saving a person's life is superior to her value of giving paramount importance to being thin although risking her life. But if so, I do not see why

---

[46] A similar problem comes up if we support an approach that focuses on the pathological origin of certain values (see Sect. 12.3). In this case, the soft paternalist jumps from origin assessment to content conclusions.

[47] I do not assert that the addressed problem cannot be resolved. It may be possible to justify soft paternalism based on the argumentative strategy presented. But any argumentation facing this challenge would have to account for the jump from form to content. No account known to me does so.

decision-making capacity assessment should matter at all.[48] Soft paternalism is about justifying a certain value content or, in other words, a normative proposition while overriding another one (the value content of, for example, an anorexic person). Are decision-making capacity and its assessment simply superfluous for soft paternalistic normative argumentations? There are three possible ways of dealing with the problems raised:

1. We could accept that decision-making capacity is actually superfluous.
2. We could enrich our conception of decision-making capacity by including some normative and in terms of content non-neutral element. (Substantivism)[49]
3. We could search for arguments supporting the claim that the assessment of values' form[50] (as in anorexia nervosa) is normatively potent. In doing so, we defend a content-neutral conception of decision-making capacity. (Proceduralism)

I think the third way might be the right one, but it requires further examination. Decision-making capacity in anorexia nervosa may be normatively relevant; however, justifying soft paternalism in the way described is clearly fallacious.

12.5 Conclusion

In this paper I gave an outline of the so-called traditional account of decision-making capacity that focuses primarily on cognitive or intellectual abilities. I defended the claim for an explicitly emotional or evaluative element in conceptions of decision-making capacity and gave some arguments in support of it. In this context, I argued that compulsive values and related emotions may compromise decision-making capacity. My illustrating psychiatric example was anorexia nervosa, a serious eating disorder that may severely change not least the self-concept and values of a person suffering from it.

I defended the thesis that decision-making capacity is a value-laden concept although it is neutral in terms of value content (it does not prescribe any certain values). In my view, a lack of decision-making capacity does not in itself justify (soft) paternalism, because this would require a 'jump' from form assessment to a content conclusion. Soft paternalism therefore cannot simply be justified by demonstrating that a person is lacking decision-making capacity.

---

[48] One might argue that decision-making capacity assessment is needed in order to disregard the paternalized person's value so that finally there is only the paternalist's value left, which then is the only 'reasonable' value at hand. In this case, a superiority claim would be superfluous. Still, I do not think that this argumentation is valid because it – in itself – does not account for why 'unreasonable' (or the like) values need not to be regarded in paternalistic argumentations. It shifts the problem to notions such as (un)reasonable.

[49] See footnote 32.

[50] Or, alternatively, the assessment of values' origin.

It is worth expanding on the notion of 'compulsory values' in future articles. Another challenge is to find possible valid arguments that fill the gap between form assessment (as in decision-making capacity assessments) and content conclusions (as in paternalistic argumentations). In this article I only gave a sketch of a broader theory. Many details still need to get clarified.

Acknowledgements I thank audiences in Hamburg, Lübeck and Bochum for extremely helpful comments. Special thanks are due to Jochen Vollmann, Adrian Viens, Thomas Schramme, Iara Cury and Michael Dunn.

References

American Psychiatric Association. 1996. *Diagnostic and statistical manual of mental disorders (DSM-IV)*, 4th ed. Washington, DC: American Psychiatric Publications.

Appelbaum, Paul S. 1998. Ought we to require emotional capacity as part of decisional competence? *Kennedy Institute of Ethics Journal* 8(4): 377–387.

Appelbaum, Paul S., and Thomas Grisso. 1995. The MacArthur Treatment Competence Study. I. Mental illness and competence to consent to treatment. *Law and Human Behavior* 19: 105–126.

Beauchamp, Tom L. 2009. The concept of paternalism in biomedical ethics. *Jahrbuch für Wissenschaft und Ethik* 14: 77–92.

Birley, J.L.T. 1991. Psychiatrists and citizens. *British Journal of Psychiatry* 159: 1–6.

Bortolotti, Lisa. 2010. *Delusions and other irrational beliefs*. Oxford: Oxford University Press.

Breden, Torsten M., and Jochen Vollmann. 2004. The cognitive based approach of capacity assessment in psychiatry: A philosophical critique of the MacCAT-T. *Health Care Analysis* 12(4): 273–283.

Charland, Louis C. 1998a. Is Mr. Spock mentally competent: Competence to consent and emotion. *Philosophy, Psychiatry, & Psychology* 5(1): 67–95.

Charland, Louis C. 1998b. Appreciation and emotion: Theoretical reflections on the MacArthur Treatment Competence Study. *Kennedy Institute Journal of Ethics* 8(4): 359–377.

Charland, Louis C. 2011. Decision-making capacity. In *The Stanford encyclopedia of philosophy* (Summer 2011 Edition), ed. Edward N. Zalta. http://plato.stanford.edu/archives/sum2011/entries/decision-capacity/.

Craigie, Jillian. 2009. Competence, practical rationality and what a patient values. *Bioethics* 25(6): 326–333.

Culver, Charles M., and Bernard Gert. 2004. Competence. In *The philosophy of psychiatry: A companion*, ed. Jennifer Radden, 258–270. Oxford: Oxford University Press.

Dworkin, Gerald. 1972. Paternalism. *The Monist* 56(1): 64–84.

Dworkin, Gerald. 2005. Moral paternalism. *Law and Philosophy* 24(3): 305–319.

Feinberg, Joel. 1986. *Harm to self. The moral limits of the criminal law*. Oxford: Oxford University Press.

Grisso, Thomas, and Paul S. Appelbaum. 1995. The MacArthur Treatment Competence Study. III. Abilities of patients to consent to psychiatric and medical treatment. *Law and Human Behavior* 19: 149–174.

Grisso, Thomas, and Paul S. Appelbaum. 1998. *Assessing competence to consent to treatment. A guide for physicians and other health professionals*. Oxford: Oxford University Press.

Grisso, Thomas, Paul S. Appelbaum, E.P. Mulvey, and K. Fletcher. 1995. The MacArthur Treatment Competence Study. II. Measures and abilities related to competence to consent to treatment. *Law and Human Behavior* 19: 126–148.

Hope, Tony, Jacinta Tan, Anne Stewart, and Ray Fitzpatrick. 2011. Anorexia nervosa and the language of authenticity. *Hastings Center Report* 41(6): 19–29.

Nelson, Robert M., Tom Beauchamp, Victoria A. Miller, William Reynolds, Richard F. Ittenbach, and Mary Frances Luce. 2011. The concept of voluntary consent. *The American Journal of Bioethics* 11(8): 6–16.

Tan, Jacinta, Tony Hope, and Anne Stewart. 2003a. Anorexia nervosa and personal identity: The accounts of patients and their parents. *International Journal of Law and Psychiatry* 26(5): 533–548.

Tan, Jacinta, Tony Hope, and Anne Stewart. 2003b. Competence to refuse treatment in anorexia nervosa. *International Journal of Law and Psychiatry* 26(6): 697–707.

Tan, Jacinta, Tony Hope, Anne Stewart, and Ray Fitzpatrick. 2003c. Control and compulsory treatment in anorexia nervosa: The views of patients and parents. *International Journal of Law and Psychiatry* 26(6): 627–645.

Tan, Jacinta, Anne Stewart, Ray Fitzpatrick, and Tony Hope. 2006a. Competence to make treatment decisions in anorexia nervosa: Thinking processes and values. *Philosophy, Psychiatry, & Psychology* 13(4): 267–282.

Tan, Jacinta, Anne Stewart, Ray Fitzpatrick, and Tony Hope. 2006b. Studying penguins to understand birds. *Philosophy, Psychiatry, & Psychology* 13(4): 299–301.

Tan, Jacinta, Anne Stewart, and Tony Hope. 2010. Decision-making as a broader concept. *Philosophy, Psychiatry, & Psychology* 16(4): 345–349.

Vollmann, Jochen. 2006. 'But I don't feel it': Values and emotions in the assessment of competence in patients with anorexia nervosa. *Philosophy, Psychiatry, & Psychology* 13(4): 289–291.

Whiting, Demian. 2010. Continuing commentary: Does decision-making capacity require the absence of pathological values? *Philosophy, Psychiatry, & Psychology* 16(4): 341–344.

Part IV
Paternalism and Public Health

Chapter 13
Why It's Time to Stop Worrying About Paternalism in Health Policy

James Wilson

13.1 Introduction

It is a commonplace that governments can improve population health by changing the costs and benefits associated with individuals' choices—by measures such as fining drivers who do not wear seatbelts, or raising taxes on tobacco. Interventionist policies of this kind are often criticized as paternalistic, and the more autonomous the choices which are interfered with, the more problematic such policies are thought to be. Some recent influential positions such as Thaler and Sunstein (2008) aim to avoid these alleged ethical problems by deploying only 'nudges' in public health policy, where nudges alter the framing of situations, but not the substantive costs and benefits attached to choices—as when a school canteen chooses to place the fruit in a more salient position than the less healthy puddings.

This article provides a qualified defence of public health policies which override or interfere with autonomously held choices. My defence makes two main moves. First, I argue that it is a mistake to attempt to transplant claims about the wrongness of paternalism in the doctor–patient relationship into claims about the wrongness of paternalism in public health policy. Whether a policy should count as paternalistic depends both on the goals for which the policy is enacted, and on whether those affected by the policy consent to it. It will typically be the case that at most some (but not all) of the motivations and justifications for an interventionist policy are paternalistic. Only some (but not all) citizens will dissent from any given interventionist policy. So it is much more difficult to make sense of the claim that a given policy is paternalistic than is usually thought. In addition, two of the elements that

J. Wilson (✉)
Department of Philosophy, University College London, London, UK
e-mail: james.wilson@ucl.ac.uk

© Springer International Publishing Switzerland 2015
T. Schramme (ed.), *New Perspectives on Paternalism and Health Care*, Library
of Ethics and Applied Philosophy 35, DOI 10.1007/978-3-319-17960-5_13

make paternalism problematic at an individual level—interference with liberty and lack of individual consent—are endemic to public policy contexts in general and so cannot be used to support the claim that paternalism in particular is wrong.

Second, the arguments against paternalistically justified policies are nowhere near as strong as are usually thought. Arguments against paternalism come in two main types: (i) anti-paternalist arguments which claim that avoiding self-regarding harm which is adequately voluntary never provides a reason in favour of a policy regardless of the magnitude of the harms it avoids and (ii) arguments that policies which interfere with people's autonomous choices will very rarely if ever be of net benefit to those whose lives are interfered with.

I argue that non-paternalistic interference with liberty presents a dilemma for anti-paternalists who would object to interventionist public health policies. Either the anti-paternalist must hold that it is never legitimate to interfere with personal sovereignty for non-paternalistic reasons, or that it is sometimes legitimate. If non-paternalistic interference is never legitimate, then governing would be impossible, as the fact that any single individual would suffer a minor infraction of liberty would give him a veto against a government policy being enacted. If, however, non-paternalistic interference in an individual's sovereign zone is sometimes legitimate, then anti-paternalism—perhaps surprisingly—will do little to rule out interventionist public health policies.

I then argue that policies which interfere with adequately autonomous choices can often be of net benefit to those whose choices are interfered with without their consent. First, I argue that health has as strong a claim as any good to be an uncontroversial good for states to promote, so Millian arguments from individuality which claim that the state will impose the wrong values if it attempts to benefit people are implausible when it comes to health. Whilst rankings of health relative to other goods will be controversial, it is impossible for states to avoid taking controversial stances on how to rank values, and so it is a mistake to think that states should aim to do so. Second, the types of public health policy which most plausibly raise worries about making citizens' lives worse are those that both interfere with choices which are significant for that person's ability to author her own life and do so in coercive ways. However, there are many public health interventions which would improve health, while interfering only with choices of mild significance, and in ways that are not coercive. I conclude that even when we give due weight to liberty and to the importance of autonomous choice many public health policies which interfere with autonomous choices will be justifiable.

13.2 Paternalism, Coercion and Government Action

The definition of paternalism has been the subject of a large literature.[1] It is not my intention to make a substantive contribution to this literature here, and the points I am going to make do not turn on any controversial edge cases. For our purposes, we

---

[1] Influential attempts at a definition include Dworkin (1972), Gert and Culver (1976), Feinberg (1986) and Shiffrin (2000).

shall assume that the recent definition put forward by Gerald Dworkin (2010) is broadly correct. Paternalism, on this account, has three features: first, it involves an interference with either the liberty or autonomy of the person subjected to the paternalism. Second, the interference is done without the consent of the person interfered with. Third, the interference is undertaken in order to benefit the person interfered with.[2]

It is standard to distinguish between soft and hard paternalism. Soft paternalism involves interference with a person's choices where those choices are reasonably believed to be less than adequately voluntary (for example, interfering with an addict's ability to get hold of his drug of choice). Hard paternalism involves interference with choices which are known to be adequately voluntary: for example, preventing someone from taking her own life, even if she has thought long and hard about the decision, and her choice is autonomously made.

The soft/hard paternalism distinction refers to the voluntariness of the choices interfered with, and has nothing to do with the coerciveness or otherwise of the means employed to interfere with these choices.[3] So it is possible to interfere in a soft-paternalistic but very coercive way, or a hard paternalistic but non-coercive way. An example of very coercive soft paternalism would be arresting anyone who has been diagnosed as a problem gambler if they set foot in a casino. An example of noncoercive hard paternalism might be providing someone who endorses their current identity as a smoker with a very large cash reward if they quit. Our interest in this article is in the justifiability of interference with autonomous choices. If (as I argue) interference with autonomous choices is often legitimate in public health policy, *a fortiori* will interference with non-autonomous choices be.

On the account of paternalism I shall be using, paternalism is defined by its aim, not by its consequences. It is essential that the interference aims to benefit the person interfered with, not that it succeeds in so doing: interference with someone without his consent which leaves him worse off could still count as paternalism.

Much discussion of paternalism (particularly in the medical ethics literature) focuses on simple cases where one individual acts paternalistically towards another. In such cases, we are typically dealing with an individual whose values (and likely choices given the available options) are either reasonably clear or could easily be clarified by asking him or her. And it is also a situation in which it would be relatively easy for the paternaliser to tailor her actions to what the paternalized believes

---

[2] 'I suggest the following conditions as an analysis of X acts paternalistically towards Y by doing (omitting) Z: 1. Z (or its omission) interferes with the liberty or autonomy of Y. 2. X does so without the consent of Y. 3. X does so just because Z will improve the welfare of Y (where this includes preventing his welfare from diminishing), or in some way promote the interests, values, or good of Y' (Dworkin 2010). Notice that there are some such as Shiffrin (2000: 216) who define paternalism in a broader way to encompass taking over or controlling 'what is properly within the agent's own legitimate domain of judgment or action', whether or not this is done for that person's benefit.

[3] It is worth noting that lawyers and economists sometimes use the soft/hard paternalism distinction in a different way—to distinguish between coercive and noncoercive interferences. Where I refer to this latter distinction, I draw it simply in terms of coercive and noncoercive interventions.

would benefit him or her. Suppose that a patient in a hospice believes that it would be beneficial for her to have CPR should she go into cardiac arrest, despite indications that she would be very unlikely to receive any significant medical benefit from this. It will usually be perfectly feasible to mark the patient for attempted resuscitation. Paternalism in this kind of context seems particularly objectionable because it is easy to ask the individual what she wants and to tailor the treatment to her preferences, but the paternaliser either does not bother to ask, or asks and then overrides the paternalized's preference. It is not hard to see why paternalism like this is thought to involve a wrongful disrespect to the person's status as a competent agent.

However, in public policy contexts the status of paternalism is much more complex, for two kinds of reasons. First, the very idea of paternalistic policies is problematic. Second, public policy interventions are by their nature blunt instruments compared to the precision and subtlety possible in one-to-one relationships. States routinely coerce their citizens without their individualized consent in ways that would be deeply problematic in an interpersonal context, so the problem with paternalistically justified policies cannot simply be that they involve coercion without individual consent.

13.2.1 The Very Idea of Paternalistic Policies

Paternalism requires that the interference with choice be done in order to benefit the recipient. It follows that it is not acts of interference with choice on their own which are paternalistic, but rather acts of interference in conjunction with the end of benefiting the persons interfered with. So, in order to be able to describe a public policy intervention as paternalistic we would need to know its goal.[4]

When we ask what the goal of a public policy intervention is, we could take ourselves to be asking a question about the psychological states of the legislators or officials who shaped it, or we could take ourselves to be asking a question about what the best normative justification would be for the policy in its current form.

If we adopt a psychological reading, there will not usually be a single goal of a given policy intervention. Policy makers will typically have a variety of motives for advocating a particular intervention. For example, a minister may have several of the following reasons for introducing legislation to increase the rate of taxation on alcohol: wanting to reduce levels of violence in society, providing more tax revenue for a government, complying with a WHO recommendation, making a name for himself, as well as the paternalistic motivation of trying to benefit people without their consent by making it more expensive for them to drink.

If, alternatively, we adopt a normative reading of the goal of a public policy intervention then we face a problem of circularity. If the goal of a public policy intervention depends on what the most plausible normative justification of the policy would

---

[4] I use public policy intervention in a broad sense to encompass both legislation and softer means of influence such as tax incentives or public health advertising campaigns.

be, then our judgements about the justifiability of paternalism will determine the types of policies we describe as paternalistic.

As Peter de Marneffe puts it, 'if paternalistic justifications are illegitimate, as some antipaternalists surely believe, then the "best rationale" will never be paternalistic, and therefore no policy will ever be paternalistic according to this account'[5] (2006: 73).

Quite separately from the question of whether a normative or a psychological reading of paternalism is to be preferred, each account faces parallel problems from pluralism. Any real-world public policy intervention will have been formed by multiple psychological intentions. It is also the case that any real world public policy intervention will have multiple plausible normative justifications. Hence both accounts need to provide an account of when either an intention of benefiting people without their consent, or a justifying reason of benefiting people without their consent, is sufficient to render a policy paternalistic.

Two extreme views might be to say that (i) a policy is paternalistic if *any* of the motivations or justifications which explain its shape are paternalistic, or (ii) a policy is paternalistic only if *all* the motivations or justifications which explain its shape are paternalistic. However, both of these positions are obviously inadequate: (i) would allow that a particular policy is paternalistic even if paternalistic intentions or justifications played only a very small part in its genesis and shape. However, (ii) would in practice mean that no policy would be found to be paternalistic, as there will always be a possible non-paternalistic justification or motivation for policies which have a large paternalistic commitment. However, it is far from clear that it would be sensible to say something along the lines of 'a policy is paternalistic if *most* of the motivation behind it is paternalistic', or 'a policy is paternalistic if its main normative justification is paternalistic' given (i) the great difficulties involved in counting psychological motivations, and (ii) the fact that our interpretation of what the main normative justification for a law is will be heavily influenced by our prior normative commitments.

De Marneffe argues that given these difficulties, we should adopt an account of paternalistic policies which combines both psychological and justificatory elements: on this account it is a necessary condition for a policy's being paternalistic towards A that the government 'has this policy only because those in the relevant political process believe or once believed that this policy will benefit A in some way' and that the policy 'cannot be fully justified without counting its benefits to A in its favor' (2006: 73–74). However, (as de Marneffe admits) even this approach is not wholly satisfactory: the definition may be both too narrow and too broad. For example, it may be too narrow, in as much as it would entail that a policy which is motivated by avowedly paternalistic intentions would not count as paternalistic, if a sufficient *non*-paternalistic justification of the policy should be available. And it may be broad, in as much as it classes a policy as paternalistic whenever paternalis-

---

[5] We see this kind of reasoning in action when judges who are convinced that laws should not employ hard paternalism interpret laws requiring motorcyclists to wear a helmet as having an implicit harm-to-others justification such as avoiding extra medical care costs.

tic motivations and justifications play a *necessary* role in it. But it will often be the case that non-paternalistic motivations and justifications such as harm reduction also play a necessary role in the genesis and justification of a policy which is rightly described as paternalistic according to de Marneffe's criteria. So if we define paternalistic and non-paternalistic in parallel ways, then such policies would simultaneously be paternalistic and nonpaternalistic.[6]

I take it that this shows that there are significant difficulties that the opponent of paternalism in public policy must overcome even to state what it is that he is objecting to. And I shall assume that, following Husak, insofar as there is a cogent case against paternalism in public policy, it must be a case against types of *justification* for policies, rather than against policies per se. If this is the case then, as Husak argues, an antipaternalist should not say of any particular law that it is wrong because it is paternalistic, but rather make the more restricted claim that the law 'is unjustified in so far as it exists for paternalistic reasons' (Husak 2003: 391).

13.2.2   The Unavoidable Coerciveness of States

Paternalism, as we have seen, has three elements: interference, lack of consent and aiming to benefit. We have seen how difficult it is to make good on the claim that a particular public policy *is* paternalistic. We shall now argue that many public policy interventions unavoidably share significant features of what makes paternalism problematic at an individual level whether or not the policy aims to benefit citizens without their consent.

It will usually not be feasible to gain individualized consent from all citizens affected by a government policy. It would generally be both too expensive and too burdensome on the electorate to get each to vote every time a minor rule or policy were changed.[7]

Even if a government could consult each affected person and get their consent or dissent, it is unclear how it should interpret the results of any such referendum. If we wanted a strict analogy with the case of individual medical treatment, we would have to offer each citizen an individualized veto, and say that even one dissent would be enough to render the policy a no-go.

If we were to take individualized consent this seriously, policy making would be completely stymied: we would not achieve unanimity *either* for keeping the status

---

[6] De Marneffe could instead define nonpaternalistic in such a way that it encompasses all and only those policies which are not paternalistic in his sense. But without further explanation this would be an odd move: if both paternalistic and nonpaternalistic reasons would be necessary to justify a given policy adequately, why then conclude that we should categorise the policy solely as a paternalistic one?

[7] It is noticeable that governments typically and sensibly reserve referenda for more weighty changes. Indeed, if a government had to call a referendum every time it wanted to make a minor regulatory change then this would collapse the distinction between representative and direct democracy.

quo, *or* any proposed reform to the status quo.[8] For any proposed policy, at least some would object, and hence it would turn out that *no* government policies were legitimate.

It will not usually be the case that governments can allow those who dissent from a policy to opt out of the policy. Some policies are impossible to exclude people from (e.g. a defence policy), while others will simply be prohibitively expensive to tailor to individuals (e.g. in the case of a policy of fluoridating the water, we could build separate non-fluoridated water pipes to the houses of the dissenters, but this would be too expensive to be practicable). Even where it would be possible to grant individualized exceptions to a policy, it will usually be unfair and may be self-defeating to do so.[9]

In short, given the bluntness of public policy instruments governments will inevitably interfere in their citizens' lives in a myriad of ways, forcing them to acquiesce in public policy interventions whether or not they individually consent to do so.[10] So two elements of what is found problematic about paternalism in an individual healthcare context—interference and lack of individual consent—are endemic to public policy more generally.

I shall assume that this type of state interference is in general legitimate in a well run democracy.[11] Given that non-paternalistic policies may be coercive and infringe liberty to exactly the same extent as paternalistic policies, it follows that those who want to defend the wrongness of justifying policies paternalistically need to show that there is something wrong about the paternalistically justified policy over and above its infringement of liberty.

The literature reveals two basic lines of argument against paternalistically justified policies. On the first line of argument (exemplified by Joel Feinberg), which I shall call *anti-paternalism*, it is never a reason in favour of a policy that it is probably necessary to prevent self-regarding harm which is adequately voluntary.[12] Anti-paternalism thus refuses to put avoidance of self-regarding harm into the balance when we are considering which policies to adopt.

The second line of argument does not rule out reduction of self-regarding harm as counting in favour of a policy in principle, but argues that for reasons connected

---

[8] As O'Neill (2002: 163) puts it, 'Neither the status quo, nor any single route away from it, is likely to receive consent from all: unanimous consent to public policies is unachievable in real world situations'.

[9] Maintaining goods such as clean air or residential amenity requires the great majority of citizens to show consideration for others. Allowing some to obtain the benefits of the policy without contributing to the sacrifices that the policy requires would amount to an official sanctioning of free-riding. Such policies may also be self-defeating, as many citizens are willing to moderate their behaviour to maintain communal goods only if they believe that others are similarly motivated.

[10] It is worth remembering that even those elements of what the government does which merely offer options (say a free healthcare service which people can attend if they want to) are paid for through taxation which is coercively extracted.

[11] Not everyone would agree with this. For a well known view which draws the conclusion of the *illegitimacy* of state sponsored coercion see Wolff (1970).

[12] I borrow this way of formulating antipaternalism from Shafer-Landau (2005).

either to the limits of government competence, or to the nature of the human good, governments will do worse in promoting the wellbeing of their citizens if they adopt policies which are justified on paternalistic grounds than if they refrain from so doing. In a nutshell, on the second line of argument paternalism does not work. As we shall see, neither line of argument is convincing.

13.3 Against Anti-paternalism

Anti-paternalists argue that the salient moral difference between morally permissible public policies—which will unavoidably interfere with the lives of some individuals without their individual consent—and morally impermissible paternalism is that the interference in the case of paternalism wrongfully trespasses on areas of the individual's life where the individual should be sovereign. Anti-paternalism involves two kinds of claims: first a negative claim, namely that it does not count in favour of a policy it is probably necessary to prevent self-regarding harm which is adequately voluntary. Second, a positive claim that so doing is positively wrongful, because it involves a kind of disrespect or wrongful usurpation of the decision-making authority of the person.

> By and large, a person will be better able to achieve his own good by making his own decisions, but even when the opposite is true, others may not intervene, for autonomy is even more important than personal well-being. The life that a person threatens by his own rashness is after all *his* life; it *belongs* to him and to no one else. For that reason alone, he must be the one to decide— for better or worse—what is to be done with it in that private realm where the interests of others are not directly involved.[13] (Feinberg 1986: 59)

Feinberg takes his anti-paternalism to be a position about the *criminal* law. For example, he indicates that he is not against paternalistically justified taxation on smoking: taxing an activity, he argues, is both less coercive and less morally condemnatory than criminalizing those who perform it, and so does not automatically fall into the category of wrongful interferences in an individual's sovereign zone.[14] It follows that Feinberg would not have an in principle objection to hard paternalistic justifications for policies which *did not* criminalize self-harming behaviour. Presumably, he would suggest (as I would) that when considering such policies

---

[13] Similarly, Darwall (2006: 267–268) explains the wrongfulness of paternalism as follows: 'The objectionable character of paternalism . . . is not primarily that those who seek to benefit us against our wishes are likely to be wrong about what really benefits us. It is not simply misdirected care or even negligently misdirected care. It is, rather, primarily a failure of respect, a failure to recognize the authority that persons have to demand, within certain limits, that they be allowed to make their own choices for themselves.'

[14] 'I object to criminalization of smoking because it is supported only by a paternalistic liberty-limiting principle that I find invalid, but I do not oppose taxing end cigarette use, even though it too is coercive in a proper sense, and its rationale would be equally paternalistic.' (Feinberg 1984: 23)

the relevant question is whether the interference with liberty required to reduce self-regarding harm in a particular case is proportional to the good done.

Given that criminalization plays only a rather small part in the public health policies recommended by public health practitioners, Feinbergian antipaternalism is eminently compatible with most robustly interventionist policies. Hence, if anti-paternalists object to such policies they would have to be more radical anti-paternalists than Feinberg: that is, they would have claim that paternalistically justified policies are wrongful even where they are not implemented via the criminal law. It is this wider view that I shall concentrate on—though given Feinberg's pre-eminent position as the theorist of anti-paternalism, I shall continue to draw heavily on his work.

Anti-paternalists are committed to the claim that personal sovereignty *always* takes precedence over the reduction of self-regarding harm even when the harm to be avoided is very large, and the interference trivial. As Feinberg puts it, 'sovereignty is an all or nothing concept; one is entitled to absolute control of whatever is within one's domain however trivial it may be' (1986: 55).

The negative claim of the anti-paternalist position is that hard paternalistic reasons do not count in favour of a public policy. This leaves under-described the situation we face with water fluoridation, where many consent to the policy and think that it is a sensible way of regulating public behaviour, but some find the policy objectionably paternalistic, in that it interferes with their self-regarding behaviour for their benefit without their consent. One way of interpreting the anti-paternalist position would be to take Feinberg's claim about 'absolute control' within one's domain literally, and to take it that this should give the individual an absolute veto against any policy which would end up with behaviour in his or her sovereign domain being interfered with. On such a view, if even a *single* person objects to a policy's interference into his sovereign zone, then the policy treats him wrongfully, and so should not be implemented. However, this interpretation of antipaternalism would allow a single person to stymie the rest of society's attempts to regulate their behaviour. This seems grossly disproportional, and moreover is incompatible with the basic idea of personal sovereignty. It is just as disrespectful to persons' moral standing to deny them the authority to bind their wills communally by making agreements on certain curtailments to liberty as it would be to interfere with their private behaviour for their own benefit.

Feinberg, however, denies that we should interpret the zone of personal sovereign control in this way. He argues that a policy to which the majority agrees and a minority object will not usually count as paternalistic in his sense:

> When most of the people subject to a coercive rule approve of the rule, and it is legislated (interpreted, applied by courts, defended in argument, understood to function) *for their sakes*, and not for the purpose of imposing safety or prudence on the unwilling minority ('against their will'), then the rationale of the rule is *not* paternalistic. In that case we can attribute to it as its 'purpose' the *enablement* of the majority to achieve a collective good, and not, except incidentally as an unintended byproduct, the enforcement of prudence on the minority (1986: 20).

Feinberg's argument presupposes that one's zone of personal sovereignty prevents one from being subjected to paternalistic interference, but it does not prevent one from being subjected to an interference which would have the same effect on one's ability to lead one's life, but is aimed to allow the majority to achieve a collective good. This assumption is somewhat odd, and threatens to completely undermine the ringing claims about the importance of personal sovereignty. As Kalle Grill (2009: 149) argues, Feinberg does not appear to consider 'the restriction of the options of the minority to be in itself a moral obstacle to enactment of policy once the majority has consented'. Thus, Feinberg's position amounts to 'accept[ing] that societies with majorities bent on zealous self-regulation may impose strict health regimes on all citizens' (Grill 2009:149). This is, to say the least, a rather odd result for a position which is supposed to be archetypally liberal.

We saw earlier that it is very difficult to make good on the claim that any particular policy *is* paternalistic. Any anti-paternalist position which imposes a total ban on hard paternalism, but allows equal interferences with liberty when they are justified by reasons other than paternalistic ones, leaves itself wide open to a strategy of redefining the goals of policies in such a way that interference with a minority is an unfortunate side effect of the pursuit of the common good. It would be better, I suggest, to adopt an approach to policy which does not pick out paternalism as a particular evil to be avoided, but instead conceptualizes unwanted intrusions into liberty as the basic category. The question then becomes one of how significant the choices are which are interfered with, how coercive the means of interference are, what proportion of people object, and what other worthwhile goals (such as improvement of population health, or greater equality) are served by the policy. We now pass on to consider whether public health policies which intervene in order to improve population health can have an appropriate balance of harms and benefits.

13.4 Interventionist Policies Can Have an Appropriate Balance of Harms and Benefits

There are various ways in which a policy can fail to benefit its recipients. Most straightforwardly, the policy might be framed in a way which fails to take account of important systemic effects, and due to unexpected interactions with other policies, leaves its recipients worse off. However, this is a general problem which applies to all policies—whether or not they involve interferences with liberty in order to improve population health.[15] An effective argument against interventionist public health policies would need to show that there are reasons for thinking that

---

[15] I have written about this in (Wilson 2009).

interventionist public health policies are either more likely to fail to achieve their ends than other policies, or more likely to have undesirable ends.

There are two main ways of attempting to do this. First, we can argue that interventionist policies are likely to get it wrong about what sorts of weightings of values would make a person's life go better. Second, we can argue that even if the weighting of values implicit in an interventionist policy is a *better* weighting of values than the one the person would have chosen for themselves, the fact that this weighting of values is *imposed* nonetheless makes the life that contains this superior weighting of values less good than the life with a freely chosen inferior weighting of values.

13.4.1   Will Interventionist Policies Get the Weighting of Values Wrong?

Will interventionist public policies get the weighting of values wrong? Mill certainly thought so: 'But the strongest of all the arguments against the interference of the public with purely personal conduct, is that when it does interfere, the odds are that it interferes wrongly, and in the wrong place' (1977: 283). Mill bases his argument here on the importance of individuality: the best life is a self-chosen one in which one develops one's own particular talents and inclinations. On this view, it is difficult for governments to benefit citizens by assigning what seem (from the government's perspective) to be sensible weights to values, given that what makes each individual's life go well depends on features unique to that individual. Doing so will be likely to lead to rankings of values which make individual citizens' lives go worse.

This argument has two presuppositions: first the empirical claim that values *do* differ between people in this way; and second, the normative claim that people's individual valuations are a better guide to what would make their life go well than their government's judgements. Both are contestable assumptions in the case of health.

We can distinguish between on the one hand, controversial values, and on the other, controversial rankings of values. I shall argue that it is reasonable for governments to take health to be an uncontroversial value to be promoted; and that whilst the relative weightings of health as related to other goods will be controversial, it is not feasible for governments to avoid taking stances which incorporate controversial rankings of health against other goods.

The status of health as an uncontroversial good is easily established. Health is plausibly an intrinsic good, as Hurley (2007) argues. But even for those who do not agree with this, health has a value as an all purpose enabler for any number of other ends we might value for their own sake.[16] Given this, even if someone places no

---

[16] See for example, Daniels (2008). I discuss the value of health in more depth in (Wilson 2009, 2011).

particular value on their health per se, it will very often be the case that health enables them to pursue the goals they do value even better. So it is overwhelmingly plausible to say that health is a legitimate goal of governments; or if governments cannot assume that goods like health are of value then there is little if anything that governments can legitimately do.

How we should rank the value of health against other values such as liberty will be controversial. But it is simplistic to imagine that governments will do better in helping their citizens to achieve a good life if they adopt a laissez faire approach. The idea that a laissez faire approach would allow each citizen to decide for themselves the relative weight that *they* want health to have as compared to other goods in their lives is illusory. This is because individual tradeoffs about the value of health relative to other goods are made in the context of broader choices about the structure of society. Many choices about one's health (such as to be able to get exercise by walking to and from work) are only feasible if a whole set of background conditions, like street lighting, maintained sidewalks, and proper town planning are in place.

So if a government responds to value controversy by adopting a laissez faire attitude this will by no means allow all to balance health against other goods in a way they deem optimal. Rather it will favour some rather than other forms of life and some rather than other sets of choices. Public health policy inevitably involves controversial rankings of health against other values: it is a mistake to suggest that states can avoid this. The appropriate goal is not to avoid controversy, but to do what is most justifiable.

13.4.2 Self-Authorship and the Good

A last line of argument alleges the human good is special, in that it requires the person whose life it is to play an active role in its procurement, and alleges that for this reason the good is not well suited to be brought about by states. I am willing to grant that we should place a very high value on self-authorship. However, even if we grant this, it does not follow that interventionist public health policies will in general fail to have net benefits. This is because there are many types of interventionist public health policies which do not impinge in a significant way on people's ability to author their own lives.

The types of policies which engage our concern about self-authorship and the good have two features: first, they involve choices which are highly significant to the meaning and structure of individuals' lives—choices such as what career path to follow, what religion if any to profess, and what treatment regimes to choose in the face of terminal illness. Second, they are policies which largely or completely co-opt a person's will, and back this up with the coercive force of the criminal law. However, as I shall argue many health promotion policies have neither feature.

13.5 Significant Choices

Not all choices are significant for one's ability to author one's own life. If a government passed a law on avowedly paternalistic grounds mandating that prepackaged meals could not contain more than a certain percentage of salt, few people would be worried they had thereby been deprived of the ability to author their own life, given the ready availability of salt cellars.

But what if someone did, surprisingly, place a great weight on his ability to buy heavily salted prepackaged food? We could either adopt (i) an objective account of the significance of a choice, according to which whether a choice is significant would be measurable according to an objective model of which choices are necessary for a flourishing human life; or (ii) a subjectivist account, according to which whether a choice is significant depends on the values of the individual chooser.

On an objectivist view, a person can simply be wrong about whether a choice is significant in a way that should prevent it from being interfered with. If a choice is not significant then it is not difficult to justify interfering with it. As in the salty food example, many of the choices which governments would like to interfere with to ensure better population health are plausibly of this insignificant kind.

If we take a subjectivist view then we commit ourselves to the claim that there is not a deep answer to the question of which choices are significant. If just those choices are significant which individual citizens think are significant, it will be a mistake to think that we can come up with an account of significant choices which will set down a standard for what people *ought* to recognize as significant. Given pervasive differences in conceptions of which choices are significant, policy making will inevitably have to interfere with choices which some citizens rightly think are highly significant. If it is inevitable that policy making does so, it cannot be true to say that interfering with significant choices is invariably unjustifiable.[17]

In the absence of any objective measure of significance, perhaps the best that governments can do is to take the minimization of interference with choices which individuals regard as significant as one goal (amongst others) of public policy. Many interventionist health promotion policies will involve interference with choices which are thought by the vast majority of people to be insignificant.

13.6 Coercion

The other kinds of policies which are most problematic for self-authorship are those which largely or completely co-opt a person's will, and back this up with the coercive force of the criminal law. An obvious example of this would be a law which required every citizen to eat five portions of fruit and vegetables a day, on pain of imprisonment. Again, it is important to notice that not all interventionist policies

---

[17] I am again assuming that the anarchist position is false.

Table 13.1 Four types of interventionist policy	1. Coercive policies which overrule choices of major significance
	2. Non-coercive policies which interfere with choices of major significance
	3. Coercive policies which overrule choices of minor significance
	4. Non-coercive policies which interfere with choices of minor significance

need be coercive in this sense. Much of the public health policy that states will want to pursue involves manipulating choice architecture in ways that still give the individual a wide range of leeway to make her own choices and her own mind up.

Suppose that a government passes legislation imposing walkability requirements on new towns and cities, on the grounds that this will improve the lives of the people living there by making it more likely that they will walk rather than drive. Or a government introduces legislation to dissuade people from smoking by increasing the tax on cigarettes. In such cases, the committed car driver and the committed smoker are free to continue to pursue their vision of a valuable life; it is just that the government has made it somewhat more costly for them to pursue this way of life than a healthier one. As Joseph Raz (1986) argues, what matters for self-authorship is to be able to make choices about the important decisions, and to have *enough* choices, rather than to have as many choices as possible about every single minor detail. Where interventionist public health policy leaves the conditions of self-authorship intact, the importance of self-authorship does not rule out interventionist public health policies.

To sum up the discussion of this section: given a commitment to minimizing coercion, and a commitment to minimizing interference with significant choices, there are four types of paternalistic policy (Table 13.1). Interventionist policies which are both coercive and which overrule highly significant choices (1) are most difficult to justify.[18] Coercive interventionist policies which overrule choices of minor significance (2), and non-coercive policies which overrule choices of major significance (3) will be somewhat easier to justify, and lastly non-coercive policies which overrule choices of minor significance (4) will be the easiest to justify.

13.7 Conclusion

This article has had two main themes. The first theme has been that it is unhelpful to take the question 'is the policy paternalistic?' to be a fundamental question in public health ethics, for several reasons. First, we face severe problems in making good on

---

[18] There will be cases where policies which both coercive and which overrule highly significant choices are justifiable: for instance social distancing measures to prevent the spread of a pandemic.

the claim that a particular policy *is* paternalistic. Second, the marks of much of what makes paternalism morally problematic on an individual level—interference and lack of individual consent—are present in a wide range of public policy contexts, whether or not the policy in question can be described as paternalistic. We do better, I suggested, to frame the *ethical* question as about what types of justifications for public policies are legitimate and to frame the *practical* question for public policy as about which infringements of liberty are justifiable, without staking too much on whether those infringements of liberty are paternalistic or not.

The second major theme of the article has been that interventionist approaches to public health can be legitimate, and that policies which infringe liberties in various ways can make people's lives better. Health has as strong a claim as any good to be an uncontroversial good for states to promote. And whilst rankings of health relative to other goods will be controversial, it is impossible for states to avoid taking controversial stances on the value of health.

I have argued that there are at least four *pro tanto* principles which should guide a state's policies to improve citizens' health. First, it is obvious that the size of benefit to be gained or size of harm to be avoided by interfering matters. Other things being equal, the greater the expected benefit and the greater the expected harm to be avoided, the stronger the argument in favour of intervention. Second, the extent to which the population regulated endorses or consents to the policy matters. Other things being equal, the greater percentage of the affected population who endorse an intervention (and the more enthusiastically they do so) the stronger the reason in favour of the policy. Third, autonomously chosen lives matter. Other things being equal, the more significant a choice is, the more important it is that a person has the opportunity to make a genuine or authentic choice and the more problematic it is to interfere with their choice. Fourth, liberty matters. Other things being equal, the more coercive a policy is, the more problematic it is.

Some cases of potential interventions for health will be clear cut: where we have a policy which will bring a great benefit, which is supported by the vast majority of people, and involves only a mild interference with choices which are not generally thought to be significant, the intervention will be easy to justify. Where we have a policy which will bring only a small benefit, and which is opposed by the vast majority of people, and involves a coercive interference with significant choices, then the intervention will definitely not be justified.

The interesting cases will be those closer to the middle.[19] Further normative work will help to clarify types of cases, but it is important to notice that we should not expect that normative reasoning will be able to give us definitive once-and-for-all answers to these questions. This is because—if what I have argued is correct— which interferences are justifiable depends (among other things) on the level of

---

[19] One such interesting middle case is the regulation of phase one clinical trials. In Edwards and Wilson (2012), we argue that the significance of terminally ill patients' autonomous choices about their treatment is sufficiently great that interference is illegitimate—even if there are good reasons for thinking that a given research project the terminally ill person wishes to participate in will do him or her more harm than good.

general consent to the policy, and the significance of the choices interfered with. The level of consent will obviously vary relative to culture and time; and unless we can redeem the Herculean task of specifying an adequate objectivist account of which choices are significant, we will also have to take account of local differences in which choices are believed by particular communities to be significant. So while further normative theorizing will be necessary, it seems likely that it will not be sufficient to tell us which policies to adopt.[20]

Acknowledgements I thank audiences in London and Keele for comments on earlier versions of this article, in particular Kalle Grill, Dan Brock, Dan Wikler, Jonathan Wolff, Annabelle Lever, Jonathan Hughes, Giovanni De Grandis and Harald Schmidt. Thanks also to this journal's editors and anonymous reviewers.

This article first appeared in Public Health Ethics 4 (3), 2011, 269–279, and is reprinted by kind permission of Oxford University Press.

Funding My work was undertaken at UCL/UCLH who received a proportion of funding from the Department of Health's NIHR Biomedical Research Centres funding scheme.

References

Daniels, N. 2008. *Just health: Meeting health needs fairly*. Cambridge: Cambridge University Press.

Darwall, S. 2006. The value of autonomy and autonomy of the will. *Ethics* 116: 263–284.

de Marneffe, P. 2006. Avoiding paternalism. *Philosophy and Public Affairs* 34: 68–94.

Dworkin, G. 1972. Paternalism. *The Monist* 56: 64–84.

Dworkin, G. 2010. Paternalism. In *The Stanford encyclopedia of philosophy* (Summer 2010 Edition), ed. E.N. Zalta. Available from: http://plato.stanford.edu/archives/sum2010/entries/paternalism/. Accessed 6 Oct 2011.

Edwards, S.J., and J. Wilson. 2012. Hard paternalism, fairness and clinical research: Why not? *Bioethics* 26(2): 68–75.

Feinberg, J. 1984. *The moral limits of the criminal law*, Harm to others, vol. 1. New York: Oxford University Press.

Feinberg, J. 1986. *The moral limits of the criminal law*, Harm to self, vol. 2. Oxford: Oxford University Press.

Gert, B., and C. Culver. 1976. Paternalistic behavior. *Philosophy and Public Affairs* 6: 45–57.

Grill, K. 2009. Liberalism, altruism and group consent. *Public Health Ethics* 2: 146–157.

Hurley, S. 2007. 'What' and the 'how' of distributive justice and health. In *Egalitarianism: New essays on the nature and value of equality*, ed. Nils Holtug and Kasper Lippert-Rasmussen, 308–334. Oxford: Oxford University Press.

Husak, D. 2003. Legal paternalism. In *The oxford handbook of practical ethics*, ed. Hugh LaFollette, 387–412. Oxford: Oxford University Press.

Mill, J.S. 1977 [1859]. On Liberty. In *Collected works*, vol. 5, ed. J.M. Robson, 213–310. Toronto and Buffalo: University of Toronto Press.

O'Neill, O. 2002. *Autonomy and trust in bioethics*. Cambridge: Cambridge University Press.

Raz, J. 1986. *The morality of freedom*. Oxford: Oxford University Press.

---

[20] I examine the difficulties of drawing public policy conclusions from philosophical work in much more detail in Wilson (2009, 2012).

Shafer-Landau, R. 2005. Liberalism and paternalism. *Legal Theory* 11: 169–191.

Shiffrin, S. 2000. Paternalism, unconscionability doctrine, and accommodation. *Philosophy and Public Affairs* 29: 205–250.

Thaler, R., and P. Sunstein. 2008. *Nudge: Improving decisions about health, wealth, and happiness*. New Haven: Yale University Press.

Wilson, J. 2009. Towards a normative framework for public health ethics and policy. *Public Health Ethics* 2: 184–194.

Wilson, J. 2011. Health inequities. In *Public health ethics: Key concepts and issues in policy and practice*, ed. Angus Dawson, 211–230. Cambridge: Cambridge University Press.

Wilson, J. 2012. On the value of the intellectual commons. In *New frontiers in the philosophy of intellectual property*, ed. Annabelle Lever, 122–139. Cambridge: Cambridge University Press.

Wolff, R.P. 1970. *In defense of anarchism*. California: University of California Press.

Chapter 14
Individual Responsibility and Paternalism in Health Law

Stefan Huster

14.1 Introduction

Health is a special good. Health problems have an existential dimension that other social deficits do not have in the same way: As bad as it may be to have no work or to have to eke out a living on a limited income, the experience of a serious or even life-threatening disease and its associated pain, impairments, and mental stress is incomparably worse. Health is additionally a conditional – or in the Kantian sense of the word – transcendental good, because it represents the basis for many of life's other needs. As they say, "Health isn't everything, but you don't have anything if you don't have your health." To be able to deal with a constraint on lifestyle caused by health problems – as much as it is medically possible – is of central importance in a competitive meritocracy: In this way equality of opportunity and a leveled playing field will be produced that initially legitimizes the stratified results of the competitive process (Daniels 1985). This might also explain why – similarly to education – there is massive political opposition against the social stratification of health.

S. Huster (✉)
Public Law, Social and Health Law and Philosophy of Law,
Ruhr University Bochum, Bochum, Germany
e-mail: stefan.huster@rub.de

© Springer International Publishing Switzerland 2015

221

T. Schramme (ed.), *New Perspectives on Paternalism and Health Care*, Library
of Ethics and Applied Philosophy 35, DOI 10.1007/978-3-319-17960-5_14

14.2   Care and Prevention

14.2.1   Attention for the Healthcare System

There are therefore good reasons to make the welfare state additionally and even primarily obligated to maintain the health of its citizens. In this context attention is paid, at least in Germany, almost exclusively to the guarantee of medical benefits. Even the jurisprudence of the German Federal Constitutional Court gives the guarantee of medical benefits an especially high level of ranking. In this sense the Constitutional Court had for a long time ruled in a cautious way, saying that no claim to the supply of special health services may be derived from Basic Law and the principle of the welfare state; the objective legal duty of the state in setting itself to protecting and promoting the right to life and physical integrity (Art. 2 Sect. 2 Basic Law – GG [*Grundgesetz*]) is, in light of the margin of appreciation of the responsible state agencies, concerned solely with meeting those public power provisions for the protection of the Basic Law that are not completely inapplicable or inaccessible. New decisions however have increased the constitutional requirements by a considerable amount – if an entitlement to benefits of the insured under statutory health insurance (GKV [*gesetzliche Krankenversicherung*]) is to be derived from the guarantee to personal freedoms (Art. 2 Sect. 1 GG) in connection to the welfare state principle as well as from Art. 2 Sect. 2. GG, and if no universally recognized treatment conforming to medical standards is available to this person for a life-threatening or consistently deadly illness, and there is a not altogether distant prospect for a cure or for the application of the medical treatment of her choice to make a perceivably positive impact on the course of the disease (Federal Constitutional Court 2005). An additional increase arises from the fact that the Constitutional Court now deems it expressly possible for these requirements to also apply to methods, for which the Combined Federal Joint Committee, as the central control agency, has already decided against being added to treatments covered by the GKV (Federal Constitutional Court 2007). The agencies legally appointed to make decisions can do as they like – in any case the "maximum value of life" in the form of claims directly drawn from the Basic Law seems, for a life-threatening illness, to break new ground (Federal Constitutional Court 2005). In the meantime the legislature has adopted these provisions in § 2 sect. 1a of the Fifth Book of the Social Law Code [*Fünftes Buch Sozialgesetzbuch*] (SGB V).

14.2.2   The Significance of Other Factors for Health and Illness

In light of opportunity costs (cf. Huster and Kliemt 2009), this focus of the attention on the healthcare system and the parallel intensification of the constitutionally legal standards will not be welcomed without some objections if it is brought to mind that the healthcare system is of no central importance either for the health of the populace – traditionally formulated: "public health" – or for the social distribution of health – "health justice". As a matter of fact, numerous studies indicate that, in addition to access to healthcare and genetic make-up, social factors (in the broadest sense) are substantially responsible for the state of health of a population: namely, the conditions in the workplace and in the residential environment, individual health behaviors, as well as distinct social-structural factors (cf. Mielck 2000; Richter and Hurrelmann 2006).

These factors might also substantially explain social health gradations. For it can be shown that these health-related factors are distinct to a large extent in a class-specific manner: Members of the lower social classes are – as the respective studies show – more often exposed to harmful work environments and also suffer from more noise and air pollution in their residential environments; they eat less healthily, are more likely to use nicotine and consume excessive amounts of alcohol, and more often neglect bodily exercise. Indeed, they suffer just as much in their private and work lives as they do in the public (political) sphere from meager opportunities for control and self-determination. In any case, the social health inequalities in Germany can surely not, at the very least, be explained exclusively by the difference between public and private health insurance or by social entrance barriers to healthcare benefits.

14.2.3 The Attention Deficit of Public Health Policy

In light of the circumstances that the social health factors are of substantial importance to both public health and the distribution of health, the protagonists of a public health policy related to these factors frantically ask the question why this political landscape enjoys significantly less attention in comparison to the supply of healthcare. This attention deficit is also expressed in law. For sure there are entire areas of law that additionally, or even primarily, serve the protection of health; environmental law and the right to safety at the workplace are just a few. Yet in view of combatting health inequalities the balance fails miserably: The idea that statutory health insurance benefits should not only improve general health, but also reduce the social inequalities of health-related opportunities in particular can apparently only be found in § 20 sect. 1 sentence 2 SGB V.

A whole range of issues, to which the "political economy" of health policy above all belongs, are likely responsible for this attention deficit. Like any policy that promotes public goods in the public interest, public health policy has a problem of political awareness and enforcement. Unlike in the "medical system", here there is no lobby for whom the expansion of resources is a top priority (Burris 1997). Additionally the public health policy lacks the technical fascination that emanates from modern medicine. Also, prevention first of all costs money; it is indeed debatable whether in this way mid and long term costs can be saved (Beske 2002). Due to the reference population of public health measures, the motivation for an individual citizen to advocate these measure politically is, in the end, not very pronounced. While healthcare services refer clearly to the individual and thereby attract citizens' attention, the so-called prevention paradox bleeds through in the area of public health: On a statistical level even a minute change in behavior or the environment has perceivable effects on public health; on the individual level however these effects are very small and are only of marginal benefit to the individual (Rose 1985). It is therefore not at all surprising that they do not find a greater interest in the public sphere.

These characteristics are reinforced through the structure of a legal system that grants individual freedom and individual rights a normative priority. The hazards of concrete life and health concerns, which are addressed in the context of medical care, are much more easily raised as issues by means of individual rights than are the statistical effects of a public health policy. The law system thereby exacerbates the asymmetrical attitude between concrete and mere statistical dangers in life, which already plays a large role in our moral intuitions (Huster 2008). There is a similar situation for distribution effects: While the socially stratified impacts of many health policy control mechanisms in the care system – such as practice fees or the increase of out-of-pocket payments – are apparent and are debated rigorously enough, the effects of general social policies – as relevant to health as they may be – remain for the most part vague. For this reason it is likely no accident at all that Public Health in institutions such as the EU and the WHO already plays a bigger role, since these institutions are said to have a technocratic understanding of policy, and in them the health policy discussion – already because of lack of authority – is not determined by distribution battles in healthcare systems.

14.2.4 The Paternalist Objection

The biggest political and legal oppositions to a decisive Public Health Policy might indeed arise from one certain aspect that in the following will be discussed more closely: If and insofar as this policy is related to lifestyle-oriented health factors, it inevitably appears to contain a paternalistic characteristic that is opposed to freedom. As a consequence, if the policy attempts to influence the health-related daily habits of citizens, then it could be encompassed in an ultimate control-state health-dictatorship of "healthism". In doing so the question of which roles a citizen's personal responsibility should play for their own health would be discussed completely differently inside of and outside of the healthcare system.

14.3 Personal Responsibility in Healthcare

14.3.1 Personal Responsibility in Health Insurance Law

In healthcare law the principle of personal responsibility has long lived a shadowy existence. Although § 1 SGB V – in addition to the principle of solidarity – emphasizes the personal responsibility of the insured for their own health, this has so far had little practical significance for the policyholder's entitlement to the insurer's performance. That is to say, health insurance is generally geared towards the final principle, and hence any entitlement to benefits is not dependent upon the cause of a disease (Muckel 2009). Altogether this means that § 52 sect. 1 SGB V only contains restrictions on the entitlement of benefits for self-inflicted health problems if the policyholder intentionally caught the disease. This is however typically not the case, even for instances of extremely health-adverse behavior, since an afflicted person hopes that in her case the adverse health effects of the behavior will not occur; therefore intentional self-infliction is out of the question. This norm therefore has a very small area of applicability.

In 2007 however legislators implemented an attempted reinforcement of the role of personal responsibility with the Competition Reinforcement Law [*GKV-Wettbewerbsstärkungsgesetz*]: According to the newly added section 2 of § 52 SGB V, insured persons are to take on a share of treatment costs if they "inflict themselves with an illness due to a procedure that was not medically necessary, such as cosmetic surgery, tattoos, or piercings." A strict reading of this phrasing meant that the insured were obligated to co-pay even for health problems resulting from otherwise innocuous practices – e.g. the widespread and socially inconspicuous practice of piercing one's earlobes in order to wear earrings – simply because they were listed as examples ("such as"). But for the legislators this was too much, and since 01/07/2008 it has been "clarified" in the Further Development of Care Act [*Pflege-Weiterentwicklungsgesetz*] that the co-payment requirement is limited to the three explicitly listed measures – in that the phrasing "measures [...] such as" has been deleted from § 52 sect. 2 SGB V. But now a considerable problem concerning equality arises: Why are the health problems resulting from cosmetic operations, tattoos and certain body piercings treated separately? Why does the regulation not also include health consequences of comparable practices like branding, cutting, and of the aforementioned earrings (Höfling 2009; Wienke 2009)? On the other hand the question is raised how the health insurance provider can trace an illness to the items listed in § 52 sect. 2 SGB V. In addition the legislature has in the meantime introduced to § 294a sect. 2 SGB V a corresponding notification requirement for doctors and hospitals practicing under statutory healthcare; this is obviously a problematic ruling in light of the relationship of confidentiality between the doctor and the patient.

The legislature has also, through the GKV Competition Reinforcement Law [*GKV-Wettbewerbsstärkungsgesetz*], reinforced the significance of prevention-related personal responsibility. This has been achieved by linking the reduction of payment for the chronically ill from 2 to 1 % of annual income to the condition that the affected persons regularly have certain checkups (§ 62 sect. 1 SGB V). The amount of health insurance subsidy for dental care has already been dependent for a long time on whether the insured person has regular dental checkups (§ 55 sect. 1 SGB V).

14.3.2   Opposition to the Criterion

Anyone who has voluntarily decided to take a certain risk can in principle not complain that she bears the consequential costs, or at least shares in on them, if this risk is realized. Unlike cases of "brute luck," when an undeserved risk occurs the personal responsibility of the affected in the constellation of "option luck" can justify that the community of solidarity leaves her to cope with the resulting problems (for terminology cf. Dworkin 1981). In principle this goes for healthcare as well. That the idea of personal responsibility has previously been waived in health insurance law indeed shows that it can be especially problematic in this area. Even the discussion on criteria of prioritization and rationing has not been able to warm towards the criterion of personal responsibility or of personal accountability (for discussion cf. Alber et al. 2009; Buyx 2005).

The biggest problem – it may thereby be shown – is not that it is difficult or impossible to trace an illness back to a specific action in singular cases. Substantively this is true, but it does not prevent the law from assigning responsibility from the outset; even in civil law abstract endangerment offences are known that were not necessarily the result of concrete harms. Therefore it is not clear why in health insurance law risk-increasing behavior should not lead to financial consequences, if not to an exclusion of benefits. The differentiation of the ceilings on patient co-payment in § 62 sect. 1 of SGB V is obviously based upon this idea, because even here one cannot always definitively say, in singular cases, that a disease's course would have been better if the insured had had regular medical checkups. Yet, the facts determining increased risk and cost have to be chosen non-arbitrarily – as is shown in the discussion of § 52 sect. 2 SGB V. In addition it would have to be calculated without bias, which costs, but also which savings, to the community are associated with a self-inflicted disease – at least when there is a real attempt to balance individual and collective responsibility to bear costs and not a concealed effort for a health-related paternalism.

The real problems however might lie somewhere else. The example of § 52 sect. 2 SGB V shows that the consideration of health behaviors in the healthcare system can be connected to an intricate investigation into the personal sphere of someone's life. In addition, it can be debated to which extent any given health-adverse behavior really is voluntary. While some degree of responsibility can usually be attributed to select decisions about health risks together with their consequences – such as getting one's ears pierced or doing some especially accident-prone sport – there is quite regularly a much longer and more complicated genesis for lifestyles that are relevant for many chronic illnesses, such as behavior regarding nutrition and exercise. At this point it will be relevant how politics and law address these circumstances.

14.4   Personal Responsibility and Public Health

While approaches to ascribe a stronger role to personal responsibility within the healthcare system have been criticized as going "against solidarity" (Wienke 2009), because they have too strongly emphasized individual responsibility in relation to collective responsibility, the discussion about public health policy often takes a different course: Insofar as this policy concentrates on preventing behaviors, it threatens not to take citizen autonomy and personal responsibility seriously enough.

14.4.1 Health Inequality and Justice

For this reason, the above objection is important, on the one hand, in answering the question of whether state and politics are obligated to take action. This is often assumed, especially in light of social health inequalities, which are found offensive. Insofar as health effects and difference arise from how individuals lead their lives, however, the welfare state apparently has no reason to be obligated to take action for them: There is neither a normative problem nor is there justification for collective responsibility if individual citizens or even entire classes of people ruin their health through their own voluntary actions. In a liberal society it is fundamentally appropriate to ascribe citizens responsibility for the (financial) consequences of their free choices. Even egalitarian theories of justice agree that a just distribution of goods, while "endowment-intensive", is nonetheless "ambition sensitive." It must balance the undeserved advantages and disadvantages that arise from varying starting positions and environmental conditions that cannot individually be influenced, but not from the differing consequences of autonomous decisions (Dworkin 1981). Legal freedom then indeed leads to actual inequality.

This objection does not apply to the whole of public health policy, since social differences in health cannot be exclusively nor likely even primarily written off as the effects of health behaviors. But insofar as this policy deals with behavior, the objection is convincing, at least at the first glance, that there appears to be no inequality that requires correction.

14.4.2 Justification of the Influence of Health Behaviors

Even worse, according to the above-mentioned premise state action would not only be uncalled-for, but even subject to massive normative objections. This poses the rather serious question of what kind of legitimacy the public authority has to make an impact on health-relevant lifestyle choices of citizens.

The first argument is also the most obvious, in that local communities have an interest in healthy lifestyles, since this reduces costs (cf. Händeler 2008). This is a morally innocuous argument, because it is based on the well-known principle that external costs are to be internalized (Eichenhofer 2003). But to the extent that the costs of the social systems and especially healthcare are included in this point of view, it is first not clear whether and in which circumstances this argument is applicable. As is well known, it could turn out to be the case that smoking has no net cost for social systems, or that it even saves on costs. Second, influencing lifestyles appears to be the wrong approach to take; it would be consistent that those affected shared directly in the treatment costs of their illnesses. As soon as one looks beyond the above costs and takes into consideration the days missed from work and other consequential social costs, one also has to bear in mind that this approach is rather alien to a liberal social order. For independent of their health habits, citizens have no obligation to develop themselves into the most productive members of the community as possible. Anyone who wastes their talents instead of honing them and using them for the good of the community may very well typically receive a low income, but they are not additionally charged with a penalty.

A second principally unsuspicious yet also not very wide-reaching argument gears to the fact that risky behaviors do not only affect the health of the individual but can also affect the well-being of others. It is especially by this means that the legal campaign against smoking in public has been justified. The resulting jurisdiction showed, however, that an absolute ban on smoking in order to protect nonsmokers is complicated: It reaches its limits if nonsmokers are not affected, or if they are only marginally affected, or if they consent to their being affected. The German Constitutional Court had to accept strong criticism (Federal Constitutional Court 2008) insofar that it nonetheless upheld strict nonsmoking laws (cf. Gröschner 2008). Even within the court it was debated whether "a path of educational paternalism is proscribed that could be expanded to other areas and then have a suffocating effect" (Federal Constitutional Court 2008). Moreover there are a multitude of habits that are hazardous to health that do not have any indirect effects on others: Anyone who eats poorly or does not get enough exercise is not harming the well-being of others in any relevant sense.

To protect and improve the general health of the population is a third argument that can account for the pertinent policy measures, but it is normatively in a fragile condition. For historical reasons this reference to "the people's health" (*Volksgesundheit*) has had already conceptually a hard time in Germany (cf. Frenzel 2007). Despite this, however, the question actually arises whether such an object of legal protection can exist in a liberal society. This argument can be reduced to those two arguments already mentioned above, insofar as it is concerned with the resultant costs of a disease or with harm to others (for instance through contagious diseases). Yet, over and above this interpretation, health actually does appear to be a private good, and accordingly "public health" may be interpreted as a summative term for individual states of health, which itself has no transcending normative substance: A person is not part of a "body politic", for whose state of health he or she might be responsible.

That leaves a fourth line of reasoning, which is that the concern here is the health and well-being of every individual citizen. But this leads the public authority onto a political minefield: To urge, educate or coerce citizens to lead more healthy lifestyles in their own interest conflicts with the principle that the liberal state has to stay neutral towards varying ways of life and life decisions (Huster 2002). The state may step in to regulate insofar as the interests of others or of the general public are affected, but according to prevailing opinion the state does not possess any further authority to make value judgments. Therefore the state is principally denied perfectionist or paternalistic policy. How far this principle has also penetrated constitutional assessments is demonstrated in the German Constitutional Court deeming it necessary to emphasize its legal justification for banning smoking in restaurants: "Smokers are not here patronized in any manner whatsoever, and in particular no protection against self-harm is being forced upon them. The state nonsmoking protection laws [*Landesnichtraucherschutzgesetze*] are aimed neither at preventing addiction nor at protecting the individual from herself. Its goal is rather to protect against the harm of secondhand smoking. The concern is not the protection of the health of the smoker, but of the other people who in these situations are not themselves smoking" (Federal Constitutional Court 2008).

There are two reasons to doubt this construct of the Constitutional Court. On the one hand it is not clear how the protection of nonsmokers shall be able to justify such far-reaching and comprehensive bans that the Court deems admissible (cf. Federal Constitutional Court 2008). If the protection of "other people who in these situations are not themselves smoking" were really the only concern, it would make more sense to have them avoid places where people smoke. Secondly, these nonsmoker protection laws put together a combination of measures that tackle smoking as such, so it is highly unlikely that they lack any prevention policy impetus. Therefore the prohibition of paternalistic policy seems either to tie the hands of state health policy or to force it to hypocrisy.

14.4.3   The Perspective of Public Health

For public health policy apologists these results are not acceptable, especially in light of social health inequalities (for discussion cf. Callahan 2000; Wikler 2005). They may well emphasize that the social health stratification can be attributed only in part to the lifestyles in various social classes. Nonetheless, the policy would lose a highly important field of practical application if it had to ignore lifestyle-related health factors and inequalities. There is the danger in this case of relying too heavily on the principle of personal responsibility, leading to "victim blaming" the already disadvantaged, and dismissing the public authority from any responsibility to promote useful social structures for health-conscious choices. Experience up to now also shows that the habits of life in any case are extremely resistant to change if they are addressed in isolation in the form of warnings and information campaigns; simply appealing to the individual responsibility of citizens does not lead very far if the social framework is not changed.

These aspects point out – and in normative terms this is the crucial point – that the cultural and social environment may influence individual behaviors to varying degrees. Aspects of lifestyle such as nutrition and exercise are usually already laid out during childhood, reinforced by social and media influences, and very often have the characteristic of addiction. These are the typical elements that should caution us from ascribing responsibility for these behaviors and their consequences to the individual alone; here there are obviously "causes of causes" in play. That particular behaviors are not as independent as is often claimed can be easily seen in that many citizens indeed themselves want to change their health-adverse habits, but do not succeed in doing so. In this situation any policy that makes it easier to allow the citizen to enact their will (cf. Frankfurt 1971) may very well be seen as promoting freedom (Bloche 2005).

This reveals a fundamental tension between the perspective of public health argument as an empirical social science on the one hand, and on the other hand the perspective of law and social philosophy. While law and theories of justice assume individual autonomy and its power of legitimation, a public health argument is for reasons of principle alien to these perspectives: The argument does not pay attention to the individual and the freedom of her expression of will, but rather abstracts from this and attaches class-specific health behaviors to a statistical level, even though the concrete members of this class vary. This suggests the idea that it is not the individuals themselves who are responsible for these regularities and differences between social units, but rather the social causal factors.

Now the tension between the two perspectives – a participant perspective on the on hand, in which we perceive ourselves as free and responsible actors and also impute this attitude to others, and on the other hand the observer's perspective, which emphasizes the social and cultural background of our decisions and our actions – does not have to end in an irresolvable contrast. Indeed, we are accustomed to respecting individual choices in other areas of life without thereby having to claim that individuals exist independent of these background factors. Even so, for every area of interest the community must nonetheless regularly negotiate and decide upon how responsibility is to be divided between the individual and society (cf. Roemer 1993).

14.5   Legal Consequences

14.5.1   Relevant Aspects of Decision-Making

Against this background the observer's and member's perspectives have to be balanced out, even in law. This might mean firstly that the law has to protect the individual's freedom to make decisions from health policy activism. It does not follow from the recognition of influencing social factors that individual decisions would not have to be respected. It is because the thesis of social ontology regarding the social and cultural conditionality of behaviors is correct that society has the mandate to cultivate these conditions; but in normative terms society has no right to stop this behavior. This distinction is well known from the social-philosophical discussion between liberal and communitarian theories (cf. for example Forst 1994). In light of the fact that highly personal aspects of lifestyle are involved, a health prevention and health promotion policy is only possible in a liberal order in the forms of information, advice, encouragements, and in maintaining a healthy environment, and typically not in the form of legally binding laws and bans: Given enough information and free will, anyone who choses an unhealthy lifestyle must not be hindered from doing so.

On the other hand, the legal system also has to recognize that the attribution of individual responsibility can and must be made with consideration of empirical context. Here it will likely come to some sort of gradation of the kind in which policy interventions are made more possible the more it is suspected that risky health behaviors are based on social background factors, and do not represent the best interests of the individual.

14.5.2   The Relationship Between Prospective Responsibility and Retrospective Responsibility

The last mentioned point is also important to the relation between healthcare and public health policy. The retroactive allocation of responsibility for an existing illness – for instance in the form of co-payments – becomes more problematic the less we are able to plausibly base the corresponding health-adverse behavior on the will of individuals, and the more we have to trace it back to social factors. Whether and to what extent this is the case depends especially upon which efforts society has undertaken to strengthen the opportunity for taking over prospective responsibility for one's own health (for terminology see Marckmann and Gallwitz 2007). Given the known empirical relationships, it is not legitimate to push the costs of disease onto citizens who had no realistic opportunities to live healthier lives due to a poor education and poor living conditions.

But this can be applied equally as well the other way around: The critics of state prevention efforts who emphasize the principle of autonomy cannot then also obstruct the attachment of a greater weight to personal responsibility in the context of health care. Exactly those people who criticize the supposed "healthism" of public health policy from a civil rights perspective, and not from a liberal perspective, should consider that this approach might reinforce social differences.

Translated by Andrew Fassett

References

Alber, K., H. Kliemt, and E. Nagel. 2009. Selbstverantwortung als Kriterium kaum operationalisierbar [Personal responsibility as a barely operable criterion]. *Deutsches Ärzteblatt* 106: A1361–A1363.

Beske, F. 2002. Prävention: Vor Illusionen wird gewarnt [Prevention: Sounding warning against illusion]. *Deutsches Ärzteblatt* 99: A1209–A1210.

Bloche, M.G. 2005. Obesity and the struggle within ourselves. *The Georgetown Law Journal* 93: 1335–1359.

Federal Constitutional Court [Bundesverfassungsgericht]. 2005. Ruling of 6.12.2005. 1 BvR 347/98.

Federal Constitutional Court [Bundesverfassungsgericht]. 2007. Ruling of 29.11.2007. 1 BvR 2496/07.

Federal Constitutional Court [Bundesverfassungsgericht]. 2008. Judgement of 30.7.2008. 1 BvR 3262/07, 402/08 und 906/08.

Burris, S. 1997. The invisibility of public health. *American Journal of Public Health* 87: 1607–1610.

Buyx, A. 2005. Eigenverantwortung als Verteilungskriterium im Gesundheitswesen [Personal responsiblity as a criteria of distribution in healthcare]. *Ethik in der Medizin* 17: 269–283.

Callahan, D. (ed.). 2000. *Promoting healthy behavior: How much freedom? Whose responsibility?* Washington DC: Georgetown University Press.

Daniels, N. 1985. *Just health care*. Cambridge: Cambridge University Press.

Dworkin, R. 1981. What is equality? Part 2: Equality of resources. *Philosophy and Public Affairs* 10: 283–345.

Eichenhofer, E. 2003. Wahl des Lebensstils – Auswirkungen in der sozialen Sicherheit [Lifestyle choices – the impact on social security]. *Sozialgerichtsbarkeit* 50: 705–712.

Forst, R. 1994. Kontexte der Gerechtigkeit [Contexts of justice]. Frankfurt/M: Suhrkamp.

Frankfurt, H.G. 1971. Freedom of the will and the concept of a person. *Journal of Philosophy* 68: 5–20.

Frenzel, E.M. 2007. Die Volksgesundheit in der Grundrechtsdogmatik [Public health in the dogmatics of basic law]. *Öffentliche Verwaltung* 60: 243–248.

Gröschner, R. 2008. Vom Ersatzgesetzgeber zum Ersatzerzieher [From surrogate legislator to surrogate educator]. *Zeitschrift für Gesetzgebung* 23(4): 400–412.

Händeler, E. 2008. Die echte Gesundheitsreform [The real health reform]. *Die BKK* 96: 382–387.

Höfling, W. 2009. Recht auf Selbstbestimmung versus Pflicht zur Gesundheit [The right to self-determination versus the duty to health]. *Zeitschrift für Evidenz, Fortbildung und Qualität im Gesundheitswesen* 103: 286–292.

Huster, S. 2002. *Die ethische Neutralität des Staates* [The ethical neutrality of the state]. Tübingen: Mohr/Siebeck.

Huster, S. 2008. Gesundheitsgerechtigkeit: Public Health im Sozialstaat [Health justice: Public health in the welfare state]. *Juristenzeitung* 63: 859–867.

Huster, S., and H. Kliemt. 2009. Opportunitätskosten und Jurisprudenz [Opportunity costs and jurisprudence]. *Archiv für Rechts- und Sozialphilosophie* 95: 241–251.

Marckmann, G., and B. Gallwitz. 2007. Gesundheitliche Eigenverantwortung beim Typ-2-Diabetes [Personal responsibility for health in case of type-2 diabetes]. *Zeitschrift für Medizinische Ethik* 53: 103–116.

Mielck, A. 2000. *Soziale Ungleichheit und Gesundheit in Deutschland* [Social inequality and health in Germany]. Bern: Hans Huber

Muckel, S. 2009. *Sozialrecht* [Social law], 3 Aufl. München: Beck.

Richter, M., and K. Hurrelmann, (eds.). 2006. *Gesundheitliche Ungleichheit. Grundlagen, Probleme, Perspektiven* [Health inequalities. Principles, problems, and perspectives]. Wiesbaden: VS Verlag für Sozialwissenschaften.

Roemer, J. 1993. A pragmatic theory of responsibility for the egalitarian planners. *Philosophy and Public Affairs* 22: 146–166.

Rose, G. 1985. Sick individuals and sick populations. *International Journal of Epidemiology* 14: 32–38.

Wienke, A. 2009. Eigenverantwortung der Patienten/Kunden: Wohin führt der Rechtsgedanke des § 52 Abs. 2 SGB V? [Patient/clients personal responsibility: Where does the legal concept of § 52 paragraph 2 SGB V lead to?] In *Die Verbesserung des Menschen* [The Enhancement of Humans], ed. A. Wienke, W.H. Eberbach, H.-J. Kramer, K. Janke, 169–177. Berlin: Springer.

Wikler, D. 2005. Personal and social responsibility for health. In *Public health, ethics, and equity*, ed. S. Anand, F. Peter, and A. Sen, 109–134. Oxford: Oxford University Press.

Chapter 15
Can Social Costs Justify Public Health Paternalism?

Jessica Flanigan

Proponents of coercive public health interventions often justify seemingly paternalistic polices on the grounds that unhealthy citizens do not merely harm themselves with their unhealthy choices—they also harm their fellow citizens, who must then pay for medical services. For example, one justification for motorcycle helmet laws is that helmetless riders are more likely to be injured in ways that require expensive emergency services. Cigarette taxes are justified on the grounds that smokers will ultimately impose disproportionate burdens on the public health system relative to non-smokers, so smokers should bear some costs to offset those predictable burdens.

I argue that the provision of public health care cannot be used to justify coercive policies, including sin taxes.[1] That is, states cannot intervene to promote citizens' health on the grounds that unhealthy behaviors are burdensome to public healthcare providers and taxpayers. I consider two justifications for a public healthcare system to illustrate this point. Many political philosophers and medical ethicists claim that healthcare is a right and that people are entitled to access health care services even if their lifestyle choices have caused them to have very expensive healthcare needs.

---

[1] Some readers may find it controversial that I characterize taxes as coercive rather than as mere disincentives. Taxes are coercive on some accounts of coercion because they are backed by threats of force (Huemer 2012). Other accounts of coercion are moralized. They characterize coercion as a violation of one's entitlements. If people have entitlements not to be forced to pay penalties for certain things (such as eating fatty foods) then taxes are coercive on these accounts as well. I think that people are entitled to eat what they like without being forced to pay a penalty for particular food choices, and if I am right about that then my view that public health taxes are coercive is compatible with this account as well. In any case, I think that insofar as public health taxes on particular behaviors are effective it is because they penalize that behavior, and on most accounts of coercion taxes that express disapproval of behavior and sanction it are coercive. I am grateful to Thomas Schramme for prompting me to clarify this point.

J. Flanigan (✉)
Leadership Studies and Philosophy, Politics, Economics, and Law,
University of Richmond, Richmond, VA, USA
e-mail: jessica.flanigan@gmail.com; flanigan@richmond.edu

© Springer International Publishing Switzerland 2015
T. Schramme (ed.), *New Perspectives on Paternalism and Health Care*, Library
of Ethics and Applied Philosophy 35, DOI 10.1007/978-3-319-17960-5_15

I am sympathetic to this justification for health care, but if healthcare is an entitlement, then it is wrong to make access to that entitlement contingent on lifestyle choices or to demand that a person make certain lifestyle sacrifices so that she may exercise her right to access healthcare.

On the other hand, some may claim that states can permissibly demand that public healthcare recipients refrain from certain risky behaviors or pay a tax or penalty on the grounds that healthcare is actually a benefit that the state gives its citizens, even if no one has a claim to health care. On this interpretation of the justification for healthcare patients are not entitled to most medical services but rather medical services are beneficently provided by taxpayers and public officials. Like other elective benefits, states can therefore place conditions on their use. Yet if healthcare is simply a benefit that states provide their citizens, then public health paternalism is still not warranted because citizens retain the right to refuse the benefit and the lifestyle conditions that go along with it.

Therefore, whether citizens do or do not have rights to healthcare, paternalists cannot cite the public provision of healthcare as a justification for coercive policies. More generally, this argument casts doubt on justifications for paternalism that rely on indirect harms to others. I proceed as follows. I describe the argument that public costs can justify paternalism in Sect. 15.1. In the rest of the essay I will show that this particular argument for paternalism does not succeed. In Sect. 15.2, I consider whether a state's provision of healthcare can justify paternalism when healthcare is understood as a right. I draw on an analogy to Feinberg's famous 'cabin in the woods' case to show that if citizens are entitled to healthcare, then the public provision of healthcare cannot justify coercive paternalism. In Sect. 15.3 I consider whether public officials can enforce coercive policies on the grounds that public health paternalism is necessary for states to beneficently provide public goods that make all citizens better off, such as healthcare. Since healthcare is an excludable good (meaning that states are not required to provide it to everyone if they provide it to some), states cannot cite that coercive policies are necessary for the provision of a public benefit because citizens who do not wish to be subject to paternalistic policies can opt out of the benefit. Together, these arguments suggest that proponents of coercive paternalism cannot justify interference on the grounds that unhealthy choices burden the public healthcare system. I discuss the implications of this argument against coercive public health paternalism in Sect. 15.4.

15.1 The Social Costs of Unhealthy Choices

Proponents of coercive public health policies sometimes cite the fact that citizens' unhealthy choices will ultimately prove costly to taxpayers because they will burden public services (Finkelstein et al. 2008; Manning et al. 1989). Motivated by a desire to keep long-term health care costs as low as possible, public officials enforce policies like 'sin taxes' on unhealthy foods, smoking bans, and penalties for employers whose workers are overweight.

Call this the 'social cost' justification for seemingly paternalistic policies. If a choice affects others then it has externalities. Economists and ethicists generally agree that government intervention is often warranted to protect people from being forced to bear the negative externalities of their compatriots' poor choices. For example, if my neighbor owns a factory that pollutes my air, his factory has negative externalities and a third party like the state can permissibly intervene to protect me from his harmful behavior. Internalities refer to the costs that people impose on their future selves. Concerns about negative internalities have long been used to justify paternalistic policies like sin taxes and mandatory savings programs (Gruber and Köszegi 2001; O'Donoghue and Rabin 2003, 2006).

Still paternalism remains difficult to justify for a variety of reasons: it is *pro-tanto* disrespectful, it expresses an offensive attitude towards people, and paternalistic policy makers risk imposing values on citizens that citizens do not endorse or accept. For these reasons, advocates of public health paternalism have bolstered their case by re-framing the justification for seemingly paternalistic policies in terms of social costs. The idea is that overweight people and smokers make bad choices that taxpayers must then subsidize by paying for their health care. By pointing to the public health costs of unhealthy decisions, policy-makers recast a seemingly self-regarding choice as an unfair burden on taxpayers. In this way, seemingly paternalistic policies that would otherwise strike liberals as misguided, offensive, or morally objectionable are presented instead as a way of protecting the public from rising costs. What follows are examples of some cases where paternalistic policies have been re-framed and justified as ways of minimizing the social costs of poor choices.

15.1.1 Smoking and Alcohol

The most common forms of coercive public health paternalism are cigarette and alcohol taxes. European countries and the United States tax cigarettes and alcohol at higher rates than other consumer goods. Most smoking bans are limited to public places—only Bhutan has banned the sale of tobacco completely within its borders (Parameswaran 2012). Smoking bans that allow for smoking on one's private property are insulated from charges of paternalism because smoking in public plausibly harms others by diminishing the air quality and exposing people to the dangers of secondhand smoke. There is some evidence that cigarette taxes effectively reduce cigarette consumption and that higher taxes are associated with larger decreases in consumption (Meier and Licari 1997).

Part of the justification for cigarette and alcohol taxes is paternalistic—policy makers aim to reduce consumption for the sake of consumers' health.[2] Paternalistic

---

[2]The motivation behind cigarette and alcohol taxes is complicated by the fact that alcohol taxes may also effectively prevent drunk driving fatalities and cigarette taxes may reduce the public harm of second-hand smoke and smoking by pregnant women. Insofar as consumption taxes aim

bans are gaining in popularity—some European countries are following Bhutan and moving towards more comprehensive bans. Iceland (Pidd 2011), Australia (Sami 2014), the UK (Siddique 2014), Brazil (WHO Bulletin 2009), New Zealand (Thompson and Wade 2012), and Singapore (Berrick 2013) have all recently considered policies that aim to ban tobacco or some forms of tobacco completely. Insofar as concerns about medical costs of smoking motivate states to adopt these restrictions, it is unclear that smoking and alcohol bans can be justified on these grounds.[3] The health care costs of smokers are not higher than the health care costs of nonsmokers, largely because smokers are more likely to die earlier of smoking related ailments (Barendregt et al. 1997).

15.1.2 Unhealthy Food

Food bans have also been proposed as a way of promoting citizens' health by discouraging obesity. In 2011, Denmark passed a tariff on saturated fats but the tax was recalled in 2012 after it was deemed over-inclusive of foods that were not unhealthy, harmful to Danish businesses, and ineffective because many citizens simply bought fatty foods from neighboring countries (Strom 2012). In addition to the Danish taxes, policy makers in Spain, Romania, and Germany have proposed similar taxes on fatty foods (Outlawing Obesity 2010). In the UK, the European country with the most obese people on average, public health researchers recently proposed steep taxes on soda to curb the nation's obesity epidemic (Cheng 2012). In these cases, strain on the public healthcare system is cited as a reason for taxing food.

Politicians and public health experts in the United States also support targeted taxation as a way of preventing obesity. Thirty-three US states currently tax soda (Brownell et al. 2009). Like food taxes in Europe, the justification for soda taxes is to protect taxpayers from bearing the costs of soda consumption. For example, former US Senator Bill Frist argued in favor of expanding this policy by taxing sugar more generally, on the grounds that obesity costs US taxpayers $150 billion each year (Frist 2012). In 2012 the American Medical Association issued a policy statement that suggested higher taxes on sugary drinks as a way of cutting obesity rates and generating revenue to offset the public health costs of obesity and financing anti-obesity public health campaigns. This recommendation was based in part on an influential 2010 essay published in the New England Journal of Medicine, which argued that beverage taxes should be seen as routine as alcohol and cigarette taxes

---

to mitigate harms to others, they are not paternalistic and the following anti-paternalist arguments will not apply. For non-paternalist justifications for these taxes see (Grossman 1989; Ringel and Evans 2001).

[3] The American Lung Association, for example, states that "Smoking cost the United States over $193 billion in 2004, including $97 billion in lost productivity and $96 billion in direct health care expenditures, or an average of $4,260 per adult smoke" in support of its advocacy for further political efforts to discourage smoking.

because of their potential to partly fund public institutions while also conferring public health benefits (Brownell et al. 2009).

In addition to sin taxes and food bans, other anti-obesity policies have recently been passed in the name of the public's health. Japan currently offers universal health care to all citizens, and most Japanese citizens are covered either through a national health service or through their employers. In response to rapidly increasing health care costs Japan's Ministry of Health now requires companies to measure the waistlines of at least 80 % of their employees and to reduce the proportion of overweight employees in that sample over time (Onishi 2008). Companies that do not meet the Ministry's weight loss goals face up steep financial penalties and employees who remain overweight are required to attend counseling.

There is some evidence to support the claim that obesity has significant social costs, but it is not decisive. The director of the CDC's division of nutrition, physical activity, and obesity explicitly linked rising medical costs to the need for preventative public health policies (Graham 2012). Some economists estimate that obesity significantly contributes to rising health care costs (Baicker and Finkelstein 2011; Finkelstein et al. 2004, 2005, 2008). One estimate states that total annual medical expenditures in the United States would be 9 % lower were it not for health problems associated with obesity or overweight patients (Finkelstein et al. 2008). However, the empirical assumption that underlies the social cost justification for seemingly paternalistic policies may not be justified. Economists have pointed out that while obesity prevention may save lives, prevention does not necessarily cut healthcare costs because people who do *not* die of obesity-related diseases consume more health services in their extra years of life before dying of other costly age-related diseases later (van Baal et al. 2008).

15.1.3   Safety and Cost

Finally, other health policies like seatbelt laws and helmet laws are sometimes justified based on social costs. For example, the US Centers for Disease Control (CDC) cite the cost of crash-related injuries and deaths as a reason for primary enforcement seatbelt laws, which significantly increase rates of seatbelt use (Centers for Disease Control and Prevention 2011). Researchers support social cost justifications for seatbelt enforcement policies outside of the US as well (Esperato et al. 2012; Harris and Olukoga 2005). Policy makers also justify helmet laws on the grounds that motorcyclists who refuse to wear helmets are more costly.

Social cost justifications for these policies rely on an empirical assumption that unhealthy choices contribute to rising health care costs. Strictly considering healthcare expenses though, coercively preventing unhealthy behavior may be more costly on balance than permitting unhealthy behavior.[4] But whether the empirical assumption

---

[4] The authors of this study note that these finding should not necessarily undermine support for preventative health policies. They write, "the aim of prevention is to spare people from avoidable

behind the social cost justification succeeds or fails, the provision of public healthcare services remains rhetorically linked to public concern about unhealthy decisions. If only because social cost arguments have such intuitive appeal for policy makers and their constituents, they merit a closer look. In the next sections I will show that the argument does not hold up to further scrutiny whether the public provision of health care is understood as an entitlement or as a benefit. Either way, the fact that taxpayers pay for their fellow citizens' medical care cannot justify targeted restrictions on unhealthy citizens' conduct.

15.2 Paternalism and the Right to Healthcare

Can public officials justify coercive prohibitions of unhealthy choices on the grounds that the public will ultimately bear some of the costs associated with those choices? An argument in favor of coercive paternalistic public health policies can be stated like this. Citizens have a right to healthcare, meaning that other citizens have a duty to provide it in the form of a national health service. Coercion can rightly be used to compel citizens to meet their obligations to one another. Therefore, coercion can rightly be used to compel citizens to support public health services. Unhealthy citizens' voluntary behavior undermines the states' ability to provide health services, so states can coerce citizens to be more health-conscious for the sake of preserving affordable health services for everyone.

For the sake of argument, assume for now that citizens do have a right to health care, that others have a duty to provide it, and that some coercion can be used to compel citizens to do so.[5] So for example, if each citizen is entitled to some public services like healthcare then states can permissibly require that people pay taxes to provide those services. Interpretations of this duty may vary. Some countries like the United States universally guarantee emergency medical services but not routine care or non-emergency treatment. Most developed countries provide all citizens with other medical services as well, usually through a program like the UK's National Health Service or through subsidized private health insurance.[6]

---

misery and death not to save money on the healthcare system. In countries with low mortality, elimination of fatal diseases by successful prevention increases healthcare spending because of the medical expenses during added life years" (Bonneux et al. 1998).

[5] This is a consensus view, supported by several moral considerations. See for example (Buchanan 1984; Daniels 2013),

[6] Another assumption we will need to grant is that those who are subject to coercive paternalism act voluntarily, and therefore are especially liable to interference. This is plausible (though contested) for alcohol and cigarette addicts. When cigarette taxes increase consumption decreases, so it does seem that cigarette consumption is a voluntary choice at least in some cases. Similarly, the choice to ride without a seatbelt or a helmet is generally voluntary. Whether the choice to eat unhealthy foods is voluntary or not is contested as well, but largely because unhealthy foods are currently cheaper than healthy options in some places. Insofar as being obese or overweight is cast as a voluntary choice however (an assumption that underlies Japan's recent anti-obesity reforms) this

Where the argument fails is in the next step. The fact that taxes can permissibly be used to fund health services, which (we are assuming for now) citizens are obligated to provide to everyone, does not entail that any other kind of coercion is permissible for the sake of public healthcare. In particular, it doesn't follow that those who are more likely to exercise their right to healthcare should be preemptively penalized for doing so.

Consider an analogy to other rights. Assume that all citizens have rights to freedom of the press, and that the states' system of taxation funds protections for this right in the form of a legal system that protects journalist's rights to print news and to access information. Now imagine that the state, recognizing that some people produced and consumed more media than others, implemented a tax on all forms of news and required that journalists who were likely to use the legal system in exercising their rights preemptively pay a fee to offset the cost to the public.

Or, say that all citizens have rights to police protection from violent offenders. Some people may put themselves in a position where they need to exercise these rights more than others. People who move to dangerous neighborhoods or those who cultivate abrasive personalities may be disproportionately more likely to later require police services than people who choose to live in isolated rural houses and those who avoid conflict at all costs. Yet if citizens have a right to publically provided police protection against violent threats, it would be wrong to preemptively penalize residents of dangerous neighborhoods or contentious people in anticipation of their future use of police services.

Similarly, if citizens have rights to publically subsidized medical services then states should not preemptively penalize those who might exercise those rights in the future, even if their subsequent need for health services is a result of their own actions. To see why, imagine penalties were retroactive rather than preemptive. Intuitively, it would be unjustified to enforce a legal system where citizens who exercise their rights to particular public services and protections were penalized if their need for that service was a result of their own behavior. Under such a system, journalists who required the legal system to exercise their rights of freedom of the press would be required to pay for the disproportionate burdens they placed on the courts. Residents of dangerous areas who called emergency services would be required to pay a fine for each police visit. Helmetless riders would be asked to pay special fees when ambulances brought them to the emergency room, and smokers would be required to pay back-taxes in proportion to all the cigarettes they smoked when they present themselves at the hospital with lung cancer.

In these scenarios, if press freedom or police protection or emergency services or medical treatments are rights, then states cannot demand that some people pay a fee when they attempt to exercise those rights. Consider what such a fee would entail. If a person does not pay the fee for a public service then public officials could either turn her away or provide the service anyhow. If access to public services like medical treatment and police protection are positive rights and there is a duty that others

---

assumption fails because some people are obese for reasons that are unrelated to their dietary choices or exercise frequency.

provide them, then it is impermissible to turn someone away when they attempt to exercise their right to services. Even those who failed to pay the fee would nevertheless be entitled to access the service.

Yet the same reasoning applies to preemptive fees that force people whose actions make them likely to exercise their rights to public services to pay in advance for the increased costs they will impose on the public. Failing to pay a fee either at the time of service or preemptively in the form of a tax does not diminish one's entitlement to healthcare. Therefore, insofar as the intuitive rejection of a system retroactive fees for public services is justified, it is also justified to reject preemptive penalties. Coercive paternalism that aims to offset the costs of providing healthcare to unhealthy citizens amounts to taxing people in advance for subsequently exercising their right to healthcare, but if healthcare is a right, then unhealthy citizens do not make themselves liable to preemptive or retroactive penalties or coercive restrictions simply because they tried or will try to exercise that right. [7]

In other words, say that all citizens have a right to healthcare whether their health needs are a result of their unhealthy choices or a result of factors beyond their control. If so, then neither voluntarily nor involuntarily needy citizens are liable to be uniquely subject to coercion by the state for exercising that right. Another example. If a cyclist has an unconditional right to emergency services, public officials would fail to recognize the unconditional nature of his right if they made the provision of those services contingent on his behavior. Since a cyclist has a right to medical services whether he wears a helmet or not, states cannot require that he wear a helmet on the grounds that they may later be asked to provide him with medical services, which he has a right to even if he doesn't wear a helmet.[8]

Another way to illustrate this point is to think of healthcare as an entitlement; that is, as something that citizens are owed. If I owe you something, I am not therefore entitled to restrict your conduct or penalize you as a way of discouraging you from asking me to pay what I owe you in the future. If states owe citizens healthcare, that debt does not therefore entitle states to limit citizens' conduct on the grounds that some citizens are likely to use the services that they are entitled to use. This logic applies to political institutions more generally as well. We do not require that our compatriots provide us with a justification for receiving the benefits of citizenship, whether those benefits include healthcare, use of the legal system, or public safety.

One may reply that coercive paternalism, especially in the form of taxes, is no different from other taxes for the sake of public healthcare. Yet even if coercive paternalistic policies are called taxes, they are effectively penalties. The distinction between a tax and a penalty is contested and it likely varies by country, but in

---

[7] This arrangement is even harder to justify though because not all unhealthy people will ultimately consume health services, though all would be subject to unilateral penalties if unhealthy choices were taxed.

[8] This point was recently affirmed by the German Supreme Court, which found that a cyclist's failure to wear a helmet did not make her liable to pay an additional portion of her medical expenses (Reuters 2014). I am grateful to Thomas Schramme for informing me about this case.

general a tax qualifies as punitive if its purpose is to change or constrain certain behavior and if it expressed disapproval of that behavior. Sin taxes aim to shape behavior and they express disapproval, sometimes intentionally, so even if paternalistic fees on food and cigarettes are called taxes they are penalties as well. We should therefore ask whether those subject to paternalistic penalties are liable to be subject to coercive paternalism. I have argued here that if healthcare is a right, then people do not make themselves liable to preemptive penalties on the grounds that they will exercise their right to healthcare in the future.

15.3 Healthcare Benefits and Conditional Aid

One may reply to the foregoing argument by maintaining that rights to healthcare are not absolute, or that people are not entitled to access healthcare under any circumstances. For example, I might have rights of freedom of association but I would waive those rights if my actions caused my associates to take out restraining orders. People are entitled to migrate, but these rights can be limited to prevent fugitives from escaping punishment. Similarly, perhaps positive rights like healthcare are not absolute. People may be entitled to public services unless they behave in ways that makes the provision of those services so costly that they waive their rights. For example, one may suggest that students are entitled to an education unless they are willfully disruptive in classrooms, and if so then school officials can permissibly expel some students.

I am skeptical that rights to healthcare are conditional in this way, but for now, let us assume that they are. If so, then even if people have rights to access healthcare in general, they are not entitled to access healthcare if they fail to take due care to avoid overburdening the system. Yet this argument also does not justify paternalistic policies because even if people only have rights to access healthcare if they meet certain conditions, it doesn't' follow that people are required to meet those conditions so that they may exercise that right. Instead, someone might choose to waive her right to healthcare in exchange for the opportunity to smoke or eat fatty foods.

Another option at this point in the argument is to simply deny that healthcare is a right in the first place. If healthcare is not a right then people are not entitled to receive it and states can therefore place conditions on eligibility for public services. On this account, healthcare is a public good that is provided to citizens by the state, though citizens are not *entitled to* this good. Framing healthcare like this makes it more like public parks and other community beautification efforts—a benefit that all can enjoy but no one is owed.

If healthcare is a benefit and not an entitlement the case for coercive paternalism as a way of minimizing social costs is even weaker. Since healthcare is an excludable good, states cannot claim that coercive policies are necessary for the provision of a public benefit because citizens who do not wish to be subject to paternalistic policies may reasonably opt out of the benefit.[9] People do not waive their general rights against coercive interference when others' decide to provide them with a benefit without first securing the consent of beneficiaries.[10] In other words, if we understand healthcare as a benefit then it cannot license coercing citizens for the sake of a benefit they never requested. Citizens should have opportunities to either accept restriction on their conduct so that they receive benefits or to refuse the benefit.

One objection to the idea of excluding unhealthy citizens from healthcare benefits is that other members of society may find it objectionable to watch as people suffer and die because they made unhealthy choices. This objection suggests that a decisive reason to provide healthcare to unhealthy citizens is for their fellow citizens to be spared the unpleasantness of witnessing the effects of untreated illnesses. In other words, it frames the benefit of healthcare as a benefit to the community rather than as a benefit to the recipient. We should then ask whether members of the community are entitled to coercively restrict the conduct of unhealthy citizens so that those community members can be spared from watching their unhealthy compatriots suffer from treatable illnesses (Grill 2009). One reason to doubt that community members are entitled to coerce their fellow citizens in order to create a healthier community is that this view would seemingly also license restrictions on other behaviors that are generally scorned or deemed unsavory, which risks imposing a perfectionistic system of values on vulnerable minority groups.

To be clear, I am not advocating a policy of excluding unhealthy choosers from public healthcare services. I suspect that a policy of coercing or excluding unhealthy citizens from healthcare benefits would disproportionately disadvantage citizens who are already underprivileged and socially marginalized, and would thereby exacerbate existing injustices on balance. Rather, I am suggesting that *if* healthcare services are conceived of as benefits and not as entitlements then it would be permissible to exclude certain citizens from those services. That this proposal strikes many (including me) as objectionable either for humanitarian or social justice reasons indicates that it is a mistake to understand healthcare in this way, not that more people should be excluded from receiving healthcare.

Moreover, even if healthcare services are benefits and not entitlements, it doesn't follow that public officials *ought* to exclude unhealthy citizens from accessing those benefits, even if doing so was permissible. On this understanding of healthcare, it is still praiseworthy for taxpayers and public officials to provide health services to their fellow citizens even if citizens could be permissibly excluded from the health system. What this argument is intended to show is that public officials cannot permissibly coerce unhealthy citizens so that officials can more efficiently provide a benefit that they had no obligation to provide to those citizens in the first place.

---

[9] Even if states can permissibly use coercion to provide some public goods, if the goods are excludible and very costly then coercion for the sake of a public service is extremely difficult to justify. See for example (Klosko 1987).

[10] Robert Nozick illustrates this point with the example of free books. If a person never requested or consented to receive free books, the fact that he accepted the books when they were given does not justify coercively forcing him to pay for them (Nozick 1977).

15.4 Conclusion

Together, these arguments suggest that proponents of coercive paternalism cannot justify interference on the grounds that unhealthy choices burden the public health-care system. If all citizens have unconditional rights to healthcare then it is wrong to subject some citizens to targeted penalties so that they may exercise that right. If healthcare is a right, then those who seek to exercise that right should not be punished for doing so. On the other hand, if healthcare is not a right, or if it is a right that people only have under certain conditions, then unhealthy citizens should be permitted to decline to exercise their conditional right to healthcare or to refuse public healthcare benefits in exchange for the freedom to make unhealthy choices. The fact that other members of society may find it displeasing to watch unhealthy citizens bear the costs of their poor choices is not sufficient grounds for limiting the rights of unhealthy citizens.

If governments aim to limit citizens' ability to make self-harming choices they must therefore defend coercive paternalism on its own terms. I am skeptical that any attempts to defend coercive paternalism will succeed, but I have not provided a comprehensive argument against all forms of coercive paternalism in this essay. Rather, I have cast doubt on attempts to re-frame the justification for coercive paternalism in terms of sound fiscal policy, and I have questioned attempts to defend paternalism on the grounds that unhealthy citizens are costly to taxpayers.

References

Baicker, K., and A. Finkelstein. 2011. The effects of medicaid coverage — Learning from the Oregon experiment. *New England Journal of Medicine* 365: 683–685. doi:10.1056/NEJMp1108222.

Barendregt, J.J., L. Bonneux, and P.J. van der Maas. 1997. The health care costs of smoking. *New England Journal of Medicine* 337: 1052–1057.

Berrick, A.J. 2013. The tobacco-free generation proposal. *Tobacco Control* 22(Suppl 1): i22–i26. doi:10.1136/tobaccocontrol-2012-050865.

Bonneux, L., J.J. Barendregt, W.J. Nusselder, and P.J. der Maas. 1998. Preventing fatal diseases increases healthcare costs: Cause elimination life table approach. *BMJ [British Medical Journal]* 316: 26–29.

Brownell, K.D., T. Farley, W.C. Willett, B.M. Popkin, F.J. Chaloupka, J.W. Thompson, and D.S. Ludwig. 2009. The public health and economic benefits of taxing sugar-sweetened beverages. *New England Journal of Medicine* 361: 1599–1605.

Buchanan, A.E. 1984. The right to a decent minimum of health care. *Philosophy and Public Affairs* 13: 55–78.

Bulletin, W.H.O. 2009. Brazil and tobacco use: A hard nut to crack. *World Health Organization* 87: 805–884.

Centers for Disease Control and Prevention. 2011. *CDC – Seat belt facts – Motor vehicle safety – Injury center* [WWW Document]. http://www.cdc.gov/motorvehiclesafety/seatbelts/facts.html. Accessed 26 June 14.

Cheng, M. 2012. *U.K. experts: Tax soda to fight obesity epidemic* [WWW Document]. USATODAY. COM. http://www.usatoday.com/news/health/story/2012-05-15/UK-soda-tax/54979080/1. Accessed 26 June 14.

Daniels, N. 2013. Justice and access to health care. In *The Stanford Encyclopedia of philosophy*, ed. Edward N. Zalta. http://plato.stanford.edu/archives/spr2013/entries/justice-healthcareaccess/.

Esperato, A., D. Bishai, and A.A. Hyder. 2012. Projecting the health and economic impact of road safety initiatives: A case study of a multi-country project. *Traffic Injury Prevention* 13: 82–89. doi:10.1080/15389588.2011.647138.

Finkelstein, E.A., I.C. Fiebelkorn, and G. Wang. 2004. State-level estimates of annual medical expenditures attributable to obesity*. *Obesity Research* 12: 18–24.

Finkelstein, E., I.C. Fiebelkorn, and G. Wang. 2005. The costs of obesity among full-time employees. *American Journal of Health Promotion* 20: 45–51.

Finkelstein, E.A., J.G. Trogdon, D.S. Brown, B.T. Allaire, P.S. Dellea, and S.J. Kamal-Bahl. 2008. The lifetime medical cost burden of overweight and obesity: Implications for obesity prevention. *Obesity* 16: 1843–1848. doi:10.1038/oby.2008.290.

Frist, B. 2012. How to wean America from its dangerous food addiction. *The Week*, May 22, 2012. http://theweek.com/articles/475360/how-wean-america-from-dangerous-food-addiction.

Graham, J. 2012. *Even a small slowdown in obesity's rise would save big money* [WWW Document]. NPR.org. http://www.npr.org/blogs/health/2012/05/07/152184370/even-a-small-slowdown-in-obesitys-rise-would-save-big-money. Accessed 26 June 14.

Grill, K. 2009. Liberalism, altruism and group consent. *Public Health Ethics* 2: 146–157. doi:10.1093/phe/php014.

Grossman, M. 1989. Health benefits of increases in alcohol and cigarette taxes. *British Journal of Addiction* 84: 1193–1204. doi:10.1111/j.1360-0443.1989.tb00715.x.

Gruber, J., and B. Köszegi. 2001. Is addiction "rational"? Theory and evidence. *The Quarterly Journal of Economics* 116: 1261–1303.

Harris, G.T., and I.A. Olukoga. 2005. A cost benefit analysis of an enhanced seat belt enforcement program in South Africa. *Injury Prevention* 11: 102–105. doi:10.1136/ip.2004.007179.

Huemer, M. 2012. *The problem of political authority: An examination of the right to coerce and the duty to obey*. Basingstoke: Palgrave Macmillan.

Klosko, G. 1987. Presumptive benefit, fairness, and political obligation. *Philosophy and Public Affairs* 16: 241–259. doi:10.2307/2265266.

Manning, W.G., E.B. Keeler, J.P. Newhouse, E.M. Sloss, and J. Wasserman. 1989. The taxes of sin: Do smokers and drinkers pay their way? *JAMA Journal of the American Medical Association* 261: 1604–1609. doi:10.1001/jama.1989.03420110080028.

Meier, K.J., and M.J. Licari. 1997. The effect of cigarette taxes on cigarette consumption, 1955 through 1994. *American Journal of Public Health* 87: 1126–1130.

Nozick, R. 1977. *Anarchy, state, and utopia*. New York: Basic Books.

O'Donoghue, T., and M. Rabin. 2003. Studying optimal paternalism, illustrated by a model of sin taxes. *The American Economic Review* 93: 186–191.

O'Donoghue, T., and M. Rabin. 2006. Optimal sin taxes. *Journal of Public Economics* 90: 1825–1849.

Onishi, N. 2008. Japan, seeking trim waists, measures millions. *The New York Times*, June 13, 2008, http://www.nytimes.com/2008/06/13/world/asia/13fat.html?.

Outlawing Obesity: European Governments Seek to Mandate Healthier Diets. 2010. *Spiegel* Online. http://www.spiegel.de/international/europe/outlawing-obesity-european-governments-seek-to-mandate-healthier-diets-a-671334.html. Accessed 10 Jan 10.

Parameswaran, G. 2012. *Bhutan smokers huff and puff over tobacco ban* [WWW Document]. http://www.aljazeera.com/indepth/features/2012/09/201292095920757761.html. Accessed 26 June 14.

Pidd, H. 2011. What a drag … Iceland considers prescription-only cigarettes. *The Guardian*, July 4, 2011. http://www.theguardian.com/world/2011/jul/04/iceland-considers-prescription-only-cigarettes.

Reuters. 2014. German court clears helmetless cyclists from injury blame. *Reuters.*

Ringel, J.S., and W.N. Evans. 2001. Cigarette taxes and smoking during pregnancy. *American Journal of Public Health* 91: 1851–1856. doi:10.2105/AJPH.91.11.1851.

Sami, M. 2014. *AMA set to examine supporting a permanent ban on cigarette sales to anyone born after the year 2000* [WWW Document]. http://www.abc.net.au/pm/content/2014/s4033084. htm. Accessed 26 June 14.

Siddique, H. 2014. Doctors to vote on cigarette sale ban for those born after 2000. *The Guardian,* June 23, 2014. http://www.theguardian.com/society/2014/jun/23/doctors-vote-cigarette-sale-ban-children-born-2000.

Strom, S. 2012. "Fat tax" in Denmark is repealed after criticism. *The New York Times.*

Thompson, W., and A. Wade. 2012. *Smoking ban proposed for Auckland.* New Zealand Herald, January 19, 2012. http://www.nzherald.co.nz/nz/news/article.cfm?c_id=1&objectid=10779688.

Van Baal, P.H.M., J.J. Polder, G.A. de Wit, R.T. Hoogenveen, T.L. Feenstra, H.C. Boshuizen, P.M. Engelfriet, and W.B.F. Brouwer. 2008. Lifetime medical costs of obesity: Prevention no cure for increasing health expenditure. *PLoS Medicine* 5: e29. doi:10.1371/journal. pmed.0050029.

Chapter 16
Determinants of Food Choices as Justifications for Public Health Interventions

Lorenzo del Savio

16.1 Plan of the Paper

This paper is about how we should and how we cannot employ the psychological, neurophysiological and socio-epidemiological evidence regarding food choices in the discussion about paternalism in public health.[1] There are two ways of using this wealth of data to rebut anti-paternalism and they correspond to two very different versions of liberal anti-paternalism, which we could attribute, respectively, to John Stuart Mill (1859) and Joel Feinberg (1986): utilitarian anti-paternalism and liberty-based anti-paternalism. In this paper I argue that liberty-based anti-paternalism is not relevant in the case of public interferences against unhealthy lifestyles and that evidence on the determinants of food choices cannot be employed against liberty-based paternalism but can be employed against its utilitarian version. In brief, we may want to intervene in dietary choices not so much because food choices are beyond our control but because they are often *bad*. This argument will be developed at length in Sects. 16.4 and 16.5, which together form the theoretical core of the paper. While Sect. 16.4 collects arguments from the literature, Sect. 16.5 contains contributions that, as far as I know, are new for the debate. Section 16.2 explains why the topic is relevant with reference to the epidemics of nutrition-related diseases and the kind of policy responses that might be mounted by public authorities. Section 16.3 distinguishes the two versions of anti-paternalism and develops some conceptual tools that will be used afterward. Section 16.6 sums up the main results and anticipates further issues stemming from this approach.

---

[1] Previous contributions to the topic, from the alternative perspective: Cohen (2008), Skipper (2012).

L. del Savio (✉)
Istituto Europeo di Oncologia, Kiel University, Kiel, Germany
e-mail: lorenzo.delsavio@ieo.eu; lorenzo.delsavio@iem.uni-kiel.de

© Springer International Publishing Switzerland 2015   247
T. Schramme (ed.), *New Perspectives on Paternalism and Health Care*, Library of Ethics and Applied Philosophy 35, DOI 10.1007/978-3-319-17960-5_16

16.2 The Context: Lifestyle Diseases and Public Responses to the Epidemics

According to the World Health Organization, most people live in countries where overweight kills to a greater extent than undernourishment (Lozano et al. 2012). High income countries have been experiencing the effects of the epidemic of nutrition-related diseases for decades and middle income countries with expanding economies fall in the trap as soon as their population switches to urban lifestyles, with low physical activity and plenty of energy-dense and nutrient-poor foods (Popkin 1998). Overweight and its severe version, obesity, are causally linked with type II diabetes, cardiovascular diseases and some forms of cancer: these chronic diseases are leading causes of death and disability worldwide (WCRF 2009). Also, these conditions are socially stratified, especially in small children and women. For instance, at the regional level in Italy, poverty rates are predictors of prevalence of obesity and overweight in children.[2] Although unhealthy diets used to be regarded as an issue of personal appearance, lack of self-control and affluence, epidemiological evidence shows that they are a key health issue and they are associated with poverty within countries.

The incidence (i.e., new cases per unit of time) of obesity is growing and does not seem to have peaked as yet, with the notable exception of few European countries where it has remained stable over the last years (Swinburn et al. 2011). Although interventions that have been implemented so far may have slowed down the spread of the epidemic, the public health response is as yet unsatisfactory. Yet policies against this public health challenge are not mysterious: there is a growing literature on interventions that are effective against unhealthy food choices, some of which are supported by strong evidence (WCRF 2007). Policies might widen the set of choice available to consumers decreasing the cost associated with certain food items that are known to be healthy (e.g., vegetables). Also, health campaigns might persuade people to prefer certain healthier diet, as in the 5-A-DAY vegetables guideline. Furthermore, the cost associated with certain choices could be increased through direct taxes on certain unhealthy items, as in the case of "fat taxes" in Denmark or the "soda tax" in France. More radically, some food could be prohibited, as super-size portions or added trans-fats. Aside from these *proximal* policies, any modification of the food chain impacts on nutritional choices, from labeling and safety standards to the public subsidies of particular sectors of agricultural production (Popkin 2010).

In a thematic issue of *The Lancet*, Gortmaker (2011) have argued that the main obstacle for the implementation of a coherent food policy for health must be political rather than evidential. Vested interests may partially explain this political reluctance. However, the political reluctance might be due to the fact that public health measures interfere with personal choices and with features of living environments that people value. Preventive policies are experienced as unduly paternalistic, and

---

[2] Retrieved from: http://www.epicentro.iss.it/okkioallasalute/.

this may prevent decision makers from implementing appropriate measures. Ethical analysis on paternalism could provide tools for discussion to decision-makers and the public, thus addressing the political obstacles identified by Gortmaker and colleagues.

16.3 Two Versions of Anti-paternalism: Protecting Liberty and Promoting Welfare

As discussed at length in other contributions of this volume, paternalistic actions and policy are beneficent interference with an agent's choices without his consent and for the sake of his own good (Dworkin 2010). At least bans and food taxes count as paternalistic actions in this sense, though there are some complications with the application of Dworkin's definition of paternalism to public health policy. Public policies have several distinct justifications, some of which could be non-paternalistic.[3] Also, the issue of consent in the political case is not straightforward, since political procedures do produce some kind of consensual agreement on policies and regulations that are legitimately implemented and enacted. However, I will not discuss these issues here and assume that food policy could be or has been paternalistic at least in some cases in the Dworkinian sense: other contributions to this volume deal in detail with this problem.

There are two different concerns about paternalism, one is utilitarian and consequentialist, the other is deontological and pertains to the intrinsic value of personal liberty. They focus, respectively, on the two defining features of Dworkin's paternalism: the beneficent intent and its intrusive character. Utilitarian anti-paternalism is associated with John Stuart Mill (1859) and it is a widespread working assumption of economists modelling individual behaviors. Deontological anti-paternalism has been defended by libertarian authors in the Lockean tradition (e.g., Nozick 1974) and by Feinberg (1986). The utilitarian critique of paternalism is that paternalism is self-defeating as for its beneficent intent. The critique of paternalism based on liberty is that paternalism amounts to a violation of the personal sphere of liberty for unacceptable reasons. The crucial premise of utilitarian anti-paternalism is that people are mostly the best arbiters of their own interests – a descriptive claim.[4] The crucial premise of the liberty-based anti-paternalism is that interference is a bad as such and that only extremely serious reasons (e.g., harm to others) might override the prohibition of interference: a normative claim. Let us consider the two families of anti-paternalism in turn.

---

[3] E.g. recent soda taxes were admittedly a pure accounting measure in France, see Karanikolos et al. (2013).

[4] Utilitarians might also argue that making autonomous choices is an important dimension of well being, quite independently from whether autonomous choices result in good personal outcomes.

Liberty The concern with the protection of an inviolable sphere of personal liberty cuts across most liberal thought. In particular, the protection of the personal sphere *from the power of public authority* is a central tenet of liberalism and therefore involves the case of public health quite straightforwardly. Some versions of liberalism (i.e., libertarianism) limit the appropriate role of public authority to the mere protection of individual negative rights and deny that beneficence is a proper aim of *any* legitimate political action. Interference with liberty is allowed only to protect third parties' rights and thus the case of avoiding self-harm is not contemplated. The moral foundation of liberty-based anti-paternalism descends from a suspicious attitude toward public powers and from the centrality of rights and individual self-ownership. This is explicitly the foundation provided by Feinberg:

> The life that a person threatens by his own rashness is after all his life; *it belongs to him and to no one else. For that reason alone*, he must be the one to decide—for better or worse— what is to be done with it in that private realm where the interests of others are not directly involved (Feinberg 1986, pg. 59 – italics mine).

Utilitarianism There are several versions of utilitarian anti-paternalism, and they are all based on the idea that paternalism might harm its intended beneficiaries. In this sense, it is consequentialist and utilitarian: it assesses the appropriateness of a policy from the point of view of its consequences and the latter in term of human well-being (Sen 1979). The bads deriving from interferences might be due to unintended consequences, as in self-defeating cases of prohibitionism or stigmatization effects, or might be due to the fact that autonomy as such is an important piece of our well-being and its infringement makes people worse-off *as such*. Here, I deal with a more general worry, which is dependent on two *epistemic* theses.

On the one hand, third parties (e.g., public authorities) cannot make reliable judgments regarding interests, goods, well-being, etc. of their intended beneficiaries. I will call this: "ignorance of third parties". On the other hand, individuals make these judgments reliably and act consequentially most of the times. I will call this: "wisdom of individuals". Given this epistemic asymmetry between agents and their beneficiaries, non-intervention is the best policy for the pursuit of maximal well-being. In the case of self-harming activities and unhealthy choices, apparent self-harm might reflect a personal ranking of preferences that assigns a high value to hedonistic enjoyment.

Epistemic anti-paternalism is a very natural way of thinking if we abandon the idea that it is possible and/or uncontroversial to flesh out a unique theory of what make life satisfying and good. This skepticism is the ultimate motivation of the *ignorance of third parties* thesis, and could be taken as one possible interpretation of liberal antiperfectionism. Liberals uphold individual subjectivism, rejecting any suggestion that one "true self" – with its true needs and desires – might be hidden behind contingent personal preferences (Waldron 1987). The epistemic virtues of individual choices, in particular their sensitivity to *real* needs and wants of individuals, is instead a key theme of liberal economists and is well entrenched in contemporary economic thinking, including the models of food purchasing behaviors.

Notice that the key premise of the consequentialist welfarist critique is descriptive. The thesis of epistemic asymmetry contains two propositions, ignorance of third parties and wisdom of the individual, which describe what certain kinds of agents can and cannot do and know. This is important to keep in mind for section 4, where I will mobilize descriptive results in the sciences of choice to claim that the wisdom of individual thesis often fails in nutritional decisions and that the ignorance of third parties thesis cannot be always taken for granted.

16.4 Why Liberty-Based Anti-paternalism Is Unsuitable for Health Policy

In this section, I argue that the liberty-based argument against paternalism is not conceptually independent from welfarist concerns (1); it does not apply neatly to the case of public health policy (2); and it is suspicious in the case of food choices because of their peculiar socio-economic stratification (3). The relation of these three points with previous contributions is the following: (1) takes up Arneson's critique (2005) of Feinberg's liberty-based version of anti-paternalism; (2) strengthens Wilson's (Chap. 13) observation about the endemic nature of interference in public policy arguing that public interference is not only endemic but also unavoidable, at least in food policy; (3) translates to food policy the critique of anti-paternalism that has been put forward in the so-called "stewardship model" of public health (Nuffield Council of Bioethics 2007; Dawson and Verweij 2008). The conclusion of this section is the following: the real concern of anti-paternalists in public health policy against unhealthy nutrition (2, 3) and perhaps more generally (1) must be utilitarian rather than liberty-based.

16.4.1 Liberty-Based Anti-paternalism Explained

Arneson (2005) has proposed an argument against the "absolute and doctrinaire" anti-paternalism that accuses paternalists to "subordinate the right to the good" (i.e., liberty-based anti-paternalism) and in favor of the "broadly utilitarian liberalism of John Stuart Mill". His general claims will be the point of departure for the next section, dedicated to utilitarian anti-paternalism and its application to food choices: "the conclusion we should reach is not that the welfarist consequentialist should find paternalism a generally desirable policy, but that it can be morally acceptable and even required" (Arneson 2005, pg. 1).

Any liberty-based anti-paternalist must rely on the distinction between soft and hard paternalism to account for cases in which there is overt agreement that liberty-limiting non-consensual interference for avoiding self-harm is acceptable or even *required*. These are cases concerning, for instance, the relationship between parents and children and the care of persons with acute or chronic cognitive impairments. Since liberty-based anti-paternalists cannot say that welfare interests override the protection of liberty in these cases (that is by definition what liberty-based anti-paternalists cannot do) they must point out to the defective nature of autonomy to argue that it does not deserve protection in these cases. This is where the necessity of the distinction between hard and soft paternalism comes from: choices deserve protection when they are voluntary and sufficiently autonomous; otherwise liberty can be infringed upon for beneficent reasons. Why does the distinction between voluntary autonomous agency and "defective" agency bear such an important moral weight? Arneson observes that the reason must be, at least partially, utilitarian:

Our reverence for rational agency capacity is to a large degree reverence for the potential that rational agency capacity gives the bearer in most normal circumstances of human life. This is potential to develop one's individuality in particular ways, to make something worthwhile of one's life for oneself and others, to achieve any of an enormously wide range of great goods according to our choices and the luck of circumstances (Arneson 2005, pg. 14).

That is, even for the liberty-based anti-paternalist, the moral distinction between autonomous and non-autonomous choices must be informed by considerations based on interests and human goods.

This however suggests that liberty-based concerns are not independent from utilitarian considerations. Liberty-based anti-paternalists deny that the rights of x can be violated in the name of his good, yet they are forced to admit that interference is permissible and even required if x is cognitively impaired or if x's autonomy is otherwise curtailed. They would argue that the right that protects the sphere of personal sovereignty does not cover cases of deficient autonomy. Why? Arneson answers that the reason must be due – at least partially – to the instrumental value of autonomous choices in the pursuit of personal interests and goods. Moreover, though *bad* voluntary choices are neither a conceptual nor an empirical impossibility, autonomy is overall *reliable*: this explains why the distinction between hard and soft paternalism is morally relevant.

16.4.2   Interference Is Endemic and Unavoidable

Wilson (Chap. 13) has argued that interference with liberty without consent cannot be the moral problem *as such* of paternalist policy, since *by its nature* public authority acts non-consensually and coercively. Interference is *endemic* to public policy.

Given that non-paternalistic policies may be coercive and infringe liberty to exactly the same extent as paternalistic policies, it follows that those who want to defend the wrongness of justifying policies paternalistically need to show that there is something wrong about the paternalistically justified policy over and above its infringement of liberty (Wilson (Chap. 13), pg. 4).

Wilson's argument can be strengthened by the observation of libertarian paternalists that interference is unavoidable (Thaler and Sunstein 2008). Unavoidability and endemism are two different concepts. Perhaps it is always very problematic that public policy is endemically coercive and so endemism cannot count in favor of relaxation of the justification requirements for paternalistic actions. Yet public regulation is not only endemically coercive, it is unavoidably so: without the layers of restrictions, bans, taxes, etc. the very choices that the anti-paternalist wants to protect would not be possible. The picture of individual agents making unfettered choice *upon which* public authorities put restrictions is descriptively naïve. Food choices in urban settings are paradigmatic here. The price, availability and nature of the foods displayed in supermarkets are not simply the outset of a gigantic set of unfettered private exchanges between producers and consumers, but depend for its smooth functioning on restrictions and regulations enforced by public authorities that make the whole system viable, from agricultural policy to safety standards (Popkin 2010). Any change in the latter variables results in modification of prices, sometimes in the disappearance of certain products: why is this less worrisome than bans or taxes introduced for paternalist purposes? Again, it is not that since interference is widespread then the burden of justification becomes lighter. Rather, since interference is unavoidable, the anti-paternalist must explain why it is so worrisome in the case of paternalism. The benchmark of free-choice that is sought by deontological anti-paternalism is very elusive: this brings again the focus on utilitarian anti-paternalism and suggests instead doing away with its liberty-based version.

16.4.3 The Ideology of Good Deliberators

Lifestyles in general and dietary choices in particular are sharply stratified socially: poorer people tend to make poorer choices. This social dependency casts doubts on liberty-based anti-paternalism while letting untouched the utilitarian version. In brief, single-minded insistence on non-interference is at least suspicious because it favors, from the welfare point of view, good deliberators. As Arneson (2005) puts it:

Anti-paternalism, most especially hard anti-paternalism but definitely Feinberg's soft paternalist compromise variety (i.e., liberty-based anti-paternalism), looks to be an ideology of the good choosers, a doctrine that would operate to the advantage of the already better off at the expense of the worse off, the needy and vulnerable (pg. 12).

This will not convince as such the libertarian anti-paternalist for whom freedom is valuable whatever people make out of it: liberty-based anti-paternalists would point out that a right is not suspended if it does not get used for the best purposes. Although this is correct, if bad and good deliberators are not a random sample in the population but disproportionately represents, respectively, disadvantaged and advantaged people, non intervention might prohibit measures that could potentially address important social issues, including the remedy of some injustices. This is not as yet conclusive, and must be completed by a full-blown moral assessment of health inequalities that depends on lifestyles. Yet liberty-based anti-paternalism appears in light of social epidemiology an "ideology of the good deliberators" in the case of food choices.

It is very suggestive that the same critique does not apply to utilitarian anti-paternalism critique. Remember that paternalism might be self-defeating because it might damage the allegedly bad deliberators that it purports to protect. If bad deliberators are more common about disadvantaged people, interference is suspicious because it might damage these worst-off groups to a greater extent. The verdict is turned upside down: paternalism looks like an ideology of the "good deliberators" from the utilitarian point of view.

Powers, Faden and Shagai (2012) have argued that a proper understanding of the relevance of the Millian tradition in public health would focus much less on the harm principle and much more on the skeptic utilitarianism that characterizes Mill's broader contributions: liberty is honored by Mill because of the "essential role that the value of self-determination plays in human well-being". The next section is dedicated to that tradition and its relevance for public health policies against diet-associated diseases.

16.5   On the Real Moral Significance of the Determinants of Food Choices

Dietary choices are influenced by cognitive biases, poor environments and lack of information. Some liberty-based anti-paternalists would argue that interference is thus admissible because cognitive biases, poor environments and lack of information subtract to the autonomous character of choices. In reply to Resnik's (2010) observation that in the case of public health restrictions on diet "at stake is a freedom that most of us exercise every day but often take for granted: the freedom to choose what we eat", Boddington (2010) says that "it is quite right that we should not take this for granted: this is because many of us are not in fact straightforwardly free to choose our diets" and that environmental, social and cognitive factors are "curtailing our autonomy" (pg. 43). Skipper (2012), endorsing libertarian paternalist strategies to tackle the epidemic of obesity, has argued that "anthropologists, evolutionary biologists, neuroscientists, psychologists, sociologists and others have provided considerable empirical evidence that our capacity to choose and maintain a healthy diet is severely limited" (pg. 182). There are two problems with these arguments: while they rely on important empirical findings, (1) they address the wrong kind of anti-paternalism, as I explained in the previous section, and (2) they depend on a conception of what it takes to make autonomous choices, which is unduly demanding Here I will systematically analyze the literature on food choices and explain how it can be employed in discussing anti-paternalism in its utilitarian version. Determinants of dietary choices are morally relevant because they often cause decisions to be bad, i.e., poor means to pursue personal aims.

Behavioral psychologists and empirical economists have provided extensive evidence to the effect that human behavior is poorly modeled by the assumptions of rationality of traditional economic models: this is the point of departure of libertarian paternalists (Thaler and Sunstein 2003, 2008). The idea of rationality is very simple. Agents are endowed with a set of coherent and complete preferences over certain states, i.e., they assign to each state an ordinal utility. This is a comparative judgment about how much well-being they expect to gain from each state. Completeness requires that agents are always able to tell if a state A is superior, inferior, or equal to the state A in term of expected utility. Coherence requires transitivity, e.g., if A is superior (equal) to B, and B superior (equal) to C, then A is superior (equal) to C. Also, agents are endowed with the relevant information that permits to connect decisions and states, e.g., "if you do g, then A will ensue with probability p": i.e., they have a reliable picture of the causal structure of the world. Given these assumptions, behavior is modeled as if it was a systematic effort to maximize expected utilities.

In the case of food choices as in other fields, a more realistic depiction of the cognitive machinery underlying decision-making highlights its potential dependency on factors that are rationally irrelevant. For instance, if an agent makes two different choices, let us say on the evaluation of a paper, depending on whether his blood sugar level is above or under a certain threshold, this must be rationally irrelevant (Kahneman 2011). The information that we gain from this kind of evidence is morally important because it highlights cases where the autonomy of the agent is not instrumentally effective in the pursuit of his goals: agents *fail* to behave rationally. The worries of epistemic anti-paternalists should be smoothened by this kind of evidence if it was available for food choices.

In the rest of this section, I will review several pieces of evidence regarding the determination of food choices. In each case, I will argue that the liberty-based interpretation is implausible, assess the utilitarian verdict and defend its normative assumptions when it entails that choices are bad.

The obesogenic environment In the last decade, there has been an extensive attention to the notion of "obesogenic environment" (White 2007). The details of urban planning are thought to influence food choices and patterns of physical activity: absence of cycle lanes and pedestrian areas, high prevalence of car use and the presence of retailers of cheap high-calories low-nutrients food are recognized risk-factors for obesity. Here we are concerned only with the latter: the geography of retailers. Notice that the geography of retailers is not as such an irrelevant factor for rational agency: driving for 1 hour to the next fresh vegetable markets because only fast food is available at the back door can substantially decrease welfare. In this case, if there is a problem with the geography of food shops, it is not one of *unwise personal behavior* that might be addressed with paternalist restrictions. Indeed, restrictions could even damage the purported beneficiaries. The real issue concerns the geography of retailers as such, not the resulting individual decisions about food. In the context of utilitarian anti-paternalism, the relevant question about the obesogenic environment is whether the geography of food retailers can be considered rationally irrelevant, perhaps because it underlies a gross "framing effect" not dissimilar to the case of the display of food in canteens discussed by Thaler and Sunstein (2008), where people choose what to eat on the basis of what they see first.

Notice however that in food canteens switching from one to the other menu (e.g., from vegetables to fatty foods) is costless, so the framing-dependent choice cannot have any rational explanation. In large scale contexts as those pertaining urban geography it is hard to imagine fully costless alternatives. Scarcity of healthy food (i.e., "food deserts") is a rationally relevant factor of choice because transportation is expensive. Abundance of fast food is a rationally relevant factor because it is less time-consuming to get there rather than looking for the sparse vegetarian restaurants. And so on. Choices are free and *prima facie* optimal: neither versions of anti-paternalism can be criticized on the basis of the literature regarding the obesogenic environment.

Sub-optimal opportunity set Anand and Gray (2009) have constructed a more complex model to explain why the latter verdict about the obesogenic environment might not be the end of the story. They start observing that the relative abundance of ready-to-eat unhealthy foods might indeed be optimal from the point of view of welfare for the reasons depicted above, i.e., time sparing and cost. Then they argue that the evolution of the "opportunity set" guided by demand, i.e., the evolution of market offer of food, can be nonetheless sub-optimal because individuals have second order preferences about opportunity sets on the top of punctual preferences about consumption. For instance, I might be delighted in discovering that I can eat a sandwich at the bar because I have a meeting in 10 min and yet regret the fact that sandwiches are always available there because of my gluttony that I would like to keep at bay. In this case, the environment (i.e., the opportunity set) might indeed be a factor that renders unwise the choice of food: the dynamics of opportunity set dependent on first order preferences frustrates second-order interests and intervention is justifiable if second order preferences take priority.

The next four cases of determinants of diets are taken from Skipper (2012), who discusses the significance of these factors within a liberty-based framework. That choices are constrained and determined by these factors cannot be morally interesting since the alternative is a metaphysically oddity. In a deterministic world, external influences and determination *as such* cannot curtail the autonomous nature of decisions: otherwise there would not be *any* autonomous choice.[5] There are of course cases in which determination excludes autonomy, e.g., straightforward addiction, but I will point out why this is implausible in the examples discussed below.

Neural correlates of feeding behavior Skipper (2012) reports the behavioral and neurological study by Page et al. (2011) that shows impaired inhibition of desire for food after lunch in obese people. The level of blood sugar regulates appetite through

---

[5] This metaphysical position is called "compatibilism", i.e., the thesis that we can make sense of autonomy, personal responsibility and liberty in a deterministic world. A good theory of these concepts will tell apart when these concepts do not apply to a certain action because of specific liberty-limiting determination. I do not need such theory here because I will explain informally why determinants of food-choices cannot be considered liberty-limiting in *any* plausible theory of these concepts.

a stress response mechanism that is repressed when sugar level is normal and it is instead triggered by hypoglycemia. Page and colleagues manipulated sugar level and evaluated appetite in obese and normal subjects, screening their pattern of neuronal activation. They discovered that the repression of appetite seems to depend on the pre-frontal cortex, which is associated with the repression of impulses. The key result of their study is the observation of poor activation of the pre-frontal cortex in obese subjects with normal blood sugar levels and their increased appetite, as evaluated by personal rating, in comparison with healthy subjects. It is not surprising that sugar level influences food choices with the mediation of appetite response, yet it is noteworthy that the mechanism is impaired in obese subjects. This cannot be morally relevant because it shows *determination*: in this respect healthy and obese subject do not differ at all. Instead, the crucial fact is that the impairment of the pathway that regulates appetite is an irrelevant factor for the decision on whether or not consuming fat and sugary foods. Moreover, it is associated to obesity, a condition that would require even higher restraints on diet: as a consequence, food choices of obese subjects are very likely to be *bad*, indeed very bad for their health.

Mindless eating Wansink and Cheney's (2005) have studied environmental framing effects on the evaluation of portion size and consumption. In a series of experiments, they have shown that visual clues such as the tidiness of dining tables and the size of serving plates influence the quantity of ingested food. Tidy tables and bigger serving plates determine significant increases of consumption, especially among men. Skipper (2012) argues that "what is important here is not so much the explanation of how the cues influenced the subjects, but rather that the cues had influence at all [...] managing our consumption of food is not firmly in our control" (pg. 184). Although in this case it is indeed surprising that visual cues modify the amount of food-intake, the problem again is that "we are not *good* at judging portion sizes or tracking our consumption in distracting environments" (*ibidem*) rather than an issue of poor *control*. What would count for *full* control? Perhaps ignoring visual clues will improve control because the size of the serving plate and the tidiness of the tables are obviously irrelevant as for whether the next piece of sandwich will be good health-wise or otherwise, but again this is a case where "our reverence" for control is instrumental and related with the role of autonomy in personal well-being.

Obesity is contagious Christakis and Fowler (2007) have shown that being related biologically (e.g., siblings) or socially (e.g., friendship) with obese people increases the chance of obesity. These correlations are difficult to interpret since they can be due to factors other than influence at the social level, e.g., genetics. In case of social groups, the effects can go from obesity to social ties as well, since people might select for peers with similar Body Mass Index. However, even if causation runs from social ties with obese people to obesity, we ought to discard the hypothesis the food choice are bad *simply* as a result of this contagion. Perhaps there are habits regarding foods that are specific to certain groups and some individuals adapt their behaviors to gain the advantages of membership. This would not be irrational if indeed adaptation is a condition of membership and advantages of membership are big enough.

Evidence that would suggest irrationality must consist in showing that living in a group of obese people renders obesity unwisely *acceptable*, i.e., by masking the information about the adverse effects of unhealthy diets. This however would count as lack of information rather than poor control, an issue that I discuss below. In this case, there is neither poor control nor irrationality, at least at the first glance.

The evolution of obesity Skipper (2012) presents the theory of mismatch between evolutionary-wired eating habits and the food plentitude of some modern societies.[6] The mismatch theory of obesity maintains the "pleistocenic" mind of hunter-gatherers is not able to limit properly food-intake in a sedentary condition where food is readily available and physical exercise scarce. In this case, even Skipper must present his argument in a utilitarian form. Given the evolutionary mismatch, food choices are very likely to be *bad*. It would be indeed implausible to argue otherwise: any capability of decision-making ultimately relies on cognitive resources that have been molded by our evolution, but that cannot detract from their autonomous nature. The mismatch theory does not show that human beings are "simply without the sort of decision-making autonomy that advocates of paternalism are accused of violating" (pg. 182). They are simply very bad choosers since human techno-sociological developments have outstripped the process of adaptation.

Contradictory inter-temporal discounts Skipper's list of factors that explain the quality of food choices is incomplete. Experimental economists have elaborated models of decision-making in case of procrastination and poor investments for the future that might be applied to the case of food (Pampel et al. 2010; O'Donoghue and Rabin 1999, 2003; Sassi 2010).

In general, procrastination might be modeled as a clash between short-term rewards (costs) and long term costs (benefits). Individuals are often faced with the option of foregoing an immediate benefit for the sake of a future reward or with the option of gathering a present advantage by risking future losses.

An important feature of these *inter-temporal* choices is that it is generally rational to discount future benefits and losses, and the more they are distant in time, the more it is rational to discount them. The idea is simply that the chance of gathering (paying) future benefits (losses) grows smaller and smaller as benefits are more distant in time. It is however experimentally proven that several people do not discount future utilities in this way. Rather, they discount future benefits more heavily the closer they are in time. In other words, they prefer to wait for 1 day in 1 month than in 1 week. This present-bias accounts for procrastination.

In the literature, hyperbolic discounting is considered irrational because it is dynamically inconsistent. Take the decision of indulging in wine-drinking instead of jogging at time T. Now consider two earlier time points, T1 and T2, respectively further and closer in time with T. In exponential discounting, it does not matter whether I am at T1 or at T2: I will always discount the benefits of jogging with a fixed rate. Instead, in hyperbolic models, I will discount at T2 with a rate r greater

---

[6] For the introduction of the concept of "mismatch": Pani (2000).

of the rate q at which I would discount at T1. In general, the closer to T, the more I would discount. These differences might give inconsistent instructions as for how I should maximize my well-being. Perhaps from the perspective of T1 I should go jogging whilst from the perspective of T2 I should drink wine, hence the inconsistency. The instruction that is held at the point of decision will win, but that will be wrong from the point of view of an earlier (and future) self. Again, this is a reason to drop the anti-paternalist utilitarian worries and favor one solution over the other. In particular, there is at least one reason to favor the verdicts of the selves that are farther away in time: their verdict is temporally *prevalent*. That is, more often than not, and indeed always if not in the time gap immediately surrounding the decision, the alternative that is actually chosen is considered inferior by the agent.

Poor information Depending on risk aversion, two individuals might make opposite choices when balancing benefits and risks of unhealthy diets: this is unproblematic. Yet they might make opposite choices even because they lack relevant information, e.g., they underestimate risks.[7] In the case of lack of knowledge, individuals might make bad choices simply because they cannot connect appropriately their actions with the expected outcomes. These cases suggest intervening by means of informational campaign, yet if these were to fail or to be too expensive, policymakers should be unimpressed by utilitarian anti-paternalism because the wisdom of individuals thesis must be rejected any time *relevant* causal information is missing.

Bad preferences Sen (1999) famously discussed cases where preferences and/or their ranking are sub-optimal and cannot represent personal interests. This might happen when a person has adapted to dire circumstances and thus is content with very poor options. This shows the need of a rough list of bads in any theory of welfare, beyond antiperfectionist doubts. However, this is not going to be decisive in cases of food choices. Assume that chronic diseases belong to the list of uncontroversial bads. Still the link between food choices and the outset of chronic disease is too tenuous to justify intervention on these bases as instead Gostin (2004) argues. Although the population-wide effect of unhealthy diets is vast and although the personal risk is substantially augmented by unhealthy lifestyles, the individual risk due to lifestyle is too low to be incontrovertibly regarded as irrational *as such*.

Socio-economic determination Purchasing behaviors are dependent on price: people look at their budget when deliberating about their food choices. This is an important mechanism that underlies socio-economic disparities in healthy behaviors (Pampel et al. 2010). It would be wrong however to argue that choices are *bad* as a consequence of this kind of determination. Selecting cheaper foods when personal budget is meager is scarcely irrational, at least if the empirical law of diminishing returns of income on welfare is true.

---

[7] For evidence regarding knowledge of risk of cancer related with food see for instance Robb et al. (2009).

For people with smaller budgets, sparing on food is rational because alternative allocations of money are in general more useful than in the case of richer people. I hasten to notice that this does not show that there is no problem with economic determination of food behaviors. Perhaps the underlying distribution of incomes and wealth is independently bad, maybe because (among other things) it does not allow some individuals to purchase healthy items. Yet it would be wrong to argue that the moral problem of economic determination is that choices are constrained or even rationally bad: they are neither.

More generally, empirical findings regarding food choices are important for the moral evaluation of public health policy because they give reasons to believe that people could make bad choices in terms of their overall welfare when making nutritional choices. In the literature, this evidence has been used instead to show that people are not in control of their actions: I have argued that this is philosophically misleading since it is based on unacceptable conceptions of autonomy. Also, I have argued above that the version of anti-paternalism based on liberty that they target is not convincing in the case of lifestyle prevention campaigns. Nonetheless, moral philosophers evaluating preventive health policy should not ignore these empirical findings: they single out cases in which, by letting people acting as they prefer, we may let them frustrate their interests.

16.6   Conclusions

In this paper, I distinguished liberty-based and utilitarian versions of anti-paternalism; I explained why liberty-based anti-paternalism is unconvincing in the case of health policy; and I argued that the science of food choices shows that diets are likely to be bad, thus rebutting utilitarian anti-paternalism. The last result depends on a systematic analysis of the determinants of food choices that I carried out in Sect. 16.5. In general, literature on determination does not show that liberty is curtailed, but it gives reasons to believe that most people are bad nutritional deliberators. The result is a moderate defense of paternalist interventions: the utilitarian approach might require intervention when choices are clearly bad. Also, I argued that decision-makers will, in one way or another, influence personal choices about nutrition, thus they ought to strive to avoid failures of food choices.

I conclude with a *proviso*. While population-wide restrictive measures can be sometimes recommendable *all things considered*, the main problem of these policies from the point of view of the utilitarian approach defended in this paper is the heterogeneity of populations. Lack of information and irrationality are not universal features of human beings, thus restrictions can damage rational and well-informed agents. Heterogeneity suggests caution, involvement of target populations in decision-making and the development of further arguments for intervention based on the social goods attached to healthier personal nutrition. In the end, whether better personal nutrition is better for everyone will be crucial to justifying liberty-limiting public health measures.

References

Ananad, P., and A. Gray. 2009. Obesity as market failure: Could a "deliberative economy overcome the problems of paternalism?". *KYKLOS* 62(2): 182–190.

Arneson, R.J. 2005. Joel Feinberg and the justification of hard paternalism. *Legal Theory* 11: 259–284.

Boddington, P. 2010. Dietary choices, health, and freedom: Hidden fats, hidden choices, hidden constraints. *The American Journal of Bioethics* 10(3): 43–44.

Christakis, N., and J. Fowler. 2007. The spread of obesity in a large social network over 32 years. *The New England Journal of Medicine* 357: 370–379.

Cohen, D.A. 2008. Neurophysiological pathways to obesity: Below awareness and beyond individual control. *Diabetes* 57: 1768–1773.

Dawson, A., and M. Verweij. 2008. The Steward of the Millian State. *Public Health Ethics* 1(3): 193–195.

Dworkin, G. 2010. Paternalism. In *The Stanford encyclopedia of philosophy* (Summer 2010 Edition), ed. E.N. Zalta. Available from: http://plato.stanford.edu/archives/sum2010/entries/paternalism/. Accessed 1 Mar 2012.

Feinberg, J. 1986. *The moral limits of the criminal law: Volume 2. Harm to self*. Oxford: Oxford University Press.

Gortmaker, S. 2011. Changing the future of obesity: Science, policy, and action. *Lancet* 378: 838–847.

Gostin, L. 2004. Health of the people: The highest law? *Journal of Law, Medicine and Ethics* 32(3): 509–515.

Kahneman, D. 2011. *Thinking, fast and slow*. New York: Farrar, Straus, and Giroux.

Karanikolos, M., P. Mladovsky, J. Cylus, et al. 2013. Health in Europe: Financial crisis, austerity, and health in Europe. *The Lancet* 6736(13): 1–9.

Lozano, R., et al. 2012. Global and regional mortality from 235 causes of death for 20 age groups in 1990 and 2010: A systematic analysis for the Global Burden of Disease study. *The Lancet* 390(9859): 2095–2128.

Mill, J.S. 1859. On liberty. In *Collected works*, vol. 5, ed. J.M. Robson. Toronto and Buffalo: University of Toronto Press.

Nozick, R. 1974. *Anarchy, State, and Utopia*. New York: Basic Books.

Nuffield Council of Bioethics. 2007. *Public health: Ethical issues*. London: Nuffield Council for Bioethics.

O'Donoghue, T., and M. Rabin. 1999. Doing it now or later. *American Economic Review* 89: 103–124.

O'Donoghue, T., and M. Rabin. 2003. Studying optimal paternalism, illustrated by a model of sin taxes. *American Economic Review* 93: 186–191.

Page, K., D. Seo, R. Belfort-DeAguiar, C. Lacadie, J. Dzuria, S. Naik, S. Amamath, R. Constable, R. Sherwin, and R. Sinha. 2011. Circulating glucose levels modulate neural control of desire for high-calorie foods in humans. *The Journal of Clinical Investigation* 121: 4161–4169.

Pampel, F.C., P.M. Krueger, and J.T. Denney. 2010. Socioeconomic disparities in health behaviors. *Annual Review of Sociology* 36: 349–370.

Pani, L. 2000. Is there an evolutionary mismatch between the normal physiology of the human dopaminergic system and current environmental conditions in industrialized countries? *Molecular Psychiatry* 5(5): 467–475.

Popkin, B. 1998. The nutrition transitions and its health implications in lower-income countries. *Public Health Nutrition* 1: 5–21.

Popkin, B. 2010. Agricultural policies, food and public health. *EMBO Reports* 12(1): 11–18.

Powers, M., R. Faden, and Y. Saghai. 2012. Liberty, mill and the framework of public health ethics. *Public Health Ethics* 5(1): 6–15.

Resnik, D. 2010. Trans fat bans and human freedom. *The American Journal of Bioethics* 10(3): 27–32.

Robb, K., et al. 2009. Public awareness of cancer in Britain: A population based survey of adults. *British Journal of Cancer* 101: s18–s23.

Sassi, F. 2010. *Obesity and the economic of prevention*. Cheltenham: Edward Elgar Publishing in association with OECD.

Sen, A. 1979. Utilitarianism and welfarism. *Journal of Philosophy* 76(9): 463–489.

Sen, A. 1999. *Freedom as liberty*. Oxford: Oxford University Press.

Skipper, R.A. 2012. Obesity: Towards a system of libertarian paternalistic public health interventions. *Public Health Ethics* 5(2): 181–191.

Swinburn, B., et al. 2011. The global obesity pandemic: Shaped by global drivers and local environments. *Lancet* 378: 804–814.

Thaler, R., and C. Sunstein. 2003. Libertarian paternalism. *The American Economic Review* 93: 175–179.

Thaler, R., and P. Sunstein. 2008. *Nudge: Improving decisions about health, wealth, and happiness*. New Haven: Yale University Press.

Waldron, J. 1987. Theoretical foundations of liberalism. *The Philosophical Quarterly* 37(147): 127–150.

Wansink, B., and M. Cheney. 2005. Super bowls: Serving bowl size and food consumption. *JAMA* 294: 1727–1728.

White, M. 2007. Food access and obesity. *Obesity Reviews* 8(Suppl 1): 99–107.

World Cancer Research Fund, American Institute for Cancer Research. 2007. *Food, nutrition, physical activity, and the prevention of cancer: A global perspective*. Washington DC: AICR.

World Cancer Research Fund, American Institute for Cancer Research. 2009. *Policy and action for cancer prevention*. Washington DC: AICR.

Part V
Paternalism and Reproductive Medicine

Chapter 17
Selecting Embryos with Disabilities? A Different Approach to Defend a "Soft" Paternalism in Reproductive Medicine

Diana Aurenque

17.1 Introduction

Thirteen years ago, a deaf lesbian couple made for an intense bioethical discussion in the US (Spriggs 2002). Sharon Duchesneau and Candy McCullough used a friend's sperm donation very well knowing that deafness had been passed down in his family for five generations. This indicates that they wanted to have a deaf child. They claimed that deafness is not a disability but a cultural identity that requires a special form of communication. This case prompted yet another debate concerning both the boundaries of reproductive autonomy and the normative power of parental desires within the realm of reproductive decisions (Savulescu 2002, p. 771).

Since the 1990s, parallel to the development of the Human Genome Project and preimplantation genetic diagnosis (PGD) in the US, ethicists and philosophers have been emphasizing the ethical necessity to limit the reproductive freedom of future parents. At that time, they were concerned with the extent to which genetic counseling, in which parents receive important genetic information about their future offspring, should respect the maxims of nondirectiveness and value neutrality. They were concerned that parents who had knowledge of this genetic information may pursue eugenic goals, or even worse, use this information to harm the child (Green 1997, p. 141). On the one hand, these concerns were prompted by the hypothetical case discussed in scientific literature of a deaf couple who also wanted to have a deaf child and on the other hand by a real case of a short-statured couple who also wanted to have a child of short stature (Green 1997). At the focus of the discussion was the question as to whether physicians are obliged to give future parents genetic information about their offspring that has no clinical benefit and could possibly be abused by them. Many were of the opinion that they are allowed to exercise their

D. Aurenque (✉)
Department of Philosophy, Universidad de Santiago de Chile (USACH), Santiago, Chile
e-mail: diana.aurenque@usach.cl

© Springer International Publishing Switzerland 2015   265
T. Schramme (ed.), *New Perspectives on Paternalism and Health Care*, Library
of Ethics and Applied Philosophy 35, DOI 10.1007/978-3-319-17960-5_17

reproductive autonomy only as long as they do not harm their offspring as a result of their decisions.

In actuality, the discussion about limiting parental autonomy with regard to PGD has become the focus yet again. This was prompted by the results of a survey of clinics from 2008 that offer IVF in the US. The survey showed that about 3 % of PGD is being used "to select an embryo for the presence of a disability" (Baruch et al. 2008). Deaf and short-statured parents seem to be utilizing PGD for exactly this purpose. The precise number of PGD that is being used for this purpose is still unknown due to the lack of data on this matter (in the US PGD is not yet reported to a central data repository). However, one thing is certain: PGD not only allows prospective parents to have a child without a severe disability or a deadly genetic disease, but it has also been used by parents who desire that their offspring have their own disabilities. Many ethicists are against such a use of PGD. They state that although procreative autonomy should be respected it is also necessary to put some limits to it. This paper is a call for a "soft" paternalism inasmuch as this concept involves actions taken for the benefit of an individual who is known to be incompetent (the embryo) (Dworkin 2010).

Regarding the question whether the use of PGD for the purpose of selecting for a specific disability is ethically acceptable we find at least two possible answers: A way to defend this use of PGD is to make reference to the reproductive autonomy and reproductive rights of parents. According to this, parents are free to weigh their moral commitments to their own account in selecting for or against an embryo with disabilities. Against this case it has been argued that such use of PGD would be unethical because parents are not allowed to intentionally harm their future child. By selecting for a disability parents disrespect both the welfare of the child and the principle "primum nil nocere". Considering these two alternatives we are facing a typical situation in which autonomy and paternalism appear as two fundamentally opposite principles in medical ethics. But: Do we necessarily have to disrespect autonomy in order to defend paternalistic attitudes? In this paper I would like to argue that paternalism is sometimes appropriate to promote autonomy. However this is only possible because autonomy and paternalism are not necessarily mutually excluding principles. Contrary to common opinion on this issue, I will show that paternalism can be understood as more than the opposite principle of autonomy. For this purpose, I will present an interpretation of the concept of "paternalism" from Emmanuel Levinas' understanding of "paternity" with the aim of elucidating the relation between "paternalism" and the concept of autonomy. Finally, I will argue that by accepting a "soft" paternalism in medicine it is possible to make better and more ethically acceptable choices concerning PGD.

17.2   Explanation of the Problem: What Is PGD?

Preimplantation genetic diagnosis is an invasive procedure that is performed on embryos via in vitro fertilization (IVF) or intracytoplasmic sperm injection (ICSI) before they are transferred to a women's uterus. For prenatal diagnosis, cells are produced from different stages in the embryo's development in order to test these for genetic diseases or conditions. In Germany, the use of PGD via blastocyst biopsy[1] has been allowed since 2011 but is strictly monitored.[2] According to the most recent regulation, only couples who carry a serious genetic disease themselves or whose pregnancy would likely result in stillbirth or miscarriage are allowed to use PGD at licensed centers after an ethics committee has consented.

Unlike Germany, the spectrum of indications for PGD in the US, as well as in other countries, is much more liberal. While PGD is admissible in Germany solely for the purpose of testing embryos for monogenic diseases (such as cystic fibrosis or Duchenne muscular dystrophy) or chromosomal abnormalities (mainly autosomal monosomy and trisomy), it is used in other countries for identifying certain desired genetic features. Thus, HLA typing can be performed in order to determine the compatibility of a bone marrow donation from a future child to save a critically ill child (so-called "donor siblings") (Sparrow and Cram 2010). Another aim of PGD is to determine the sex of the embryo not for medical reasons but for social or familial reasons (known as "sex balancing"). Nonetheless, using PGD for those last purposes is a hotly debated topic. While PGD is increasingly accepted by the public as a means of avoiding serious genetic diseases in future offspring, critics call the ethical legitimacy of sex balancing into question. More controversial still is the implementation of PGD with the objective of sorting out the healthy embryos and only transferring the genetically affected embryos into the uterus.

A great number of people are intuitively against the preference to intentionally bring a deaf child into the world. However, if one would like to make a rational and normatively binding argument against the acceptance of parental decisions and therefore restrict their reproductive autonomy, defending "soft" paternalism proves to be a daunting task. There are at least three clear arguments that complicate such an effort: (1) the verifiable loss of the normative force of terms such as "disease" and "disability" in modern medicine, (2) the subjective understanding of the child's well-being and quality of life and (3) the central position of the principle of respect for autonomy in contemporary medical ethics.

---

[1] In a blastocyst biopsy, the embryonic cells are in the eight-cell stage and are pluripotent. At this stage, it can be clearly distinguished between cells that belong to the constitution of the embryo itself (embryoblast) and other cells that form the outer layer of the blastocyst and develop into a large part of the placenta (trophoblast).

[2] The PGD is defined by the German Medical Association (*Bundesärztekammer*) as follows: "Preimplantation genetic diagnosis (PGD) is the (invasive) diagnostics performed on cultured embryos created by way of in-vitro fertilization (IVF) before transferring embryos into the uterus while taking into consideration changes in the genetic material, which could lead to serious illness." (German Medical Association 2011)

17.3 The Difficulty of Justifying Paternalistic Attitudes in Reproductive Medicine

17.3.1 The Intuitive Rejection of the Selection of Genetically Affected Embryos Using PGD

The intuitive rejection of the selection of genetically affected embryos using PGD is mostly due to the fact that diseases and disabilities are not regarded as value-neutral but as negative aspects. In determining that "The patient is sick" the predicate serves, on the one hand, as the description of a factual condition, or is empirically provable, and serves, on the other hand, as one with normative content. Being sick is regarded as a negative condition, as something "bad". Since the very beginning, medicine regarded it its responsibility to help sick people in times of need. In the background, however, there is the guiding traditional notion in medicine that health possesses an undoubtedly higher value than disease. Accordingly, medically combating and preventing diseases and treating disabilities are ethically necessary. In modern medicine, however, not all medical services exclusively relate to preventing and curing diseases and disabilities or restoring patients' health. There are also other tasks that belong to the medical field such as enhancing the abilities of healthy people, performing aesthetic (not reconstructive) surgery and even abortions. For this reason, the question is raised in current debates as to how modern medicine can still hold onto a notion of illness.

In these debates, the loss of objectivity in the notion of illness plays an important role. As a result of the development of human genetics, the prediction of genetic disease is possible without observing symptoms. The increasingly earlier diagnosis of diseases makes it possible for patients to consider themselves "sick" before even feeling sick. Increasing uncertainty as to where a disease begins (whether it begins in the genetic disposition or in the concrete discomfort of the patient) provides evidence for the fact that diseases are not objectively measurable phenomena but are rather dependent upon technical alternatives and subjective interpretations.

The prevailing scientific model of medicine since the nineteenth century and its understanding of disease as localized, somatic disorders have proved to be a reductionist notion today. If people with certain biological or anatomical deviations can still live well and do not perceive these as disorders, they do not feel sick. For this reason, contemporary medicine is making efforts to respect first and foremost the subjective suffering of the person concerned. If this idea is applied to the treatment or prevention of a disability, then it is also necessary to take the subjective experience of people with disabilities into consideration. If people do not consider their deafness or dwarfism as something they suffer from, as the social model of disability proposes, then there is no reason to have it medically treated. Under this interpretation deafness or dwarfism are not truly harmful. Many members of the hearing-impaired and of short stature communities are convinced that diseases and disabilities are social, historical and cultural constructs. For this reason, they perceive their suffering not as a result of their impairment but as a result of a dictatorship of "normality", which is also a social construct that is solely responsible for their discrimination and social exclusion.

If one takes the meaning of the subjective assessment of suffering seriously, it can be more important than the biological and anatomical understanding of the condition in some cases. If, however, medical treatment is supposed to focus on the subjective assessment of suffering, in some cases the physician's responsibility to cure diseases and alleviate disabilities might play a secondary role.

17.3.2 The Subjective Understanding of Quality of Life and Welfare of the Child

A strong argument for limiting reproductive decisions is based on the idea of a child's welfare. The argument being made here is in favor of a "soft" paternalism that aims to protect the interests of the future child who is neither capable of consenting to nor declining medical treatment. Ron Green, for example, is of this opinion and argues that parents do not have the right to genetically harm their offspring. By this he means: "In the absence of adequate justifying reasons, a child is morally wronged when he/she is knowingly, deliberately, or negligently brought into being with a health status likely to result in significantly greater disability or suffering, or significantly reduced life options relative to the other children with whom he/she will grow up." (Green 1997) From this point of view, parents are not allowed to predetermine whether their offspring should be deaf or short-statured, because this would not only deliberately reduce their quality of life but also restrict their conception of life itself. Joel Feinberg developed a criterion to legitimize medical interventions for children that claims that children have the right to an open future. According to this criterion, medical interventions should not unnecessarily restrict children's future conceptions of life. Even though these strong arguments are for the tenability of a "soft" paternalism in reproductive decisions, they are lacking several aspects. For the most part, short-statured and deaf parents defend their preferences for a child with the same disability with the argument that they are indeed acting in favor of the child's welfare. One short-statured parent, for example, raised a question with a genuine concern about her daughter's not being of short stature: "What is life going to be like for her, when her parents are different than she is?" (Sanghavi 2006) With the guarantee that their children will have the same disability, parents are, according to their own argument, only limiting a certain conception of their children's life but thereby opening up others. Strictly speaking, by using PGD parents are not limiting the life options of their offspring because they are not making a hearing child deaf, but they are only choosing the embryo with is expected to be deaf. Parents are therefore not harming their child: If parents would have selected a hearing embryo they would have had a different child. The deaf and short-statured community has its doors open for these children and is willing to provide them with a place in which they will not be excluded or discriminated against. Furthermore, one could argue that a life without any restrictions is also impossible (Mills 2003). There was no thinker who made this clearer than Martin Heidegger. His understanding of the human being as a "thrown projection" (Heidegger 1976) means that every existence can be, on the one hand, freely formed, but is, on the other hand, "thrown" into a world that the human being did not choose for himself. The conceptions of life are therefore not an unlimited abundance of possibilities but are determined by factual factors such as the familial biography, social status, nationality or simply fate or nature. Moreover, the parental obligation to leave options open for a child can only be relatively fulfilled. Thus, if parents can only offer a limited number of possibilities for their offspring from the very outset, then it is debatable as to how this can serve as an argument for restricting reproductive freedom of short-statured and deaf couples. Finally, the argument that parents are harming their offspring by using PGD in order to select an embryo for the presence of a disability does not withhold counter-arguments. It is important to remember the "non-identity problem" here. As Parfit and others argue (Dworkin 2010; Parfit 1984; Glover 2006; Savulescu 2002) no child can be harmed only by being brought into existence as long as he or she will have a life worth living. From this point of view, no harm is inflicted by the decision to select between possible embryos because all of them have an interest in coming into existence. Since parents do not harm their child by selecting for a disability (if they had chosen a different embryo a different child would have come into existence) it is permissible to choose either embryo, even ones with some disability.

17.3.3 The Principle of Respect for Autonomy

Even though the treatment and prevention of disease are still a central concern in medicine, the highest ethical legitimacy of medical action will not be maintained automatically as a result of these but only by respecting the autonomy of the patient (Düwell 2008), which is substantiated by informed consent. Only if patients agree to a medically indicated treatment, is the physician permitted to perform it. The respect for patient autonomy is one of the most important principles in contemporary medical ethics. It was introduced in medical ethics in the 1970s with the purpose to protect patients from paternalistic acts by the part of medical doctors. Patients should decide whether they want to consent to or decline a medically indicated treatment. Emphasizing autonomy as a principle equivalent to the traditional values beneficence and non-maleficence brings about a change in the traditional doctor-patient relationship. Patients are no longer subject to the decisions of doctors but are just as crucial to the decision-making process as the physician's medical expertise. Recognizing the patient's autonomy abolishes the traditional asymmetric relationship between doctors and patients by providing the patient with the right to play a role in medical decisions despite his or her lack of medical knowledge. Düwell summarizes this as follows: "Even though the treatment of diseases is still the main concern in medicine, the patient's will is normatively the more important reason of legitimacy." (Düwell 2008) The respect for the principle of autonomy in medical ethics has a particularly strong normative power when it comes to reproductive decisions. As a result of the harrowing experiences made during the Nazi period with their eugenics and compulsory sterilization programs, all human beings should have the right to make their own reproductive decisions without the influence of public authority or social pressure. Reproductive decisions are very personal and private. Couples should be free to consent to or to refuse genetic testing of themselves or their offspring, as long as their decisions do not harm their child. As Robert Sparrow points out, the notion of reproductive liberty often aims to emphasize two aspects: the right to be free of government interference (negative right) and the idea that reproducers should receive assistance to support their reproductive decisions (positive right) (Sparrow 2008). Nowadays, there are many authors who advocate the granting of reproductive decisions as long as they are made autonomously, individually and free of any compulsion (Agar 2004; Dworkin 1994; Harris 1998; Robertson 1995). For example, the author of *Liberal Eugenics*, Nicholas Agar, states: "Prospective parents may ask genetic engineers to introduce into their embryos combinations of genes that correspond with their particular conception of the good life. Yet they will acknowledge the right of their fellow citizens to make completely different eugenic choices." (Agar 2004) Viewed from this perspective, if physicians are not ready to abide by the wish of deaf or short-statured parents to have a deaf or short-statured child they implicitly claim that they know better what is the best for the child than the parents (in terms of preventing disability and improving welfare).

In sum, one can state that justifying a "soft" paternalism in reproductive decisions is by no means an easy task. Many people are intuitively against the desire to intentionally bring a deaf child or one of short stature into the world. Health professionals in particular are beside themselves with astonishment when confronted with parents who express such an unusual wish. In the following, however, I would like to argue that "soft" paternalistic positions do not require a specific justification because they fundamentally belong to medicine, even though they do so without the normative power of the notion of illness.

17.4 A Bridge Between Autonomy and Paternalism?

In medicine, autonomy and paternalism are usually considered to be conflicting terms. According to this view, paternalistic medical acts are the ones that primarily aim to protect the welfare of the patient, even if they are carried out against the patient's will. "Paternalism" (*pater [lat.] = father*) means to act like a father, or to treat another person like a child. In this respect, paternalistic acts seem to stand in opposition to the principle of medical ethics that aims to respect the autonomy and self-determination of the patient. As Gerald Dworkin put it, paternalistic actions "are justified solely on the grounds that the person affected would be better off, or would be less harmed, as a result of the [paternalitic] rule, policy, etc." (Dworkin

2010). In medical ethics, paternalism is founded in the assumption that physicians are more capable of making a decision about whether medical treatment is necessary because of their medical knowledge. From this perspective, patients should "provide the physician with the necessary information concerning their symptoms [...] so that the physician can determine the diagnostic and therapeutical procedures" (Klemperer 2006). Medical paternalism assumes that the physician can better identify what is best for the patient. In this context, measures are considered paternalistic if they are intended to protect the welfare of the patient even against his or her will. Paternalism rests upon terms such as "aid" and "welfare", which have been fundamental to medical ethics since the very beginning. It often seems as if these terms are more important to paternalism than the principle of autonomy, a fact that is regarded as problematic in contemporary medical ethics. In the following, I would like to reveal the ethical content of paternalism. The focus of this analysis will lie on determining the extent to which paternalism actually surrenders the autonomy of an individual in favor of an ethics of care, as is often claimed, or whether paternalism actually takes the autonomy of the patient into consideration.

In order to understand better the special relationship of "paternity", which is at the bottom of paternalism, it is helpful to refer back to Emmanuel Lévinas's analysis of *"paternité"*. Lévinas is not directly referring to paternalism – in French *paternalisme* – but to *paternité* (paternity), but from his phenomenological approach about paternity we can get another understand of the ethical substance of paternalism. For Lévinas paternity does not concern biological paternity or fatherhood but rather an ethical phenomenon. According to Lévinas we become the father of every other person for whom we bear responsibility. The term "paternity" therefore refers to a particular form of interpersonal relationships in which the ethical substance of *every* intersubjective encounter reveals itself: in Lévinas's account subjectivity is primordially ethical. Lévinas therefore brings forth from the phenomenological description of paternity its ethical significance. According to his interpretation, paternity refers to a certain intersubjective relationship that is based upon tensions between distance and closeness and insuperable asymmetries. On the one hand, paternity represents the close relation between a father and his son. Paternity proves to be a word that brings a fundamental closeness and human solidarity to bear. On the other hand, paternity expresses a fundamental difference between parties (Lévinas 2011, p. 85). Father and son are not characterized by an identity; their relation does not automatically equal a fusion. According to Lévinas, even though father and son do indeed have a relation, they preserve their differences *by* being related.

Just as every other interpersonal relationship, Lévinas regards paternity as an ethical kind of relationship. Paternity differs from other relationships in that it brings the ethical nature of every human relationship to light. For Lévinas, the constant need for the other, his or her mysterious distance and reflexive unreachability are the typical characteristics of an intersubjective relationship that reveal themselves to a great extent in paternity (Lévinas 2011).

As an ethical relationship, paternity allows for a particular implementation of freedom and responsibility. In the role of the father, the ego assumes responsibility for the other, who is in this case the child. Lévinas promotes a philosophical justification of an ethics of care that decidedly forgoes substantiation from the empathy theory. For Lévinas, ethical behavior is not founded on an empathetic ability to perceive suffering of another as one's own. Lévinas stresses again and again that foreign experience is not to be interpreted as one's own experience. In this respect, the relationship to the child, and to everyone else for that matter, can be described as a relationship to a stranger that goes beyond the theoretical accessibility of the ego. For Lévinas, a relation to the other does not naturally arise from mere existence. The ego has always been rooted in the loneliness of existence, and only a social relationship can make experiencing transcendence possible. The existence of a father or of a son is not interchangeable or even comparable. For Lévinas, "existence" is "the only thing I cannot convey; I can talk about it, but I cannot share my existence. Solitude acts as a separation that marks the actual event of being" (Lévinas 1982, p. 92). Lévinas claims that a bridge to the other can only be built by ethics and not ontology. Ethics alone demands of us the willingness to assume responsibility for the other: "Without a doubt, we can say that from the very moment in which the other looks at me, I am responsible for him, without even having to take responsibility for him at all; his responsibility is incumbent upon me" (Lévinas 1982, p. 92). However, the assumption of this responsibility presumes an insuperable asymmetry in interpersonal relationships. For Lévinas, the ego always has the majority of the responsibility when compared to everyone else (Lévinas 1982).

In paternity, the father assumes "the majority of responsibility" and thus the responsibility for his child. Assuming this responsibility means that one acknowledges the fact that the child and his or her existence cannot be reduced to that of the father. Paternity is a relationship with a stranger who obviously possesses an ego as well. However, the child is neither his father's alter ego nor an extension of his identity. Instead, the child is "outside of the father". The child is not a possession of the father; he or she is independent. The child's independence is an expression of dominance over his or her own existence and therefore evidence for his or her own autonomy.

With his ethical philosophy, Lévinas is not trying to determine moral norms. The only explicit normative obligation on his mind is based on the commandment not to kill. However, from the normative content of his ethics of care, one can conjure up other standards and obligations that could enrich his approach. The ethical demand to recognize and respect the radical otherness of the other basically means to leave room for the other's conception of life, or in other words, not to reduce these claims to the ego's point of view.

In this respect, paternalistic action means to assume responsibility for the other and to stand up for his well-being and interests. Provided that the interests of the other in the ego are always relatively accessible and never absolutely transparent, ethical action demands first and foremost reservations on the part of the ego and its egoistic actions for the benefit of allowing the free self-design of the other. Therefore, paternalistic measures that really aim for the welfare of the other do not lose sight of the autonomy of the other. The approach that the point of view of the ego should not be foisted on the viewpoint of the other means nothing more than leaving room for the self-determination of the other and preserving his or her autonomy. From

Lévinas's considerations can be deduced what an acceptable paternalism truly means: on the one hand, that knowing one's own limits and accepting the fact that we will never completely see the well-being of the other and, on the other hand, that this form of paternalism requires us to always assume responsibility for the other. This is precisely the form of paternalism that is of great significance in medical ethics today.

Lévinas's interpretation of paternity with its underlying ethical asymmetries is consistent with the traditional understanding of the doctor-patient relationship. This understanding is based on an informational asymmetry that also has important ethical implications. For patients, doctors are not simply people who have medical knowledge crucial to weighting the risks and benefits of medical treatment; they are trustworthy people precisely due to this knowledge. When a patient consents to a treatment, he or she not only has trust in the doctor's competence to perform it but also *believes* that the doctor wants nothing less than the best for him or her. The asymmetry in the doctor-patient relationship goes way beyond competences. Due to their suffering or respective illness, patients find themselves in an emergency situation. In this context, medical action is not neutral but value-laden. On the one hand, the art of medicine (*ars medica*) accepts the technical responsibility of restoring the patient's health. At the same time, however, this technical responsibility has an ethical back side: the art of medicine is concerned with restoring a suffering patient's health or combating, or at least trying to minimize, the patient's suffering. This, however, presumes from the very outset that the doctor feels a certain obligation toward the patient and readily assumes responsibility.

17.5 Conclusion

With Lévinas we found an alternative to the dominant understanding of autonomy and paternalism as opposites shows that the principle of respect for autonomy and the principle of care, which forms the basis of paternalism, are related. With Lévinas's approach I have described a theoretical ethical support of children's right to an open future.

In reproductive medicine "soft" paternalism does not premise a specific theoretical moral justification, but rather to remember the ethical meaning of the phenomenon "paternity" at the bottom of paternalism. Thus we have to admit that paternity represents first and foremost a *moral duty*. This duty tells parents that they should assume responsibility for their offspring, for his or her well-being and interests, but also for its self-determination. From the one perspective, for a deaf child to be part of deaf communities may be an essential part of their sense of well-being, but from another angle deaf communities are always under pressure exerted by hearing cultures. In this light, the decision to use PGD in order to select for the presence of a disability is morally problematic because this decision involves willingly choosing a real limitation in the life of their offspring. Inasmuch as a hearing child of deaf parents can also learn the sign language as much as children can learn other languages to select an embryo without that disability assures a maximally open future for the child.

Should physicians limit the reproductive autonomy of parents? Modern medicine has learned that the "best" does not always and does not necessarily mean combating pathologies. This, however, does not imply that a patient's autonomous decision forms the only legitimate basis of medical practice. In the case of reproductive decisions, the relationship goes beyond the dual relationship between doctors and patients. Here, it is important to take not only the parent's/mother's but also the child's interests into consideration. In this respect, the refusal to grant the parents' wish to intentionally have a child with a disability by using PGD is not founded in the conviction that this disability is bad per se. It is not a matter of dogmatically following the maxim *salus aegroti suprema lex*. It rather concerns the fact that the future child's autonomy and conception of life would be extremely limited by such an intervention. This "soft" form of paternalism is therefore not based on the traditional conception that doctors know what is best for their patients and can therefore make a better-informed decision but on the moral intuition that doctors should not assist in limiting the child life's options. The mentioned objection that the conceptions of life always go beyond subjective control and should therefore be formed within limits is actually not an adequate argument against the duty of leaving life-options open for the future child. Because of the fact that parents have always influenced or even determined their children's life opportunities, be it as a result of their upbringing, society, norms or traditions, parents cannot assume that they can do so with the help of medical professionals as well. However, this fact still does not give rise to a normative argument for or against the permissibility of reproductive freedom of short-statured or hearing-impaired parents. The normative force of an ethics of "paternity" that is not based on an empathy theory offers physicians a solid argumentative basis in favor of an obligation toward the child's well-being and inability to consent as well as the child's conception of life. Thus even when parents are not harming their child by selecting the deaf embryo over the hearing one, they are willingly choosing an embryo whose life is likely to be more limited than the hearing one. To choose that limitation would be questionable inasmuch as the hearing embryo could have a wider spectrum of possibilities to develop his/her own understanding of well being.

I have demonstrated that there are moral reasons for physicians to decline the parental requests for PGD aiming at the selection of embryos with disabilities. This does not mean, however, that they should oppose to perform the intervention under all circumstances nor that they should send families to other doctors who might be ready to consent. Paternity is, first and foremost, a moral duty, but not something that can legitimately be translated into a legal obligation. Reproductive autonomy rights are based on fundamental values that we must preserve even at the cost of allowing that parents' decisions may not maximize the future options of their child. Thus, the purpose of this paper was not to promote a particular way of regulating PGD allowing some interventions while banning others. Instead, the idea was to present Lévinas's ethical approach in order to rehabilitate paternalism to some extent. Paternalistic measures do not always and ineludibly disrespect autonomy and therefore are not *prima facie* incompatible with modern medical ethics centered around autonomy. Lévinas's account facilitates a more sophisticated and potentially more fruitful consideration of the relationship between paternalism and autonomy. Above all, I hope that this may stimulate a more diverse debate not only about the question how much paternalism modern medicine can accept but, particularly, what kind of paternalism we are willing to tolerate.

References

Agar, N. 2004. Liberal eugenics. In *Defense of human enhancement*. Oxford: Blackwell.

Baruch, S., D. Kaufman, and K.L. Hudson. 2008. Genetic testing of embryos: Practices and perspectives of US in vitro fertilization clinic. *Fertility and Sterility* 89(5): 1053–1058.

Düwell, M. 2008. *Bioethik: Methoden, Theorien und Bereiche [Bioethics: Methods, Theories, and Areas]*. Stuttgart: Metzler.

Dworkin, R. 1994. *Life's dominion*. New York: Vintage Books.

Dworkin, G. 2010. Paternalism. In *The Stanford encyclopedia of philosophy* (Summer 2010 Edition), ed. N. Zalta Edward. Available online: http://plato.stanford.edu/archives/sum2010/entries/paternalism/.

German Medical Association. 2011. *Memorandum zur Präimplantationsdiagnostik (PID)* [Memorandum on Pre-inplantazion Genetic Diagnosis (PGD)]. Available online: http://www.bundesaerztekammer.de/downloads/Memorandum-PID_Memorandum_17052011.pdf.

Glover, J. 2006. *Choosing children: Genes, disability, and design*. Oxford: Oxford University Press.

Green, R. 1997. Parental autonomy and the obligation not to harm one's child genetically. *The Journal of Law, Medicine & Ethics* 25(1): 5–15.

Harris, J. 1998. Rights and reproductive choice. In *The future of human reproduction: Ethics, choice, and regulation*, ed. J. Harris and S. Holm, 5–37. Oxford: Clarendon.

Heidegger, M. 1976. *Sein und Zeit* [Being and Time], GA 2, ed. F.W. von Herrmann. Tübingen: Max Niemeyer.

Klemperer, D. 2006. Vom Paternalismus zur Partnerschaft: Der Arztberuf im Wandel. In *Professionalisierung im Gesundheitswesen*, ed. J. Pundt, 61–75. Bern: Huber.

Lévinas, E. 1982. *Éthique et infini. Dialogues avec Philippe Nemo*. Paris: Librairie Arthème Fayard et Radio-France.

Lévinas, E. 2011. *Le temps et l'autre*. Paris: Press Universitaires de France.

Mills, C. 2003. The child's right to an open future. *Journal of Social Philosophy* 34(4): 499–509.

Parfit, D. 1984. *Reasons and persons*. Oxford: Clarendon.

Robertson, J.A. 1995. *Children of choice, freedom and the new reproductive technologies*. Princeton: Princeton University Press.

Sanghavi, D.M. 2006. Wanting babies like themselves, some parents choose genetic defects. *New York Times*, 2006. Available online: http://www.nytimes.com/2006/12/05/health/05essa.html.

Savulescu, J. 2002. Deaf lesbians, "designer disability", and the future of medicine. *British Medical Journal* 325: 771–775.

Sparrow, R. 2008. Is it "every man's right to have babies if he wants them"? Male pregnancy and the limits of reproductive liberty. *Kennedy Institute of Ethics Journal* 18(3): 275–299.

Sparrow, R., and D. Cram. 2010. Saviour embryos? Preimplantation genetic diagnosis as a therapeutic technology. *Reproductive Biomedicine Online* 20(5): 667–674.

Spriggs, M. 2002. Lesbian couple create a child who is deaf like them. *Journal of Medical Ethics* 28: 283.

Chapter 18
The Limitation of a Mother's Autonomy in Reproduction: Is the Ban on Egg Donation a Case of Indirect Paternalism?

Clemens Heyder

Considering the German Embryo Protection Act (Embryonenschutzgesetz – ESchG), it should be noted that there are some cases of assisted reproduction that are not allowed to certain people. For example, it is prohibited to procreate by using techniques such as egg donation, cloning, surrogacy and ectogenesis. In addition, by the Professional Code for Physicians in Germany it is forbidden to couples that are not married or in a long-standing relationship to use any kind of assisted reproductive technologies.

Within a liberal democratic society[1] which is characterised by plurality of moral beliefs there is a distinction between a private and a public sphere. State interventions cannot be applied to actions which are based on private moral grounds that do not violate another person's rights. Personal autonomy is valuable by itself and interference needs to be justified. Good reasons must exist for any restriction on individual liberty. This is strengthened especially by the fact that the cultural and legal background is linked to a liberal tradition.

The question is now the following: Are there any good reasons for a legal restriction? Regarding the context of reproduction there are distinctive features that cannot be ignored. Of course it could be reasonable to protect the mother from a risky medical service but while there is a third party involved, the focus of benefit should rather be placed on the prospective children. At this point it is necessary to distinguish between direct and indirect paternalism.[2] In the following I will discuss the case of

---

[1] "The liberal society is one in which to the maximum degree people are at liberty to exercise their personal autonomy." (Charlesworth 1993, 16).

[2] Paternalism is the interference of the state or a person with another person, for the purpose of benefiting (protecting from harm) the person. In the case of indirect paternalism the interference is motivated by benefiting another one as the acting person.

C. Heyder (✉)
Department of Philosophy, Bielefeld University, Bielefeld, Germany
e-mail: clemens.heyder@uni-bielefeld.de

© Springer International Publishing Switzerland 2015
T. Schramme (ed.), *New Perspectives on Paternalism and Health Care*, Library of Ethics and Applied Philosophy 35, DOI 10.1007/978-3-319-17960-5_18

egg donation[3] and analyse whether there are some valid arguments to outlaw reproductive autonomy and to figure out in which way direct or indirect paternalism could be justified.

At first I give a short outline of the meaning of reproductive autonomy and make clear how an intervention into one's individual freedom of action can be justified by the harm principle. After that I will analyse whether any argument applies against the harm principle on distinct levels so that egg donation could be legally banned for direct or indirect paternalistic reasons. Since this is not the case, I conclude at last that, if we want to accept a ban on egg donation, we need other premises than an autonomy-based ethical framework.

18.1   The Meaning of Autonomy

The term *reproductive autonomy* is very common in biomedical ethics. The academic debate does not only examine the justification of autonomy in general, but it also tries to find a definition for the term *autonomy*. Commonly, the first notion for autonomy that comes into mind is the one that stands in a Kantian way.

> Autonomy of the will is the property of the will by which it is a law to itself (independently of any property of the objects of volition). The principle of autonomy is, therefore: to choose only in such a way that the maxims of your choice are also included as universal law in the same volition.[4]

According to Kant, the principle of autonomy is the Categorial Imperative itself. In a negative sense it refers to the independence of material determinants (Kant implies even interests and passions), in a positive sense to self-legislation by a will free from heteronomy.[5] A person as a being of reason could act independent from natural causalities. Therefore, "freedom and the will's own lawgiving are both autonomy and hence reciprocal concepts".[6] Autonomy is a precondition of morality.

Autonomy as self-legislation, however, could not be understood as a mere application on special areas of life. Kantian autonomy is intended to be all-encompassing and focusses not only the individual action in its special contexts. It is valid in all places at all times as all people necessarily come to the Categorical Imperative as the basic moral law. But the matter of biomedical ethics is rather an act of individuals dealing with self-interests and prudential aspects of well-being. Kant's notion of autonomy is therefore not helpful for the questions discussed in this paper. In consequence, it seems plausible to use the term autonomy rather in the tradition of Mill, as the absence of external influences.[7] Mill's concept of autonomy defined in *On*

---

[3] Many arguments against egg donation apply as well on surrogacy because in both cases they deal with split motherhood, i.e. the distinction of genetic and biological motherhood. See fn. 32.

[4] Kant (1996, 89).

[5] Höffe (1996, 199).

[6] Kant (1996, 97).

[7] Schöne-Seifert (2007, 40), Knoepffler (2008, 145); see Beauchamp (2007).

Liberty is the origin of (Western) ethical thinking in medicine.[8] Although Mill uses the term *autonomy* just once – he mainly uses the word *moral freedom* – the connotation of its reception cannot be ignored.[9] Within the meaning of self-determination he states "that the free development of individuality is one of the leading essentials of well-being", not only as an element but also as "a necessary part and condition".[10] A typical objection is that the idea of liberty is not reconcilable with the idea of utilitarianism; when Mill purposes to find a principle to protect the individual against the tyranny of the majority.[11] For him, of course, it is not a genuine problem, since utility is a kind of justification: "I regard utility as the ultimate appeal on all ethical questions; but it must be utility in the largest sense, grounded on the permanent interests of man as a progressive being."[12] When he states liberty as sole source of improvement,[13] he gets a teleological justification of liberty as a means at least in a negative sense. Only in the absence of coercion and the subsequent evolvement of individuality it is possible to develop as a human being, at least from a naturalistic perspective such as Mill's, as well as mankind itself in its own interest.[14]

Nevertheless, Mill understands freedom not solely as a means, but also as an end in itself. Personal freedom is the highest and the strongest wish of every human being, after meeting his basic needs, e.g., shelter and nutrition, fixed in human nature.[15] Self-determination as an "intrinsic worth"[16] is a required element of one's identity, whose development and refinement is for its own sake. Finally, it is not just what a human being is doing but also what he is, similar to what is said in Aristotelian virtue ethics.[17] Thus, reason and character are the two constitutive elements of human nature, of which individual autonomy is a crucial part. This is because otherwise someone "who lets the world, or his own portion of it, choose his plan of life for him, has no need of any other faculty than the ape-like one of imitation."[18]

Assuming the improvement of mankind, Mill aims to find a principle to define the relationship between individuals and political society. In this distinction he

---

[8] Even if Beauchamp and Childress emphasize respect for autonomy as prima facie duty and has therefore another moral ground, their notion of autonomy obviously follows Mill. (Beauchamp/Childress 2001, 63 f).

[9] O'Neill (2005, 30). For example: "Mill in whose philosophy naturalism and the ideal of rational autonomy are the two deepest convictions, is particularly committed to the assumption that they are indeed reconcilable." (Skorupski 1989, 43).

[10] Mill (1996a), On Liberty, 261. "Autonomy – the freedom to make one's own decisions in one's own private domain – is in its own right a categorical human end, one of the essentials of a worthwhile life." (Skorupski 1989, 21).

[11] Mill (1996a), On Liberty, 219.

[12] Mill (1996a), On Liberty, 224.

[13] Mill (1996a), On Liberty, 272.

[14] Gräfrath (1992, 30); for discussion see Riley (2004, 167–176).

[15] Mill (1996b), Principles of Political Economy, 208.

[16] Mill (1996a), On Liberty, 261.

[17] Gräfrath (1992, 75).

[18] Mill (1996a), On Liberty, 262.

draws a line between the sphere of public and the sphere of regarding oneself and no others in which autonomy is linked to public policy, and accordingly to the sphere of morality.

> That principle is, that the sole end for which mankind are warranted, individually or collectively, in interfering with the liberty of action of any of their number, is self-protection. That the only purpose for which power can be rightfully exercised over any member of a civilized community, against his will, is to prevent harm to others.[19]

As the consequences of harm may be very direct (a slap in the face) or nondirect and possibly not apparent for the actor (buying non-fair-trade coffee), the harm principle lacks in view of the fact that there is no harm as an indirect consequence of an action. Aside the situation of a hatemonger who instigates a crowd to harm someone else, there are other kinds of non-direct harm as well. On the one hand one could be harmed mentally; maybe someone says something disturbing or insulting to a labile person. And on the other hand there could be some socially harming actions.[20] Both of these could justify an intervention to one's autonomy. Accepting this fact, the harm principle needs to be extended.[21] In this matter the *offense principle* gets its very importance in the discussion about harms to the society as a liberty-limiting principle.

A special feature of Mill's account is that it is only applicable to a human being, "arrived at the maturity of his faculties".[22] It applies neither to children nor to mentally disabled persons. He does not want to define any rules when someone is able to do his own well-considered action or could be held responsible for it. Rationality and self-consciousness are presupposed by Mill. In this meaning autonomy is not described as a capacity or a condition but rather as an ideal. He does not deal with characteristics of self-determination. The main item of his conception is a morally normative element, namely the question how society should be. Hence, autonomy is a moral right, which is not linked to any personal condition or constituted legal law but it is quite related to the status of personal sovereignty.[23]

18.2   The Meaning of Reproductive Autonomy

Even if it is now clear what is meant by *autonomy* it is still not clear what is meant by *reproductive autonomy*. If autonomy is described as a moral right to design one's own life plan we have to distinguish different approaches when we are speaking of

---

[19] Mill (1996a), On Liberty, 223.

[20] Indirect does not mean the harming consequence is not obvious but that it is, and not directly linked to the action. In such cases the consequence of an action leads to some harm not the initial action itself.

[21] It is contested whether the *offense principle* is a genuine part or rather a self-contained extension of the *harm principle*. For further discussion see Riley (2004, 176–187).

[22] Mill (1996a), On Liberty, 262.

[23] Feinberg (1989, 48).

reproductive autonomy as a moral right. What does it mean to say there is a right to reproduction? And, in which matter is it important to be unhampered by coercion and not to be restrained by others? In the following, I focus on the absence of coercion which means the direct interference by other people or the state.[24] Additionally I do not consider how this right could work and what it is claimed in a sense of positive liberty. My scope is narrowed to negative liberty. In other words, I do not concern myself with which way assisted reproduction could be provided but rather which justification could be fit to prevent someone's procreation.

At this point I shall make clear that I will not discuss moral issues of terminating pregnancy. While in this case other moral values are at stake (unborn life, dignity etc.), I will discuss whether a human being should be begotten or not and the moral problems that arise before there is any kind of life.

In a first notion of reproductive autonomy someone has to decide whether he or she wants to procreate at all and if so, it is also to be decided when. Having said this, Onora O'Neill emphasizes that autonomous reproduction not only means the absence of coercion, it also means the liberty of choice. However, she does not understand a state's duty to provide any possibilities to reproduction but its duty to enable to choose. Reproductive autonomy is more than the choice of when and how someone wants to reproduce; even the choice of the reproductive partner is covered by this right. So it is to say that a forced marriage potentially concludes a forced reproduction.[25] Thus, reproductive autonomy could mean, in a second notion, the choice of potential partners as well as the choice of the means. Based on the fact that modern medicine allows using a method other than the old-fashioned sexual way of reproduction, there is a possibility to decide for medical assisted reproduction without any medical indication, although most people have good reasons. In a third notion there is the parental predetermination of the prospective child's particular qualities. For sure, it cannot be excluded, but it is very unlikely, that there are parents who want to create a designer baby that would be pieced together like a puzzle, the more so since today (and in the near future) it cannot be realized medically.[26] More important should be noted that parents often bear in mind avoidance of any kind of suffering to their child. This is especially a reason for being concerned about the genetic makeup. There are some severe diseases which cannot be treated and not even be avoided by embryo selection after PGD. But in some cases it is possible to avoid genetic diseases by using donated gametes. A woman can use donated eggs without abstaining from the experience of pregnancy and childbirth. In a fourth meaning reproductive autonomy concerns the intended relationship. Not only could some genetic attributes be avoided, there are some cases in which one or both of the

---

[24] Of course there are a lot of external and internal influences (e.g. social, economical, religious...) on our daily decision-making process, but as I mentioned before I presuppose a sufficient capacity of self-determination.

[25] O'Neill (2005, 50).

[26] Assuming there is a strong parental desire of having a genetically related child, the world will not be crowded by supermen and wonderwomen. But even if it would be, this fact itself is not an objection.

parents are not able to reproduce. While tubal ligations or poor gamete quality can be compensated by assisted reproductive technologies sometimes infertility is an inevitable fact. In this case, the only solution to get pregnant and bear a child is through gamete or embryo donation.

In the following my purpose is to figure out aspects of reproductive autonomy and to make clear whether and how it could be justified to interfere with it. Based on the fact of the normative validity of the *harm principle* (in addition the *offense principle*) it is a minimum requirement to justify paternalistic interventions. To intervene in someone's action is only reasonable if not just the person himself is affected but it is "prejudicial to the interests of others"; social or legal punishments could only be imposed "if society is of opinion that the one or the other is requisite for its protection."[27] A distinction on different spheres should help to determine which action is against whose interest, and whether a paternalistic intervention could be justified for this reason.

18.3   Sphere of Parents

To start with, intended parents usually desire to have a genetic related child, which is commonly realised through sexual intercourse between man and woman. During the last few years new ways of making babies have been developed as a result of progressive medical technologies. Hence it is possible to fulfil people's desire for their own child even if it is not conceivable in the usual way. Infertility or sterility of one partner is a general challenge for intended parents. A couple of years ago this case entailed obviously no chance for an own child. Recently, this problem could be resolved by gamete donation (with an increasing rate of success in the last years). Either partner uses donated sperm or eggs or even both. It is possible as well to transfer donated embryos in the utero. The worst case for parents is if neither sperm nor egg are suited for reproduction and moreover if the intended mother is not able to bear a pregnancy to term, whereby a surrogate is needed. Under these circumstances, five parents could be accountable for one child: a genetic mother and father, a biological mother (the surrogate), and a social mother and father. Although this might happen very seldom and adoption seems to be the easier way, such cases cannot be ignored.

Beside these cases with medical indications the new reproductive technologies allow to overcome social infertility. It happens that some people (e.g. singles, gay couples, people who don't like to have sex, women after menopause) want to have a child but cannot conceive due to social reasons. Moreover, it is also possible to fulfil a desire for an own child of a man and two woman by mitochondria replacement. Assuming reproductive autonomy refers to a strong interest in having children, there is no moral relevant difference between biological and social infertility. Hence, all kind of infertility and every kind of desire for a child needs to be considered.

---

[27] Mill (1996a), On Liberty, 292.

Taking into account that all participating parents and the medical staff have some interest in procreation, irrespective of whether they want to take social responsibility, there is no interference by any other person. In compliance with current quality standards of good clinical practice the patients[28] will be informed in advance so that medical risks as well as social and emotional follow-up will not be a surprise. Assuming a sufficient competency of the patients there is no good reason to believe that they were coerced by external influences.[29] The decision to procreate is made by fully autonomous persons. Finally, an assisted reproduction is no action by accident but rather planned well in advance. If we acknowledge (from a liberal perspective) reproductive autonomy as a moral claim with a right to self-legislation, then by fulfilling these conditions any interference with the autonomous decision to and the action of procreation is a paternalistic intervention. That could morally not be justified by the harm principle because there is no direct harm, neither to donor/receiver and/or intended parents nor to the medical team.

One might object that there will be an unforeseen conflict between the different parents. For example, a surrogate might decide not to hand out the children to the intended social parents as agreed.[30] Thus, she destroys the plan of parenthood and limits their freedom of action. Let me emphasize that the future is unwritten and nobody can foresee all complications of life. But I am sure that a medical assisted reproduction is well planned and all involved parties are aware about the particular kinds of occurring problems when they make their decision towards this type of parenthood.

Furthermore, the interests of the prospective children will be even more affected in this case as the parental ones, which leads to the discussion about moral issues concerning the children.

18.4   Sphere of Children

As I stated before the capability of self-determination, which means to make life plans and to realise them, is a necessary requirement for granting autonomy as a moral right. Little children or babies, comatose patients, and those who are not fully

---

[28] To simplify the discussion I used the term patient according to a typical doctor-patient relationship where the patient suffers (from involuntary childlessness) and therefore seeks help by a doctor. In what way this meets a consumer oriented model of a doctor-patient relationship needs to be discussed on another stage.

[29] For sure, there are external influences in social and economic dependencies as well as social constraints controlling individual's life. Especially arguments of commercialisation and exploitation strengthen the importance of that matter. Actually women in economic distress might not be aware of medical risks that arise in the procedure of harvesting eggs and thus women in low income countries bear the main part of the burden of donation. (Berg 2008, 245; Graumann 2008, 182 f.) Albeit such cases that form a constraint on someone's autonomy could justify paternalistic interventions, there is a need for a separate consideration about the actual condition of autonomy but this cases cannot be discussed with the argument of autonomy as a moral claim.

[30] Government draft bill ESchG, BT-Drs. 11/5460, 7; Beitz (2009, 224).

conscious are not able to make extensive life plans. Of course, the moral right applies to all persons, but the realisation will be taken by surrogates. A baby may decide and communicate that it is hungry or not, but it does not know that it should not touch the hot cooktop. The baby always needs someone who takes responsibility and helps to meet the baby's interests. This is commonly a parent's duty. Birth is the beginning of an inevitable dependency, which is characterized by a strong kind of responsibility. Parents develop the child's life plan and take care for its realisation.

For sure, there will always be some problems in a parent-child-relationship and probably no child is perfectly satisfied with its situation. Nobody would ever consent to every parental decision, neither as child nor as an adult from a retrospective. However, nobody would ever be dissatisfied with his or her own life so much that he or she regrets his or her own existence and seriously wishes to never have been born. Life is still worth living, irrespective of individual tragedies or any bad luck.

One might object that conceiving a child with any harm or suffering is morally counterintuitive and there is an obligation to avoid it (at least if it is possible in a trivial manner). The German legislator argued in exactly this way[31]: It assumed a gestational bond as well as a genetical bond between the mother and the unborn child, with unforeseeable consequences for the child if those bonds would diverge and results in ambiguousness of the mother's identity (split motherhood)[32] which might jeopardise the child's development of personality. Incalculable is the risk of any mental disorder for the child (especially in puberty) when it comes to know that two mothers are responsible for his or her existence. Furthermore, this divergence raises some issues with the child's discovery of identity. The legislature identified this as a high risk for a serious disorder that could be avoided and therefore considered the ban on egg donation to be necessary as a consequence of split motherhood. One aim of the German Embryo Protection Act as it is explicitly formulated[33] is to avoid split motherhood which, again, means nothing else then avoiding any mental disorder or disability from children.

This is the point the well-being of the not yet conceived child is at stake and, therefore, the harm principle might justify to interfere with the parent's reproductive autonomy if the future child will be (indirectly) harmed by the parental action. In this case, so-called indirect paternalism, avoiding harm to other will overrule the moral claim of autonomy. To argue that it would be better to avoid any unnecessary harm can be done in two ways. Either avoiding harm is good in general or avoiding harm is good or might be better for the prospective children[34] who have at least a basic interest in not being harmed before birth.

---

[31] By passing the German Embryo Protection Act in 1990.

[32] Split motherhood (*gespaltene Mutterschaft*) is originally a term of the German juristic debate about the Embryo Protection Act. Unfortunately there is no clear translation because this argument exists in this way only within the German discussion.

[33] Keller/Günther/Kaiser, ESchG (1992), § 1 Abs. 1 Nr. 1, margin number 1; Günther/Taupitz/Kaiser, ESchG (2008), § 1 Abs. 1 Nr. 1, margin number 1.

[34] This is the way the German legislature argued.

In the first notion all people are taken into account.[35] Any person who can live a pleasant life should not be hindered to do so; otherwise he or she would be harmed. The same condition applies for unborn humans as well. If they were hindered to come into existence and therefore never experience happiness, there is no contribution to maximizing happiness or avoiding any harm, respectively. Hence, we have a moral obligation to procreate and should therefore bring children into existence as much as the maximum total utility is reached before it decreases by overpopulation.[36]

In the second notion the question is whether a child is in fact interested in not being born just for avoiding any harm. To explain this, I shall distinguish between possible and future persons concerning a concrete biological existence.[37] On the one hand there are future persons who will come into existence regardless of people's actions today. Even if a nuclear bomb will explode tomorrow some people would live at this place in a couple of years. Their very existence is independent of our actions today. These are unknown humans in an unknown future. It is only certain that they will live in the future.

On the other hand there are possible persons whose existence is uncertain and will only become reality by a concrete action. As a product of a specific sperm cell and a specific egg cell a possible person would just come into existence by a specific act of procreation. The following examples shall explain: if a pregnant woman has the chance to avoid a serious disease for her unborn child by taking some medicaments and does not, then she harms the unborn child directly, which is interested in not to be harmed as a concrete human being. On the contrary, if a woman has the chance to conceive while she is sick and the child will be born disabled, or she waits for a couple of months until she is physically recovered, no child can have any interest in that. Harm is not linked to any action during pregnancy but to existence itself. The alternative to procreation is non-existence.[38] Thus, in a contemporary mode the question could be reformulated: Can I have an interest in not to have been begotten?

For sure, there are serious diseases and impairments no one wants to live with that should therefore be avoided because of a personal feeling (and maybe by social conventions) to be disadvantaged in contrast to all others. But that doesn't mean a life with it is not worth living. It' hard to see that a person who is badly off by health has a serious wish to have never come into existence. This is at least true because an overall rating of the own life requires a minimum of positive interest in living. I do not want to say that there is nobody who sees no sense in his or her life and seriously wants to die; just as well as someone might say it would be better if he or she has

---

[35] Cf. Hare (1993).

[36] If someone does not accept the strong relation between maximizing happiness and avoiding harm, it could be said alternatively, avoiding harm means to bring no children into existence anymore.

[37] Parfit (1986, 355 f).

[38] From this point of view arises no moral obligation towards the parents of bearing a healthier and happier child. Parfit refers to the problem as that of "non-identity" (Parfit 1986, 351–356), and Kavka dubs it as "Paradox of Future Individuals". (Kavka 1982, 95).

never been born. Nonetheless, this happens just in the heat of the moment, particularly with respect to a life-affirming condition as required for summing-up.[39]

I will clearly not deny that a lot of people will come to the conclusion that a sufficiently happy or good life is absolutely not compatible with some diseases,[40] and thus, they would rather not want to come into existence than to have such a disease. But they form their opinion from an external point of view. Just because someone does not want to live (maybe it might be better said: cannot imagine living a life worth living) with a terrible disease does not mean that to live with a disease is bad and should therefore be avoided.[41] It may be understandable if parents think in this way, so they do it merely from an adult's retrospective. They try to see and feel the world from a child's point of view. They ask themselves if the child would be just as happy about his or her existence as a child without this disease. But if so we have reached the initial problem. The only alternative to this life is non-existence, not to be another person.

After all one question is left: Why is it always argued with the concept of the best interest of the child?[42] This juristic technical term is not well defined and moreover widely interpretable.[43] Before I will start the discussion I shall remark that the interest of a not yet conceived human being could only be anticipated. It is highly implausible to speak about future children's interest. These are not genuine because these are supposed by others.

The principle of the best interest standard always comes into play when patients in case of a disease can no longer decide for themselves and no possible preferences could be read from their life history. The method is sufficiently well known in rescue medicine.[44] It was developed within the clinical decision making process for incapacitated patients whose decisions are taken by a surrogate in the greatest purposes of their well-being.[45] These surrogates are usually the parents (or people with legal

---

[39] Steigleder (1998, 106). David Heyd emphasizes moreover that these two options are incommensurable. It is not possible to set non-existence as (selectable) alternative to existence. (Heyd 1992, 32).

[40] I will not concern any diseases which will lead to death directly after birth, e.g. spina bifida combined with anencephaly.

[41] Steigleder, 109. "One who has never known the pleasures of mental operation, ambulation, and social interaction surely does not suffer from the loss as much as one who has. While one who has known these capacities may prefer death to a life without them, we have no assurance that the disabled person, with no point of comparison, would agree. Life and life alone, whatever its limitations, might be of sufficient worth to him." (Robertson 1975, 254).

[42] Within the German juristic discussion this concept is expressed by the term *Kindeswohl* (literally: the children's welfare).

[43] Keller (1989, 710).

[44] If a patient is not able to interact or his actual will is not directed in advance the substituted judgment standard means to figure out his will according to affiliates. Even if this is not possible anymore, one attempt to define the best interest standard in general.

[45] President's Commission (1983, 134–136).

parental responsibility) who decide for their child[46] and act literally paternalistically. Although they do not even know better than others what is the best for their child, which in the case of newborns it is evident, they can estimate future living conditions, see the planned life course, and accordingly vary it, so that the highest possible well-being will be reached (from a parental point of view). That is the point I will emphasize. The best interest standard is less determined by clear criteria as it is regulated by the individual context. In this manner the decision principle is not the interest of the child but rather the interest of the parents.[47] The interpretation is not universal; it depends on cultural and temporal conditions.[48]

Something similar diagnoses Heta Häyry in her discussion about paternalistic interventions in medical context.[49] Assuming the restrictive action is always taken in the best interest of the patient, she points out four motives of justification behind it. Prior is obviously the real interest of the concerned person, but as already shown, this could not justify any restriction against reproductive autonomy. Second, the intervention can be justified if acting would be irrational. Since presupposing the capability of self-determination and rational decision making it can be neglected below. Additionally, it could be argued that an action is immoral or does some harm (not absolutely in a physical sense) and has to be prevented for this reason. The last two items I will discuss hereafter.

As it can be shown, there is no setback of interest of prospective children, neither of their real interests nor of the presumed best interest. This means prospective children will not be indirectly harmed by the parent's procreative decision and therefore it could not be justified by the harm principle to interfere with their reproductive autonomy. This is not a case of indirect paternalism.

To emphasize this point again, I do not want to say that it is morally good to conceive a child who comes into existence with a serious illness or disability. I would only like to say that it is not morally wrong, because the moral judgment of a possible life cannot be linked to the interests of an individual human being. But, this is not equivalent to the absence of other good reasons why conceiving such kind of life should be avoided.

18.5   Sphere of Society

So far I have showed that neither the real interests of the parents nor the anticipated interests of the prospective children can be taken into account against a moral right for procreation, whereas the interests of the society (whatever those should be) have

---

[46] President's Commission (1983, 6–8), British Medical Association Ethics Department (2013, 133 f).

[47] See Downie/Randall (1997).

[48] See Eekelaar (1994). Accordingly, the criticism of the best interest standard refers basically to the individual interpretation and indeterminacy, which is subjected to cultural and time-bound interpretations. (Dörries 2003, 124).

[49] Häyry (1991, 6).

not yet been considered. This leads to the next issue in the debate. Presupposing the validity of the harm principle as well as the offense principle, not only harm to a single person will justify interference with an autonomous action, a group of persons or the entire community could be affected as well. For example, Joel Feinberg describes a situation of a garrison of settlers under attack from Indians. In the meantime, a person kills himself, and because he can no longer lend his support he endangers the entire community.[50] Doubtless, this example is antiquated, and we could say that any action of an individual in modern complex societies is not necessarily harmful, but there is a point at which the community will be harmed significantly. Feinberg continues that it probably would have no effect if a person is addicted to drugs but if large numbers are greatly addicted to drugs, the burden would fall back notably to the community which would at least bear the actual costs.[51] Hence, there is a limit that should not be exceeded since otherwise there would be a serious public harm. Even though an action is directed against oneself and no other, and it seems the action will not lead to further consequences for anyone else, like in the case of the garrison or drug addicted person, it might happen that the community or society will be indirectly harmed by that autonomous action. In these cases indirect paternalism will be justified by the harm principle because an intervention will avoid harm to society. In the following I will make clear whether this is the case.

This now taken on the birth of a child (within the background of split motherhood), it is evident that the community would not be harmed. Even if a child will be harmed mentally, psychologically, or something else that is caused by split motherhood, there will be no serious impact on the society. And even if in all cases in which children could only come into existence by splitting motherhood, it would happen so, and there would probably be no serious public harm at all.[52] But what would happen if all future children will be begot by egg donation (or optionally will come to birth by surrogacy)? Would that be harm to the community itself?

On the one hand one might argue that possible mental disorders, which the child could suffer, have an impact on society if a larger number is affected. So it would be possible that the costs for any treatments are funded due to solidarity, and perhaps more important, it results in a reduction of social welfare (at least from an economic point of view). Although it is only a hypothetical future vision, it cannot be simply ignored. To express such a forecast, it is important to have consolidated findings on the effects of splitting motherhood. In the few studies, which are conducted so far, no significant impact on the psychological well-being of children were ascertained,[53]

---

[50] Feinberg (1989, 21 f).

[51] These are among others the costs for health care services as well as the opportunity costs.

[52] It can hardly be estimated how many cases there are in fact. Affected are women whose ovarian activity is not given, but who could meet their desire to have children by egg donation. In addition to some acquired diseases that can make this happen, there is as well a possibility of surgically removed ovaries (e.g. as a result of cancer or accident). Another case for using donated eggs is in prevention of transmission of genetic characteristics. Overall, this should be a quite manageable number. Most likely the number of children who cannot be begotten due to reproductive age is significantly greater.

[53] See Golombok et al. (2005), Murray/MacCallum/Golombok (2006).

and the expressed objections concerning any problems in discovering their own identity could not be documented. "The egg donation children were well adjusted in terms of their social and emotional development."[54] Although there is a lack to comprehensive longitudinal studies, which can verify the previous results, it must be noted that parental care is an important factor for a child's development and psychological well-being. Insofar we have to ask, at last, what weighs more and leads more into a harmonious development: blood or love?

Therefore, there is no expectable serious risk for society, because, firstly, according to the current state of research, no mental impairment is noted, especially not to that extent it was suspected, and secondly, it is highly unlikely, even if there is any mental impairment, that it occurs in such large quantities that it becomes a serious public harm.

On the other hand one might argue that the legal system is based on a society's morality. Morality constitutes society as network within their boundaries and shapes therefore the social fundament. Ignoring social conventions may thus undermine the social order to its foundations. For this reason, it is necessary that the bonds of morality become manifested in positive rights and every violation of these rights will be sanctioned.[55] According to the harm principle the limitation of reproductive autonomy seems to be reasonable in this line of argumentation. In that effect it is intended to sustain society, which on the contrary collapses when compliance with the moral order is not guaranteed, in which a traditional image of the family is a part. Splitting motherhood shapes a deep contrast to an institutionalised traditional family, built on a father-mother-child-relationship, constituted by sexual polarity and generational difference, and characterized by durability, exclusivity and commitment.[56] The practice of egg donation (as well as surrogate motherhood) runs "counter to the established principle of unambiguousness of motherhood which represented a fundamental and basic social consensus."[57] Tolerating this necessarily destabilises the existing tradition as well as the social order.

Arguing this way presupposes a homogenous moral order and requires mutual consent about morality. Divergent moral beliefs are not intended. However, there is a big problem in recognising this homogenous moral order if pluralistic moral beliefs will be acknowledged as a matter of fact. Moral plurality is more than common in modern, complex societies. A picture of moral conformity could not be drawn. Insofar there is no necessity of a consensus about moral beliefs in all areas of life. Where no metaphysical or basic moral agreement exists and social structures are characterized by moral diversity, there is no evidence of an imminent danger to

---

[54] Murray/MacCallum/Golombok (2006, 610), Revermann/Hüsing (2010, 164). Neither physical nor psychosocial development indicates any abnormalities. Ibid. 166.

[55] See Devlin (1962).

[56] Funcke/Thorn (2010, 19). There again, the formal structure of a nuclear family has not lost any binding character for family arrangement since other kinds of family life emerged. Ibid. 23.

[57] S.H. and others v. Austria, ECHR Judgment (Grand Chamber), Application no. 57813/00, 3 November 2011, § 70.

society, not even if we live in a world of moral strangers.[58] Indeed, such a pluralistic and secular society requires a conception of common values but no overall moral agreement.[59] In a liberal and democratic state this conception is built on the idea of autonomy as a moral claim and a right of self-determination derived from that. The societal conventions are based upon unlimited personal freedom with respect for other individual life plans and acknowledging the strong desire to control one's own life reciprocally.

18.6 Harming vs. Wronging

Although no harm will be done, neither to parents nor to children nor to society, it nevertheless seems to be wrong to conceive a child which will be born with (a significant risk to develop) a serious disease or mental disorder.[60] Nobody wants to suffer by pain or by reduction of opportunities of a normal life. Nobody prefers to live in constant dependence to other persons or technical assistance caused by a kind of disability. Even though it will not be interpreted as particularly bad by a person affected, and the life is still worth living, it is a circumstance from which one wishes it would not be there. However, it is not clear how purely a wish could justify a limitation of reproductive autonomy, and it is worth considering whether a moral obligation could be justified to beget only a sufficiently healthy child.

As shown before, it is not possible to protect a child's interest by avoiding his or her existence, because no harm would be done purely by existence. The concept of harm can only be taken into account if there would be a better-off state, which cannot be realised in non-existence. Since the concept of harming is not suitable to describe the current status of the child,[61] Carson Strong[62] proposes to leave this aside and discuss wronging instead. He argues that someone could be wronged without being harmed, thus, he postulates a "right to a decent minimum opportunity for development" which every human being has qua existence. In a similar manner Feinberg postulates a "right to an open future".[63] This right will be violated if someone would be conceived with awareness of a serious impairment.[64]

Now, it could be objected that this approach is not reconcilable with the harm principle (as a minimum requirement of interference with someone's autonomy) and if so, no intervention could be justified. But, if the harm principle would not only be interpreted as restriction on someone's action to protect another's freedom

---

[58] Engelhardt (1996, 7).

[59] Hart (1977, 86).

[60] Again, this is the way the German legislature argued. See fn. 32.

[61] Feinberg (1987, 99).

[62] See Strong (1998, 284–286), and for further discussion Strong (2005).

[63] Feinberg (2007).

[64] However, Strong does not establish a right against nature nor undermine the natural lottery as a point of social justice.

but also as enabling a freedom of action, meaning the free development of the individual,[65] this right could be justified from a liberal point of view.

If this right could be derived from the harm principle, it means that under some circumstances to conceive a child is (at least morally) wrong because the child's birth right is violated. However, we have to ask, in which way such a right could be realised and will take effect. It seems impossible to violate a right at the same time the right arises, namely in the act of procreation. Moreover, it is by no means clear what a sanction for violating could be. It would be paradox if the child would complain about his existence or sue his parents for conception and birth. Bringing this case to court and suing for compensation (i.e. non-existence) is a contradiction in terms.

Since the general function of rights is to protect the owner's interest, it is not clear how such a right qua existence might be established when there are no interests. Or, are there any other interests that shall be protected by this right? It seems truly much more plausible that such a right is based on a general theory of good life, which needs to be protected as it is a value. Consequently, the legitimating justification for a restriction of reproductive autonomy could rather be found on a societal level.

Steinbock and MacClamrock argue in the same line of a value-based impersonal morality, albeit they do not define a right of a child but a moral obligation of the parents deriving from a special responsibility within the parent-child-relationship. Designed as prima facie principle, the entire capability within the parental responsibility will be seen as a mission to enable the best possible life for the child. The decision on the child's initial capabilities is directly derived from the educational duty as parents. However, the challenge is not in doing the best for the child and the question is not, what should be done for the child's best possible well-being but rather could there be any best well-being at all. That decision could never take place from the child's point of view, insofar the evaluation behind is linked to an external point of view. While this will always be prejudiced, it could not make the grade of an internal view, especially since an external view is not independent from another person's interests and theories of good life.

To point out the problem, it will be possible to put another value (or more) next to autonomy and further, these values must be weighed against each other. In case of autonomy another value will weigh more,[66] the harm principle will be overruled and a limitation of a mother's reproductive autonomy might be justified. But, this challenges the initial presupposition of the harm principle. As I took it as a matter of fact at the beginning, autonomy is a very high value and the harm principle is the minimal requirement to interfere in the public sphere. Any way arguing with a supreme theory of good life is contrary to presupposed plural beliefs and therefore not reconcilable with a liberal democratic society.

---

[65] At this point liberty is not any more meant as a negative freedom but becomes a requirement of enabling a positive freedom.

[66] E.g. the value of naturalness of childmaking.

18.7 Conclusion

At last, it should be noted that to interfere with someone's reproductive autonomy is not in the best interest of the child and could thereby not be justified, at least in a liberal society. To penalise the practice of egg donation does not conform to the harm principle since there is neither harm (or offense) to the mother nor to the child nor to society, because without any harm there is no justification for direct or indirect paternalism. From a non-occurring harm no moral claim against the parents can be established. Regarding the above-mentioned elements of reproductive autonomy, limiting procreative liberty could in no way be justified. There is no moral obligation to go without a child with a serious impairment. Other moral values go further beyond an ethical framework of liberalism. It is rather a matter of responsible parenthood to provide the best possible life to an already conceived child.

If legal restrictions on assisted reproduction, like egg donation, shall be based on ethical considerations, a different way of justification is needed. Speaking with Mill the "only purpose" and "sole end" for which the liberty of a person may be legitimately constrained by others is self-protection.[67] For sure, one might argue against an ethical approach of liberalism. Assuming a non-liberal society, in which personal autonomy is not the supreme value, it could be argued for a kind of moral legalism. In this case a public morality will overrule individual interests and personal autonomy is limited by the will of community. From this point of view liberalism is a moral doctrine and autonomy is a value among others.[68] This objection seems to be plausible, particularly for a society of moral plurality. But, in consideration of autonomy as a value of public morality, most other values concern private morality. Liberalism claims to regulate only the public sphere whereas other ethical approaches rule the private sphere. Therefore liberalism should not be compared with any "particular ethical or religious doctrine, but the general morality of totalitarianism".[69] One may say that we agree to disagree about moral values and find a consensus in autonomy,[70] even if it should be the second best choice.

References

Beauchamp, Tom L. 2007. History and theory in "applied ethics". *Kennedy Institute of Ethics Journal* 17(1): 55–64.

Beauchamp, Tom L., and James F. Childress. 2001. *Principles of biomedical ethics*, 5th ed. Oxford: Oxford University Press.

---

[67] Mill (1996a), On Liberty, 223.

[68] Lee (1986, 15–17).

[69] Häyry (1991, 106).

[70] Charlesworth (1993, 166 f).

Beitz, Ulrike. 2009. *Zur Reformbedürftigkeit des Embryonenschutzgesetzes: Eine medizinisch-ethisch-rechtliche Analyse anhand moderner Fortpflanzungstechniken.* Frankfurt am Main: Peter Lang.

Berg, Giselinde. 2008. Die Eizellspende – eine Chance für wen? In *Umwege zum eigenen Kind: Ethische und rechtliche Herausforderungen an die Reproduktionsmedizin 30 Jahre nach Louise Brown*, ed. Gisela Bockenheimer-Lucius, Petra Thorn, and Christiane Wendehorst, 239–253. Göttingen: Universitätsverlag.

British Medical Association Ethics Department. 2013. *Everyday medical ethics and law.* New York: Wiley.

Charlesworth, Max. 1993. *Bioethics in a liberal society.* Cambridge: Cambridge University Press.

Devlin, Patrick. 1962. Law, democracy, and morality. *University of Pennsylvania Law Review* 110(5): 635–639.

Dörries, Andrea. 2003. Der Best-Interest Standard in der Pädiatrie – theoretische Konzeption und klinische Anwendung. In *Das Kind als Patient: Ethische Konflikte zwischen Kindeswohl und Kindeswille*, ed. Wiesemann Claudia, Andrea Dörries, Wolfslast Gabriele, and Simon Alfred, 116–130. Frankfurt am Main: Campus.

Downie, Robin S., and Fiona Randall. 1997. Parenting and the best interests of minors. *Journal of Medicine and Philosophy* 22(3): 219–231.

Eekelaar, John. 1994. The interests of the child and the child's wishes: The role of dynamic self-determinism. *International Journal of Law, Policy and the Family* 8(1): 42–61.

Engelhardt, Hugo T. 1996. *The foundations of bioethics*, 2nd ed. New York: Oxford University Press.

Feinberg, Joel. 1987. *Harm to others*, The moral limits of the criminal law, vol. 1. New York: Oxford University Press.

Feinberg, Joel. 1989. *Harm to self*, The moral limits of the criminal law, vol. 3. New York: Oxford University Press.

Feinberg, Joel. 2007. The child's right to an open future. In *Philosophy of education: An anthology*, ed. Randall R. Curren, 112–123. Malden: Blackwell.

Funcke, Dorett, and Petra Thorn. 2010. Statt einer Einleitung: Familie und Verwandtschaft zwischen Normativität und Flexibilität. In *Die gleichgeschlechtliche Familie mit Kindern: Interdisziplinäre Beiträge zu einer neuen Lebensform*, ed. Dorett Funcke and Petra Thorn, 11–33. Bielefeld: Transcript.

Golombok, Susan, et al. 2005. Families Created by Gamete Donation: Follow-Up at Age 2. *Human Reproduction* 20(1): 286–293.

Gräfrath, Bernd. 1992. *John Stuart Mill: "Über die Freiheit": Ein einführender Kommentar.* Paderborn: Schöningh.

Graumann, Sigrid. 2008. Eizellspende und Eizellhandel – Risiken und Belastungen für die betroffenen Frauen. In *Umwege zum eigenen Kind: Ethische und rechtliche Herausforderungen an die Reproduktionsmedizin 30 Jahre nach Louise Brown*, ed. Gisela Bockenheimer-Lucius, Petra Thorn, and Christiane Wendehorst, 175–183. Göttingen: Universitätsverlag.

Günther, Hans-Ludwig, Jochen Taupitz, and Peter Kaiser. 2008. *Embryonenschutzgesetz: Juristischer Kommentar mit medizinisch-naturwissenschaftlichen Einführungen.* Stuttgart: Kohlhammer.

Hare, Richard M. 1993. The abnormal child: Moral dilemmas of doctors and patients. In *Essays on bioethics*, ed. Richard M. Hare, 185–191. Oxford: Clarendon.

Hart, Herbert L.A. 1977. Immorality and treason. In *The philosophy of law*, ed. Ronald M. Dworkin, 83–88. Oxford: Oxford University Press.

Häyry, Heta. 1991. *The limits of medical paternalism.* London: Routledge.

Heyd, David. 1992. *Genethics: Moral issues in the creation of people.* Berkeley: University of California Press.

Höffe, Otfried. 1996. Immanuel Kant, 4 Aufl. München: Beck.

Kant, Immanuel. 1996. Groundwork of the metaphysics of the morals. In *Practical philosophy*, ed. Mary J. Gregor, 37–108. Cambridge: Cambridge University Press.

Kavka, Gregory S. 1982. The paradox of future individuals. *Philosophy and Public Affairs* 11(2): 93–112.

Keller, Rolf. 1989. Das Kindeswohl: Strafschutzwürdiges Rechtsgut bei künstlicher Befruchtung im heterologen System? In *Festschrift für Herbert Tröndle zum 70. Geburtstag am 24. August 1989*, ed. Hans-Heinrich Jescheck and Theo Vogler, 705–721. Berlin: de Gruyter.

Keller, Rolf, Hans-Ludwig Günther, and Peter Kaiser. 1992. *Embryonenschutzgesetz: Kommentar zum Embryonenschutzgesetz*. Stuttgart: Kohlhammer.

Knoepffler, Nikolaus. 2008. Patientenautonomie – Anspruch und Wirklichkeit am Beispiel der Sterbehilfedebatte. In *Medizin zwischen Humanität und Wettbewerb. Probleme, Trends und Perspektiven*. Beiträge des Symposiums vom 27. bis 30. September 2007 in Cadenabbia, ed. Volker Schumpelick and Bernhard Vogel, 145–156. Freiburg: Herder.

Lee, Simon. 1986. *Law and morals: Warnock, gillick and beyond*. Oxford: Oxford University Press.

Mill, John S. 1996a. On liberty. In *Essays on politics and society 1*. Collected Works of John Stuart Mill, vol. 18, ed. Robson John M. Reprinted, 213–310. London: Routledge.

Mill, John S. 1996b. *Principles of political economy: With some of their applications to social philosophy; Books I – II*. Collected Works of John Stuart Mill, vol. 2, ed. Robson John M. Reprinted, London: Routledge.

Murray, Clare, Fiona MacCallum, and Susan Golombok. 2006. Egg donation parents and their children: Follow-up at age 12 years. *Fertility and Sterility* 85: 610–618.

O'Neill, Onora. 2005. *Autonomy and trust in bioethics*, 3rd ed. Cambridge: Cambridge University Press.

Parfit, Derek. 1986. *Reasons and persons*. Reprinted. Oxford: Clarendon Press.

President's Commission for the Study of Ethical Problems in Medicine and Biomedical and Behavioral Research. 1983. *Deciding to forego life-sustaining treatment: Ethical, medical, and legal issues in treatment decisions*. Washington: U.S. Government Printing Office.

Riley, Jonathan. 2004. *Mill on liberty*. Reprinted. London: Routledge.

Revermann, Christoph, and Bärbel Hüsing. 2010. Fortpflanzungsmedizin – Rahmenbedingungen, wissenschaftlich-technische Entwicklungen und Folgen: Endbericht zum TA-Projekt. TAB-Arbeitsbericht Nr. 139.

Robertson, John A. 1975. Involuntary euthanasia of defective newborns: A legal analysis. *Stanford Law Review* 27(2): 213–269.

Schöne-Seifert, Bettina. 2007. *Grundlagen der Medizinethik*, Kröner Taschenbuch, vol. 503. Stuttgart: Kröner.

Skorupski, John. 1989. *John stuart mill*. London: Routledge.

Steigleder, Klaus. 1998. Müssen wir, dürfen wir schwere (nicht-therapierbare) genetisch bedingte Krankheiten vermeiden? In *Ethik in der Humangenetik: Die neueren Entwicklungen der genetischen Frühdiagnostik aus ethischer Perspektive*, ed. Marcus Düwell and Dieter Mieth, 91–119. Tübingen: Francke.

Strong, Carson. 1998. Cloning and infertility. *Cambridge Quarterly of Healthcare Ethics* 7(3): 279–293.

Strong, Carson. 2005. Harming by conceiving: A review of misconceptions and a new analysis. *Journal of Medicine and Philosophy* 30: 491–516.

Index

© Springer International Publishing Switzerland 2015   295
T. Schramme (ed.), *New Perspectives on Paternalism and Health Care*, Library
of Ethics and Applied Philosophy 35, DOI 10.1007/978-3-319-17960-5

Printed in the United States
By Bookmasters